P9-EIG-954

Saints and Schemers

Saints and Schemers

Opus Dei and Its Paradoxes

JOAN ESTRUCH

Translated by Elizabeth Ladd Glick

New York Oxford
OXFORD UNIVERSITY PRESS
1995

Oxford University Press

Oxford New York
Athens Auckland Bangkok
Calcutta Cape Town Dar es Salaam Delhi
Florence Hong Kong Istanbul Karachi
Kuala Lumpur Madras Madrid Melbourne
Mexico City Nairobi Paris Singapore
Taipei Tokyo Toronto

and associated companies in
Berlin Ibadan

Published by Oxford University Press, Inc.
198 Madison Avenue, New York, New York 10016

Originally published in Spain in 1993, as
L'Opus Dei i les seves paradoxes (in Catalan)

Library of Congress Cataloging-in-Publication Data
Estruch, Joan, 1943–
[L'Opus Dei i les seves paradoxes. English]
Saints and schemers : Opus Dei and its paradoxes / Joan Estruch.
p. cm.
Includes bibliographical references and index.
ISBN 0–19–508251–6
1. Opus Dei (Society) 2. Religion and sociology—Spain.
I. Title. BX819.3.068E7413 1995
306.6'67182—dc20 94–37014

1 3 5 7 9 8 6 4 2

Printed in the United States of America
on acid-free paper

Preface

It would probably be far too difficult—and of little interest to the reader, as well—to try to reconstruct the genealogy of this book by going back to the remote origins of the research on which it is based. Anyone who is acquainted with the evolution of Spanish history during the second half of the twentieth century will of necessity have come face to face on many occasions with the existence of an organization in general little known, surrounded by mystery and a great deal of controversy, called Opus Dei. No social scientist could fail to be curious about the omnipresent—or perhaps we should say omnilatent—character of this organization. And in the case of a social scientist whose field is the sociology of religion, this curiosity is even more likely to become a fascination.

But in order for this to happen some specific factor must precipitate the leap from the nebulous remote beginnings of interest in a topic to its more immediate causes. In my case nothing could be easier to explain, since the immediate source of the present study can be located very precisely, both in time and in space: a conversation with Peter Berger one sunny winter afternoon on the Rambla de Catalunya in Barcelona at the end of 1989. Anyone who has studied sociology during the second half of the twentieth century will of necessity have come across the work of Peter Berger, and anyone whose work is in the sociology of religion is probably his disciple.

At least that was the case with me. I had looked on him as my teacher for many years; for many years I had read and studied his books; and for many years I had quoted him repeatedly in my own work. I had also translated some of his books, and—according to him—I was responsible for the "loss of his virginity in Catalan." This had given rise to an exchange of letters. However, that conversation in the winter of 1989 was, in fact, our first face-to-face meeting.

In his capacity as director of the Institute for the Study of Economic Culture at Boston University, Peter Berger wanted to gather information about several original and specific aspects of our "economic culture." What sparked the process leading to the publication of these pages was the fact that the first two business schools established in Barcelona, which played such an important role in the revitalization, stimulation, and modernization of the Spanish business world, were both connected to the Catholic Church, one through the Society of Jesus and the other through Opus Dei.

In my 1984 prologue to the Catalan edition of Max Weber's *The Protestant Ethic and the Spirit of Capitalism,* I wrote that in the same manner in which one could demonstrate "the relation existing between the Jesuit ethic and the spirit of capitalism, I confess that for years I have awaited the appearance of a similar study on the ethic of Opus Dei and the spirit of capitalism" (Estruch, 1984, 10). If we regard this sentence as a synthesis of what I have called the remote origins of the research, Berger's response—"so why don't you do this study?"—was without question the immediate origin.

◆ ◆ ◆

But if in fact this constitutes the immediate origin, it is because Peter Berger's question, which might have been merely a friendly rhetorical question, immediately became a question-proposal, so much so that the classic phrase in all scholarly acknowledgments, which goes "this study would not have been possible without the help of . . . ," is in this case not at all rhetorical. The Institute for the Study of Economic Culture accepted the present research as part of its program and subsidized it from the very beginning. Once the actual research was finished, the institute invited me to spend a semester in Boston to prepare the final version of the book and to participate in the institute's activities. Thus my old intellectual debt to my mentor has multiplied notably; I owe him partly for the opportunity to complete this work, but above all I owe him for the very special privilege of being able to treat a great teacher as a real friend.

I must also thank the Universitat Autònoma de Barcelona for making its facilities freely available to me for my research. The personnel in the research office showed themselves to be a team that willingly made things easy, quite unlike the stereotypical bureaucracy that only erects obstacles. My colleagues at the Department of Sociology and the academic authorities freed me of all duties during the semester I spent in Boston.

I also wish to extend my sincerest gratitude to the Fundació Jaume Bofill of Barcelona, which, at the suggestion of its director, Jordi Porta, took part in the project and contributed materially to its realization.

For the first time in many years I have engaged in a relatively long and complex research project without the collaboration of Salvador Cardús. I must confess that it has not been easy for me to get used to working without him again. Still, he was the first—after my wife, that is—to read the first version of the manuscript, and he criticized its faults without apology, as it ought to be.

Of the many persons who participated in the research in various ways I must mention just two, although this seems unjust to the rest. For a whole year I had the help of Esther Fernández Mostaza and Josep Verdaguer, who located, read, and analyzed mountains of documents. My always fertile discussions with the latter gave rise to quite a few of the hypotheses in the second part of the study. As for the former, her capacity for work, and for work well done, smoothed the road considerably for me with respect to many questions about the origins of Opus Dei and the character of its founder.

Barcelona J. E.
September 1994

Contents

Foreword

Opus Dei, a Catholic order of priests and lay members, has attracted public attention most recently because of its activities in Latin America, particularly because of its influence in the Peruvian church and the association of some of its members with repressive regimes in Chile and Argentina. But well before this recent phase of activity, Opus Dei was known in the United States for its well-publicized and controversial efforts to recruit new members on college and university campuses and for its prominence in international economic activities. The order has been surrounded by an air of mystery, which the present volume aims to dispel, and its international spread has tended to draw attention away from its specifically Spanish value system, associated with the early years of Franco's dictatorship and the promotion of "national Catholicism"—which Estruch here documents in great detail.

In this book we learn that Opus Dei was founded in Spain not, as its own literature states, in 1927, but somewhat later, possibly as late as 1940. Its founder, José María Escrivá, was a Spanish priest interested in evangelizing the professional and intellectual sectors of Spanish society—precisely those sectors that had typically rejected traditional Catholic and Spanish values and who constituted the bedrock of support for the liberal Second Republic (1931–1936).

After the Spanish Civil War (1936–1939), the victorious Franco regime orchestrated a concerted attack on liberal Spain. As Victor Pérez Díaz notes, in the mean-spirited ambiance of post–Civil War Spain, "occupying cultural or intellectual space meant excluding all the cultural rivals of earlier times

. . . this was the purpose of the church's cultural mesogovernment."[1] By "mesogovernment" he means the cession to the church of a large bloc of Spanish life, including most of the educational structure and anything having to do with public morals, which it enforced with a vengeance.

"But if the world was there to be conquered," Pérez Díaz continues, [Catholics] had to live in a state of constant vigilance because the war had to be won again and again. Their mold must be heroic, their rhetoric one of conquest, their style almost military. Self-control, discipline, heroism, and asceticism: such were to be the moral and emotional keys to their behavior."[2] These values were precisely those of Opus Dei, whose aim was to rebuild in an explicitly Catholic guise the broken liberal intellectual and professional structure and to use those structures for the evangelization or re-evangelization of the elites. Their primary targets were the universities and the print media.

Opus Dei was eventually granted the official status of a "secular institute" of the church. Secular institutes have different aims and practices but in all of them members continue to pursue their lay activities, take vows of perfection (chastity, obedience, and poverty consistent with their secular way of life), and do not live together in a common religious life or wear a distinctive habit, as monks and nuns might do. They are "secular" in that members remain occupied with worldly affairs, but they are not to be confused with lay sodalities or associations of Catholic laypeople whose members do not take such vows.

Opus Dei's uniqueness lies in its "numerary" members, who take vows—like priests—of poverty, obedience, and chastity and live in well-appointed Opus residences but continue to pursue their secular professions. These are far outnumbered by "supernumerary" members who take no vows and live in their own homes, frequently with spouse and children.

Vatican II

The famous council of bishops convened by Pope John XXIII in 1962 was charged with modernizing the church. *Aggiorniamento*—bringing the church up to date—was the buzzword popularly associated with the council. Formerly, the priest performed the mass in Latin with his back to his parishioners, facing the altar, so that the congregation could not see what he was doing. Vatican II provided both for vernacular languages and for a new physical disposition of the service whereby the altar was brought forward and the priest now stood behind, not in front of it, facing the congregation. Escrivá was appalled, and secured a dispensation allowing Opus priests, who did their best to subvert the new layout of altars, to continue using the Latin liturgy. Yet an Opus official would declare in 1989 that "though founded a full generation before Vatican II, Opus Dei welcomed the council's repeated statements that confirm its own teaching—that the laity have as 'their special vocation . . . to

1. Victor Pérez-Díaz, *The Return of Civil Society: The Emergence of Democratic Spain* (Cambridge, Mass., 1993), p. 135.
2. Ibid., pp. 124, 137.

seek the kingdom of God by engaging in temporal affairs and directing then to God's will.'"[3]

At best it could be said that Opus, in stressing a heightened role for secular professions in Catholic life, offered a kind of traditional and conservative alternative mode of *aggiornamento*. This is particularly true insofar as Opus projected a technocratic image, particularly in Spain, where prominent members played leading roles in the economic modernization of the country in the 1950s and '60s as well as in the administrative overhaul that such reforms required. These technocrats were not ideologues in the conventional political sense. One ex-Opus member went so far as to describe the early movement as kind of counter-culture that opposed the highly conformist, conventional Catholicism promoted by the Spanish regime.[4] The same zealot—or ex-zealot, in this case—described the order as the "last remnant of that militant messianism which is endemic in the Abrahamic religions." Recent events have demonstrated that Abrahamic religions, whether of the Christian, Jewish, or Muslim variety, have no want of militant messianists. As for the Spain of the 1940s, it is far easier to see Opus as the spiritual shock troops (to use a term that has frequently been applied to Jesuits) of Franco's backward-looking, anti-modern, integralist national Catholicism.

There has been widespread confusion as to the theological orientation of the order. Some have identified Opus with other Catholic groups that fill the role which in the nineteenth century was filled by the movement called integralism, conservative Catholics who opposed "modernism" and all its works. Such groups are the Catholic equivalent of Protestant fundamentalists, but with one major distinction: they are not scriptural literalists but rather "papal fundamentalists" for whom "an uncritical adherence to papal and curial documents serves as the litmus test for orthodoxy."[5]

The notion that Opus is slavishly loyal to the papacy is reminiscent of another order founded by a Spaniard and known for its militancy: the Society of Jesus, or Jesuits. It was commonly said in Spain that the Jesuits greatly resented the intrusions of Opus into higher education, the historical bailiwick of the Society, and there is considerable evidence to support this contention. From within the church and without, the Opus was tagged with accusations formerly directed at Jesuits: "secretiveness, greed for corporate wealth and power, elitism, and propensity for manipulation."[6] It is a fascinating comment on the multifaceted nature of Roman Catholicism that its constituent institutions do in fact evolve over time and play quite different roles in distinct historical situations. Thus in Latin America, Opus has emerged as the leading opposition within the church to liberation theology, many of whose most

3. William A. Schmitt, "Apologia: Opus Dei [Letter]," *Commonweal,* October 6, 1989, p. 541.

4. Ramon Panniker, quoted by Michael Walsh, *Opus Dei: An Investigation into the Secret Society Struggling for Power within the Roman Catholic Church* (San Francisco, 1993), p. 170.

5. John A. Coleman, "Who are the Catholic Fundamentalists?" *Commonweal,* 27 January 1989, p. 46.

6. Paul Hoffmann, *Anatomy of the Vatican: An Irreverent View of the Holy See* (London, 1985), p. 231.

outstanding practitioners have been the same Jesuits—now leaders in the progressive wing of the church—who some decades earlier had fought to reserve their prerogatives in higher education against Opus pretensions in Franco's Spain.

Popular Suspicions

The public has been most concerned about two aspects of the Opus Dei's practices: first, to what extent was it a cult, trapping the innocent; and, second, to what extent did its members secretly infiltrate professional organizations, in particular education and the media?

With respect to the cultic aspects, attention was focused on recruiting, overcontrol of recruits, and enforced separation of recruits from families. Recruiting is called "fishing" in Opus Dei parlance: one does not apply for membership, but is discovered, searched out, "fished" by the order, whose operatives have a good idea of the kind of fry that will fit in. Catholic secondary schools and colleges, of course, are prime targets for fishing. In such endeavors Opus has typically worked through clubs (such as the Midtown and Rosedale Clubs in Chicago and New York, respectively) that it established and, at universities, full-fledged Opus residences (some two hundred throughout the world). In such surroundings, Opus Dei priests could cultivate the order's particular esprit de corps that can be especially attractive in a university environment. In the course of recruitment into the order, young people appear to have been told not to discuss the order with their families, for fear they would not understand, and to inform family members only after joining. This kind of behavior has aroused fears of "kidnapping" and brainwashing among the general Catholic public. In certain dioceses, such as that of Westminster, England, in 1981, bishops have issued "recommendations" to the effect that no one under eighteen years of age should be permitted to take Opus vows; before making the decisions to join, young persons should discuss the matter with parents or diocesan priests; Opus should not exert undue pressure on recruits; Opus activities should be clearly identified as such.[7]

Official church reactions such as these make it clear that the documented incidence of such activities has been great enough to oblige bishops to take action. It is also probably true that such strictures have had no effect on Opus procedures or were deliberately ignored.

With regard to the issue of secrecy—semisecrecy, some commentators say—some of the accusations may be more related to style than to substance. A Catholic friend of mine reported attending an Opus mass in Boston which, he said, "had the spirit of a secret society" even though the service was in no way secret. He went on to say that the spirit of camaraderie among Opus members was such that he felt excluded. Members of the order collectively project a mystique that lends credence to rumors, whether founded or not, to the effect that is a "secret society."

7. Walsh provides the full text in *Opus Dei*, pp. 165–66.

Objectives and Strategies

Opus Dei made a mark in the Spanish intellectual world by buying existing print media and by establishing their own, in particular the publishing house Rialp, and by its enterprises in higher education, notably the establishment of the University of Navarre at Pamplona. This characteristic thrust was extended quite naturally to Opus efforts in other countries. A 1979 Opus report listed participation of its members in 197 colleges and universities, 694 newspapers and magazines, 52 television and radio stations, 38 news agencies and 12 cinema companies, worldwide.[8] Notable Opus educational institutions include Strathmore and Kianda Colleges in Kenya, and, in Latin America, universities in Peru, Mexico, Colombia, Chile, and Argentina. It also created a series of agricultural centers for the training and evangelization of peasants in Chile, Peru, and Mexico.

The order typically relies on complex networks of Opus-owned or controlled holding companies and banks and shadowy networks of financial supporters to fund its operations. In England, Opus established a number of foundations such as the Netherhall Educational Association and the Dawliffe Educational Foundation to channel or hold Opus property and to channel money to Opus institutions. Some of this money came from abroad and some was notoriously tainted, for example, the millions of dollars channeled to the order by the Spanish holding company Rumasa before its well-publicized bankruptcy in 1983. Rumasa's president José Maríe López-Mateos is an Opus member. Similarly, a high-flying Chilean company of the Pinochet era, Larrain Crusat, supplied the order with 10 million pesos a month before it too failed. It is never clear whether a company is Opus-owned, or merely owned by a member of Opus, or, indeed, whether such a distinction has any real meaning, inasmuch as Opus trusts receive money from Opus businessmen.[9]

The Opus was also implicated in the Vatican Bank scandals of the late 1970s and early 1980s. Purportedly, a deal was cut that would have funneled Opus money through the network of banks controlled by Rumasa to the cash-strapped Vatican Bank, to the tune of one-third of the Vatican's yearly expenses.[10]

The Personal Prelature

In 1982 Pope John Paul II granted a request from the order that it be converted from a "secular institute" to a "personal prelature," that is, a religious order directly under the supervision of the pope. This made the order's head the equivalent of the bishop of a worldwide superdiocese, with a bishop's right to

8. Cited by Hoffmann, *Anatomy,* p. 240.
9. Walsh, *Opus Dei,* p. 152.
10. For details, see *ibid.,* pp. 153–58.

ordain priests. It was the first such prelature in the long history of the church. This unprecedented move evoked howls of outrage and protest from enemies of Opus within the church, who feared the order was on its way to becoming a church within a church or a parallel church. As a high church official complained to Paul Hoffmann, shortly before the personal prelature application was approved: "Some of my priests are Opus, although they won't say so. I tell them to take a new assignment, and, would you believe it, they reply they'll think it over for a few days. I know, of course, that they ask their Opus Dei superior whether they should obey me, their archbishop! Now, imagine if Opus Dei were to get what it wants! This would undermine the authority of all bishops everywhere and create a church within the church."[11]

Once Opus became a personal prelature, however, it became a kind of institutional hybrid, because in spite of official Opus statements to the contrary, its priests have increasingly received preferential treatment and the number of priests in the order has continued to grow. In common with other secular institutes, moreover, the Opus has tended, over time, increasingly to resemble a normative monastic order, in which members are encouraged to share a common life in an Opus residence that is a monastic establishment in everything but name.

The personal prelature coincided with a spurt of expansionism that vastly extended the geographical range of Opus activities. It is interesting to note that the scope of Opus evangelization, previously limited largely to the Spanish-speaking countries and those of Western Europe, broadened considerably in the 1980s: thus branches were established in Zaire and the Ivory Coast in 1980; Hong Kong in 1981; Singapore in 1982; Trinidad and Tobago in 1983; Taiwan in 1985; Cameroon in 1988; New Zealand, Macao, and Poland in 1989; Czechoslovakia and Hungary in 1990; India and Israel (!) in 1993. As of 1991, there were 76,934 lay members of Opus Dei, together with 1,459 priests. Most of these were in Europe (46,138 lay, 928 clergy) and America (26,507 lay members in North and South America combined, with 420 priests). Opus Dei membership in the United States stands at around 2500.

The international character of Opus Dei today should not, however, obscure its roots in the distinctive ideological and political milieu of the Spain of the 1940s and '50s, when its ideology was forged in close harmony with that of the Franco regime and its program of national Catholicism, while at the same time the moribund Spanish economy became a laboratory for neoliberal Opus economists. In the interplay between those two areas—the church's mesogovernment and the exercise of civil profession as a spiritual calling—emerged the unique dynamic so effectively analyzed here by Joan Estruch.

Boston
February 1995

THOMAS F. GLICK

11. Hoffmann, *Anatomy*, pp. 231–32.

Saints and Schemers

Introduction

Basic Research Perspectives

Although I wish I were mistaken, the truth is that as soon as I began this research I got the definite feeling that the publication of a sociological study about Opus Dei was going to be, almost by definition, controversial.

From the day of its founding—officially in 1928—Opus Dei seems to have had only defenders-to-the-death or diehard detractors. On one side are its extraordinarily touchy partisans, practically incapable of tolerating any kind of criticism; on the other stand ferocious critics, ready to detect signs of weakness and tacit complicity in the smallest manifestation of sympathy. One might say that the former tend to identify with the evangelist's assertion, "whoever is not with me is against me" (Luke 11:23): whoever does not sing the praises of Opus Dei is its enemy. The latter, meanwhile, tend more toward another assertion, by the same evangelist, "whoever is not against you is with you" (Luke 9:50): whoever does not denounce Opus Dei becomes its accomplice.

With Opus or against Opus—are these, then, the only alternatives? Of course not. On the contrary, the only way to undertake a sociological investigation of Opus Dei consists precisely, in my view, of attempting to transcend a dichotomy that can only serve to confirm preexisting stereotypes and prejudices. On this matter, it is better to distrust such prejudices and to withdraw from such stereotypes, to prevent them from coloring the development of the research. In other words, in a work of this nature it is sometimes difficult to avoid taking as a departure point certain questions which one believes have already been answered in advance. This is necessary, even when in the end it sometimes happens that some of those earlier answers were correct.

Throughout the pages that follow we will see again and again that this is precisely one of the most serious problems posed by a good part of the literature available on Opus Dei. For example, the various biographies of the founder of Opus, all written with unconditional sympathy and lacking the tiniest dose of criticism—to such an extreme that they cease to be biographies and become hagiographies—are countered by other works of great virulence written from a critical stance and absolutely devoid of sympathy for Opus Dei.

My research has been conducted, by contrast, from a sympathetic and a critical point of view at one and the same time. By joining these two words I am not trying to juxtapose—and merely juxtapose—two apparently contradictory notions, nor am I gettting ready for a dialectical pirouette that would distance me from both of them at once. I simply want to say that in sociology, sympathy and critique, whether they appear paradoxical or not, constitute in fact two sides of the same coin; they are the two indispensable ingredients that enable us to comprehend the reality of what is being studied. Some might object that this is penny-ante epistemology, but I grow more and more convinced that it is, in fact, epistemology: that in the world of the social sciences, knowledge implies the combination of both ingredients, simultaneously and in equal conditions and proportions.

A critical sympathy alone can lead to a blind love, but not to that mixture of perception and clairvoyance called insight, not to a lucid and deep understanding. On the other hand, apathetic or even antipathetic criticism alone can lead to a distancing, and perhaps a correct description, but never to an understanding from within. Real knowledge and understanding, clear and unbiased but still capable of preserving the profound significance of the real object under study, require the combined and simultaneous presence of criticism and sympathy.

I do not mean to claim unequivocally that I have achieved this true knowledge and comprehension, but only that this has been the intent that has presided over the whole investigation. The patient reader who reads to the end will have to personally judge to what extent I have succeeded. What is obvious, in any case, is that in no way can this work be considered definitive. Specific aspects of the history and present situation of Opus Dei have not even been touched upon; others are treated in a fragmentary and incomplete manner; there are many gaps. On some occasions the text suggests both the possibility and the utility of subsequent monographic analysis of certain questions. The available documentation is not always adequate, and it is frequently either nonexistent or "classified."

As a result my conclusions can be only provisional, and often they have a merely hypothetical character. As Max Weber said in his introduction to *Gesammelte Aufsätze zur Religionssoziologie,* in the end every scientific work is destined to be surpassed: this one, too, and by a great deal, and the sooner the better. The best that one can hope for from this study is that it will stimulate others, and that in so doing it will rapidly grow old.

The Triple Thesis of the Book

This book is structured around a triple thesis, which can be expressed as follows: in order to understand Opus Dei it is necessary to (1) relate it to the figure of its founder, (2) situate it in the historical context in which it was born and developed, and (3) place it in relation to the Jesuits.

The Figure of the Founder

The history of Opus Dei is still relatively short, and it appears to be decisively marked by the personality and work of its founder, Monsignor Escrivá de Balaguer, who was its lifetime general president until his death in 1975. Without him, and without the small group of his closest associates, neither the origins of the peculiar organization he created not its evolution over the course of the last half-century can be comprehended.

On this point there would certainly be general agreement, the only exception being, perhaps, that some in the world of sociology tend to propound an analysis of social events independently of the role played by the actors considered individually. But it is obvious that from a point of view that wants to take into account and assess the significance implied by any action for the actors involved, this is an essential element of every analysis. Of course we take for granted that it is necessary to avoid the temptation of making an interpretation centered exclusively on individual motivations and omitting the sociohistorical context. And let it be equally clear that it is not a matter of concentrating solely on the intentions of the actors, but also on the effects of their actions, and especially on the possible lack of coincidence between the intent and the effect.

This is a point on which discrepancies and disagreement may arise with writers belonging to Opus Dei. Undoubtedly they would agree that their founder ought to be singled out as a prominent figure; but it is not certain that they would share the basic hypothesis behind my project: that is, that all the work of Msgr. Escrivá de Balaguer constitutes an extraordinary and fascinating illustration of the classic Weberian axiom of the *unforeseen consequences* of an action.

From the thousands of pages that members of Opus Dei have written about Escrivá, one gets the clear impression that—whether because their emotional involvement is still very intense or because one of their major objectives was to campaign for his beatification—they wanted to make him into an individual made of a single piece, without any cracks or blind spots. By contrast, the impression that clearly emerges from my approach to Opus Dei's founder is that he was an immensely complex personality, paradoxical and even contradictory. To put it differently, the official biographies of Escrivá tend to mask the reality, whereas my perspective is necessarily—for methodological reasons, not ideological ones—a perspective that unmasks.

The Historical Context

Nor should it seem particularly daring to assert that an understanding of Opus Dei requires that it be situated in its historical context. This should be fairly obvious, and ought not to give rise to disagreements.

It so happens, however, that the specific framework in which Opus Dei arose and began to develop was very peculiar: it was the Spanish context of the thirties and forties, which were the years of the republic, the civil war, and the installation of the Franco regime. Today, it seems a particularly unpleasant framework, even to the followers of Opus Dei, who, although they were delighted with it at the time, now prefer to convince us—and certainly to convince themselves—that their delight never existed.

To accomplish this they seem to have chosen a radical and effective formula: denial, not of their delight, but of the entire past. That is, they deny the historical context, they deny their own history.

Thus in the name of the subsequent internationalization of Opus Dei, they deny their original Spanish character, their in a way even "typically" Spanish character. In the name of the relative diversity of the social positions of present-day members, this same diversity is claimed for the past as well. In the name of the no less relative plurality of its contemporary ideological options, the existence of a previous mode of thinking that was not in the least pluralistic is denied. And now that Opus Dei has realized—in spite of itself—the irreversibility of a whole series of changes introduced in the Catholic Church as a result of the Second Vatican Council, it tries to present itself today as a precursor of those changes.

One could say on this subject that the entire history of Opus Dei is a monumental and extraordinary example of an attitude that leads to modifying the past and reinterpreting it, adapting it to the circumstances and to the interests of the present. Thus the effort to situate Opus Dei in the historical context in which it arose and developed becomes a particularly attractive exercise, which is at the same time particularly complex and predictably polemical.

The Relationship Between Opus Dei and the Society of Jesus

No less interesting or complex is the question of the relationship between Opus Dei and the Society of Jesus. My assertion that to understand Opus Dei one must see it in relation to the Jesuits is the only statement in my initial formulation that might be considered somewhat surprising, and controversial as well, and in this case controversy might well be forthcoming from both sides. As Xavier Moll wrote, to assert that "without the Jesuits an Opus Dei would never have come to pass" (Moll, 1991, 2) might offend one side—which would be offended by the attribution of "such un-divine filiation"—as well as the other, who would find distasteful the attribution of "such irresponsible paternity."

All this notwithstanding, once we realize that this is one of the topics on which Opus Dei is most cautious, most silent, and most reserved, we shall also see that the Society of Jesus is manifestly the model that initially inspired Escrivá de Balaguer, and that the relations which were established between the two institutions—one nascent, the other firmly established—are the relations of confrontation and conflict. But to situate things once again in their historical context, the conflict appears to have been provoked by their *similarity,* and not by their differences. It was provoked, definitely, by competition.

Their respective evolutions led the Jesuits and Opus Dei today to occupy totally different positions in the Catholic Church. This is why the conflict between them, which has by no means disappeared, has come to be explained today, paradoxically, using reasons just the opposite of those used to explain it in the Spain of the 1940s.

The Underlying Dual Implicit Postulate

Beneath this triple thesis about Opus Dei's relationship with its founder, with the historical context, and with the Society of Jesus, there lies an implicit dual postulate. We must dwell here for a few moments on these assumptions, so that they will be clearly explicit.

In presenting the threefold reference to the founder, history, and the Jesuits as necessary "in order to achieve an understanding of Opus Dei," I am implying a "wish to understand" as a fundamental aim of the study. Although this is undoubtedly a legitimate option, it is not the only option possible.

Thus it would be perfectly conceivable to propose a sociological analysis of Opus Dei that did not give priority to this dimension of comprehension, or of comprehension "from within." It is quite possible that the directors of Opus Dei would prefer that someone who in fact is not "from within" would avoid such an effort at understanding. "There are many people," said Msgr. Escrivá, "who do not understand your path. Do not try to make them understand: you will waste your time and create opportunities for indiscretions" (*The Way,* no. 650).

In giving priority to the wish to understand we assume a specific conception of how to practice sociology, which is clearly tributary to the Weberian *verstehende Soziologie* and the *humanistic perspective* propounded by Peter Berger. Even so, and although in the case of the present study the search for an understanding of Opus Dei does not seem to me to have been a waste of time in any way, I must admit the possibility that "from within" it might be interpreted as a frankly "indiscreet" kind of sociology.

On the other hand, every choice carries with it certain consequences. To postulate a wish to "achieve an understanding of Opus Dei" as a fundamental goal of the work implicitly suggests that this understanding has to come before any considerations of its ethics, or of its economic ethics specifically. The consequences translate, then, into the fact that while at the outset the investigation was framed as a study of "the ethics of Opus Dei and the spirit of capitalism" (to express it as a simple, but graphic, formula), as the work went along the treatment of this topic kept being postponed. That is hardly to say it disappeared, but it did come to occupy less space than originally planned, while all the *prior* questions acquired considerably more weight, comparatively, becoming in the process—in the literal sense of the word—*basic* questions.

The Plan of the Study

Given this partial modification of the focus of the study, the first chapter will be an attempt to synthesize "the present state of the question" on matters relating to our knowledge of Opus Dei and to the difficulties placed in the way of any substantial increment in this knowledge. From a sociological point of view, this initial chapter tries to distinguish and present various possible models for approaching the organization created by Msgr. Escrivá de Balaguer and its historical evolution.

In accordance with the stated thesis, the next two chapters (Chapters 2 and 3) will be devoted to the figure of the founder, not with intent to invalidate the official biographies, but with the goal of correcting their linearity and counteracting their unilaterality, emphasizing the complexity of this personage as well as some of the features—which come from an attentive reading of his biographies and his published work—that have appeared to me as most indicative of his singular personality.

The next section, the longest of all, comprises a series of chapters (4–10) which, following chronological order, pursue that understanding of Opus Dei, its origins, its evolution, and its meaning that was mentioned earlier. The following periods will be analyzed in succession: the period considered as the founding period (1928–1936), with a brief extension into the three years of the Spanish civil war (1936–1939); the stage of the implantation of Opus Dei in Spain (1939–1946); and the phase of its first international expansion (1947–1958).

Not until the end of the fifties were all the elements present that permit us to speak of an economic ethic within Opus Dei. These chapters, consequently, are an attempt to identify those questions characterized in the last section as "prior" and which I consider *basic* to a comprehension of Opus Dei. In addition, so as not to interrupt the historical approach in 1958, instead of passing directly to the last part of the work I include a supplementary chapter devoted to the period after 1958 (Chapter 11), which is more descriptive in character and whose structure differs from the previous chapters.

In this last chapter of the first part, in fact, questions that have most fundamentally affected the recent life and evolution of Opus Dei have been grouped thematically: the progress of its international expansion; the diversification of its activities; the impact of the Second Vatican Council and the vicissitudes of its "Roman" insertion; and, finally, the long process of modification of its juridical status until it was established as a Personal Prelature of the Catholic Church in 1982.

The structure of the chapters covering the earlier years (1928–1958), which from our viewpoint are decisive in importance, offers a more detailed approach. Each starts with an overview of the basic events of the period analyzed, as they are presented in the texts of authors belonging to Opus Dei; in a second section those questions not totally resolved—or questions resolved for now in an unsatisfactory manner—are emphasized and then the chapter concludes with the formulation of several hypotheses pointing to possible later clarification. In each of the periods considered I have tried to situate Opus Dei in a triple context: the Spanish context of the epoch, which never ceases to be important, even when it ceases to be the only context in which Opus moves; the international context, which over the years acquires growing importance; and the ecclesiastical context, in which the conflict between Opus Dei and the Society of Jesus occupies a salient position, especially in the early years.

In regard to certain particularly decisive moments in its history I wished to go a little further than the simple formulation of hypotheses that closes each chapter. Thus three additional chapters (6, 8, and 10), based on several small monographic research studies, are inserted between the others in chronological order. These deal respectively with the following years: 1939, the year of the end of the Spanish civil war, the year of the beginning of the development of Opus, and the year of the publication of Escrivá's celebrated book *Camino* (*The Way*); 1946, the date of the founder's first trip to Rome, the establishment of his official residence in Rome, and the beginning of the process of the internationalization of Opus Dei and the preparations for its recognition as a universal institution of the Church; and 1958, the year in which Opus Dei began the process of withdrawal from the juridical status of secular institute it had been granted, and the year of the death of Pius XII and the election of John XXIII, which meant for Catholicism the end of an epoch and the beginning of a period of changes, innovations, and also crisis.

This set of events leads us, at last, to the second part of the book. Personally comforted after verifying that God, in his infinite mercy, would not permit this study to become purely an umpteenth gloss on *The Protestant Ethic and the Spirit of Capitalism,* I have tried in the final chapters, nevertheless, to render to God that which is God's and to Weber that which is Weber's.

The first chapter in this section (Chapter 12) attempts to locate the precise context in which specific parallels between the ascetic Protestantism analyzed by Weber and the asceticism of Msgr. Escrivá's followers can be established. This is followed by a chapter emphasizing the importance of the role played by the first schools specifically designed for the training of entrepreneurs, within the framework of the economic development that characterized Franco's Spain in the seventies (Chapter 13).

But it is not enough simply to point out the similarities between the ethic of the Puritans and that of the members of Opus Dei: the differences between them are no less significant. Even though in both cases we are speaking undoubtedly of an innerworldly asceticism, this takes on some unique characteristics in Opus Dei, especially with relation to the assumptions on which it is based (Chapter 14).

The analysis of these parallels and differences leads to a last puzzle, treated in the concluding chapter (Chapter 15): namely, the paradoxical combination of traditionalism and modernity that seems inherent in the movement founded by Msgr. Escrivá and which for many years has attracted the attention of all observers. The paradox is examined from the point of view of the existence of a dual ethical reasoning, which varies according to the sphere with which it is concerned. This dual reasoning, in fact, is not a phenomenon exclusive to Opus Dei, but is also quite widespread in the highest official circles of the Catholic Church today.

Research Methodology

It is nearly obligatory in introductory pages such as these to say something about methodology. I confess that I was tempted to dispense with the matter by saying that I had confined myself to following C. Wright Mills's slogan: "Never mind methodology, and let's get down to work!" (Mills, 152). However, I feel it necessary to make at least two points.

(*a*) The whole work was carried out using two sources of information: written documentation and interviews. (In general, tapes [audio and video] are not considered worthy of being elevated to the category of sources, but I used them also. Moreover, conversations and discussions with friends and colleagues, which were extremely useful to me, are not properly considered formal interviews.)

The text relies heavily on written documentation, and I have tried to pay my debts as fully as possible: quotations are numerous and are always accompanied by the appropriate bibliographical reference, presented in full in the bibliography at the end of the text. On the other hand, the content of the interviews and the useful information imparted to me during them are not explicitly referenced in the text.

Here it should be mentioned that the type of interview I have used has nothing to do with the classic opinion poll (not even in terms of qualitative techniques). The truth is it never once occurred to me to ask anyone's opinion of Opus Dei, perhaps precisely because, in contrast to what usually occurs with opinion polls, the only persons who were interviewed had of course already formed an opinion about Opus. In other words, no "results" were extracted from the interviews. They served, in the first place, to formulate working hypotheses, refine them, enrich them, and put them in context; and later on, in more advanced stages of the research, they served to contrast, confirm, or invalidate those hypotheses, and to correct and reformulate them. In my opinion, in usual sociological practice interviews are too often used as a pretext to avoid the process of what Pierre Bourdieu calls "the construction of the object." Here, on the other hand, they have been used in a systematic way—and almost exclusively—precisely for the purpose of constructing the object.

For this very reason any quantification of such interviews would make no sense at all. During the months from March 1990 until July 1991 I conducted several dozen interviews, which took several hundred hours. They were with persons who have had some relation with Opus Dei in Latin America (not in all the countries where Opus is present, but in most of them), in North America, in Australia, and, principally, in Europe (Great Britain, France, Belgium, the Scandinavian countries, Switzerland, Germany, Austria, and above all Spain and Italy). Among these persons were members of Opus Dei (both numerary members and nonnumerary members), others who used to belong to Opus, and others who never took part in it. They include priests, people from the world of education, economists, historians, doctors, jurists, theologians, and sociologists, as well as bishops, persons connected with the world of the Vatican Curia, Benedictines, Claretians and the like, directors of various secular institutes, and Jesuits.

In quite a few cases, regardless of whether their relation with Opus was direct or indirect, interviewees formally asked me to preserve their anonymity. I then made a decision never to mention any name that could not be accompanied by a reference to a written document. I regret this, basically because now my gratitude to all of them must also remain anonymous and generic. Even when we speak of thanks extended to all, if I had to distinguish some interviews as especially important I would mention—with respect to the last part of the book—those I conducted with various teachers and former students of the business schools in Barcelona run by Opus Dei and by the Jesuits, respectively. As for the first part of the book, I would mention the half dozen long interviews with persons who, independently of whether they belonged to Opus Dei, share the feature of having known from very close up, or even of having lived with, Msgr. Escrivá, in Spain and in Rome, especially during the forties and fifties.

(*b*) The basic source of the study as a whole, however, is without doubt the written documentation. To the greatest extent possible I have worked fundamentally from materials that emanate from Opus Dei itself. Of course I have also consulted and used texts written "from the outside" about Opus Dei. But the wish to understand "from the inside" led me to give maximum preference to the first source of information, which will translate here into the multitude of occasions on which I will allude to what I call the "official literature" of Opus Dei.

This is where we reach the second point I wish to make, in order to avoid any possible ambiguity. One might object that the expression "official literature" is inappropriate, insofar as the only official literature, in the strict sense, might be said to be those documents that issue from the headquarters of the organization. All right: if on any occasion the adjective "official" does not appear in quotation marks, this is an inadvertent error of omission. The quotation marks mean that the expression "official literature" designates those texts written by persons who belong to Opus Dei, or else are published by Opus Dei presses or in Opus Dei magazines. (Here we go again! Officially, according to its Statutes, Opus Dei has neither magazines nor publishing houses. But this does not prevent the existence of magazines linked to institutions of Opus and directed by members of Opus, and presses where the members of Opus publish nearly all their writings, and in whose catalogues the majority of the books are by Opus members.) One can locate other trustworthy indicators of "official literature": a systematically eulogistic tone when it comes to any question related to Opus Dei, and the absence of even a hint of criticism; constant reference to other texts by Opus members, who always cite one another and never mention work written about Opus by anyone who does not belong to it; and so on.

Some Final Observations

This, then, is the sense in which I am using the expression "official literature." The use of this type of material as the fundamental basis of the research carries with it certain consequences, and before concluding this introduction I would like to explain them in a couple of observations.

First, due to the mere circumstance of working from documents, the textual citations and bibliographical references are inevitably quite abundant in some sections, which imparts a certain unevenness to the rhythm of the discussion, slower and apparently more erudite where there are many citations, and faster where they are not necessary. Second, in order not to add even more to this possible feeling of discontinuity, I have deemed it preferable not to interrupt the discussion with footnotes. There could have been many, but I have relinquished them, incorporating essential points into the text. Bibliographical citations are listed at the end of the text.

Finally, indubitably the most important consequence of approaching Opus Dei basically from an examination of writings by its own members is that the image reflected will not be so much that of Opus as a whole as it will be of the minority who write books and articles. This minority, the creator and transmitter of the "ideology" of Opus Dei, tends to be more monolithic and less plural, more "orthodox," frequently more intransigent and less open-minded, than the majority of the regular members, who do not perform, inside the organization, any job that is representative or carries responsibility.

From the interviews one can verify the existence of some diversity in ways of thinking and acting in the bosom of Opus Dei, although not to the extent of being able to assert that, in Opus Dei, "there's a little of everything, as there is everywhere," since certain types of mentality could hardly be represented. It is, rather, only a relative diversity, but greater in any case than among the "writers," who, no matter how many times Msgr. Escrivá liked to repeat that they were "totally free," always end up saying approximately the same thing.

One example should suffice by way of illustration: in an Opus Dei girls' school (in accordance with the "official" formula one would have to say: a school that entrusts its spiritual leadership to Opus Dei clergy), a two-page leaflet was circulated with more than fifty questions that were to serve as a guide to preparing for confession. Some girls took it very seriously, some received it with anxiety, others saw it as routine, and some merely went through the motions, playing the game without believing in it, or even treating it as a joke. However, the only written evidence that exists and remains is the guide itself.

At bottom, the question we are dealing with concerns the existence of a series of possible adaptive mechanisms, in the context of the socializing forces of what Erving Goffman called "total institutions" (Goffman, 1961). The patterns of Opus Dei's functioning offer many parallels to the characteristic patterns of a "total institution." Throughout the study we will keep Goffman's analytical model before us, which in my view is more useful for an understanding of Opus Dei than is the notion of "sect"—frequently used especially in literature in English on Opus—which is not very pertinent in this case. But at the same time, because the study relies largely on written documentation, and in particular on the "official literature" of Opus Dei, it is clear that our research will reflect less of the relative plurality of positions and points of view in Opus than of the remarkable homogeneity of the ideology of its elite leaders.

Part I

Opus Dei: A Historical and Sociological Approach

1

Toward a Historical Sociology of Opus Dei: The State of the Question

The Historical Evolution of Opus Dei

Monsignor Escrivá "engendered Opus Dei just as we know it today," writes Malcolm Muggeridge in the prologue to a short illustrated biography of Opus Dei's founder (Helming, 4). With all due respect, I believe we must say this statement is false, or at least so exceedingly ambiguous that it leads to confusion.

The assertion is confusing first of all because it seems to assume that knowledge about Opus Dei is at present something simple and basic, within the reach of everyone. But that is not true: the knowledge we are given about Opus Dei continues to the present day to be very partial, very fragmentary, and very contradictory. There is a series of clichés, stereotypes, and typifications on which our knowledge of Opus Dei—like our knowledge of anything else (Berger and Luckmann, 52ff.)—is based; but in this case the stereotypes generated by the "official" literature of Opus Dei and the stereotypes found in the literature that adopts a "critical" position toward Opus Dei counteract and contradict one another. Thus it does not make much sense to speak of Opus Dei "just as we know it today," as Muggeridge does, without specifying at the same time the sources on which this knowledge is based.

Furthermore, this almost schizophrenic dichotomy is reinforced by the fact that in most cases the critical literature does not seem to ultilize and cite the "official" sources except to disprove and refute them, while the literature by authors belonging to Opus demonstrates a surprising unanimity—interpreted by its adversaries as adherence to a "party line"—in totally ignoring any study or any text that does not come from "official" circles. As might be expected, this rule has its exceptions, and some of these exceptions will be of particular interest here, both for their rarity and because I intend for the present work to be included in this group of exceptions to the rule. In fact, my intent is to take into consideration, as seriously as possible, the "official" literature, but without, in so doing, renouncing the right to take into account the most rigorous of the contributions of the literature adverse to Opus Dei.

There is a second reason why Muggeridge's statement is false, or at least excessively ambiguous and confusing: to maintain that Msgr. Escrivá "engendered Opus Dei just as we know it today" seems to indicate—leaving aside for now the question of what type of knowledge we have about it—that Opus Dei has neither changed nor evolved from the official date of its founding in 1928 to the present day. This is clearly—absolutely—not true. Opus Dei has grown and developed; it has internationalized; it has diversified its activities; on repeated occasions it has modified its juridical status in the Catholic Church, and has continued to modify its own Statutes (or Constitutions).

Over the years, it has gradually changed its language and its style. What Durkheim called "the modes of doing, thinking and feeling" of its members have evolved toward an increasing pluralism, as the "official" literature in fact tends to assert, and as its "critics" frequently try too hard to deny. And this is not the only way in which it has changed. In spite of the fact that the "official" literature still seems unwilling to admit it, even its objectives, its purpose, the conception it spreads of itself and its "mission" or "vocation," even, to a great extent, its very "spirituality," have all evolved.

At bottom, this is strictly a matter of common sense, and at first glance it would be as absurd to deny it as to be shocked by it. From the republican Madrid of the thirties and the fascist Madrid of the forties to the still Francoist but different Spain of the seventies—and many prominent members of Opus Dei have had a lot to do with this difference—and to the Spain of today, a member of the European Economic Community, the economic, social, political, cultural, and also ideological and religious transformations have been too profound not to have affected absolutely everything to the utmost degree, including Opus Dei. From the prewar world of the thirties to the world of Yalta and the cold war, through the alternation of periods of euphoria and expansion with those of recession and crisis, to the present, the changes have been remarkable. How could they not have had an effect on an institution made up not of anchorites in the desert but of individuals who expressly state that they want to live "in the midst of the world"? Finally, from the Catholic Church of Pius XII through John XXIII and the Second Vatican Council, to Paul VI and the troubled postcouncil years to the present, the ecclesiastical institution's relationship to the world has clearly been transformed and, consequently, Opus Dei has also had to change.

Nothing is being said here, then, except the elementary and the obvious; it is almost trivial and, as I have said, pure common sense. There does not seem to be any need to invoke the epistemological argument of the "principle of historical specificity" in order to be able to formulate a plain and simple statement such as : *Opus Dei is the fruit of a particular epoch and particular circum-*

stances, and over time it has evolved and adapted itself to different epochs and changing circumstances.

Why is it, then, in spite of the apparent banality of all this, that things are so much more complex, and that the question is not only problematical but constitutes one of the most difficult obstacles to surmount when one tries to approach an understanding of Opus Dei? We will try to answer this question, basing the reply on three considerations which can be schematically summarized as follows:

1. Because the "official" texts of Opus Dei frequently tend to deny any type of internal evolution, to the point of pure *reification*.
2. Because its adversaries often unwittingly fall into the same trap, citing the famous axiom, "Plus ça change, plus c'est la même chose."
3. Because finally, as a consequence of the foregoing, the truth is that at the moment we do not have a true history of Opus Dei written by one side, neither do we have a minimally trustworthy history written by the other. What we do have is a variety of approaches which are either highly fragmentary or else appear to be impressive exercises in *alternation.*

History and Reification

Reification, Berger and Luckmann write, in a paragraph that even from a terminological point of view seems exceedingly apt,

> is the apprehension of the products of human activity as if they were something other than human products—such as facts of nature, results of cosmic laws, or manifestations of divine will. Reification implies that man is capable of forgetting his own authorship of the human world and, further, that the dialectic between man, the producer, and his products is lost to consciousness. The reified world is, by definition, a dehumanized world. It is experienced by man as a strange facticity, an *opus alienum* over which he has no control, rather than as the *opus proprium* of his own productive activity. [Berger and Luckmann, 1966, 89]

Examples abound. Let us choose one at random: some declarations by Msgr. Escrivá in Venezuela, a few months before his death. "And I have to say that I did not found Opus Dei. Opus Dei was founded in spite of me. It was the will of God that has been verified and that is that" (quoted in Vázquez, 472). Vázquez de Prada says, in the opening sentence of his chapter on the founding of Opus: "On the morning of October 2, 1928, Don Josemaría 'saw' Opus Dei, just the way God wanted it, just the way it would be at the end of the centuries. On this date it was founded" (Vázquez, 113). If it were not for the fact that this type of statement is made systematically in the literature we are calling "official," commentaries like the following, which obviously come from the other side, would be discredited:

> On October 2, 1928, the religious obfuscation of Escrivá's mind had reached a temperature high enough to produce condensation. Later, at the end of the Spanish civil war, prevailing atmospheric conditions were so favorable for Escrivá's ideas that the Spain of 1970 was still suffering under the heavy clerical-authoritarian

> downpour that began on July 18, 1936. The inundation of Opus Dei members is one consequence of this downpour. [Ynfante, 12]

It would be difficult to find a more illustrative example of the radical counterpoising of mere stereotypes than this pair of statements from Vázquez de Prada and Ynfante.

Reification also has important consequences with respect to the strategy of the social actors. Insofar as it implies a loss of consciousness that the social world, however objectified it may be, is a human product, reification implies equally a lack of consciousness of the possibility of changing it (Berger and Luckmann, 1966, 89); and it thus becomes, in the hands of anyone who has the power to impose specific definitions of reality, a formidable instrument of defense and manipulation.

Thus when we say that it is obvious that Opus Dei has evolved and has been transformed in the course of its history, we are making a "commonsense" affirmation that is nevertheless denied by a good part of the "official" literature of Opus, in the name of a supposed "supernatural sense" which frequently leads to reification.

At the other extreme, change is also denied by the argument according to which *something* must be changed in order that nothing fundamental be changed, hence, "plus ça change, plus c'est la même chose," and, in the final accounting, it is always "the same dogs with different collars." The Spanish political and social situation has changed, but only to make possible the perpetuation of exploitation and inequality; the world situation has changed, but only to enable the preservation of disequilibrium and imperialism. In a similar fashion, Opus Dei would have changed its appearance, its facade, in order to ensure the preservation of its basic objective, which would be nothing other than "the will of power."

The writers who take a perspective similar to this as their point of departure are most likely to adopt a terminology that characterizes Opus Dei as a secret society, as white Masonry, or as a holy Mafia (see, for example, Le Vaillant or Ynfante). This view is based on, among other things, the "virtue of discretion," which Opus Dei has always advocated and practiced; at the same time these writers tend to forget that this "discretion" is not exclusive to Opus Dei in the religious world, but is rather a fairly generalized characteristic of many religious organizations and, in the last analysis, of every social institution (political, economic, etc.) that exercises a degree of power. One must not forget that, sociologically speaking, "there is no power that does not owe a part—and not necessarily a small part—of its efficacy to general ignorance about the mechanisms on which it is based" (Bourdieu, 28).

Whether you start from the axiom that Opus Dei has not evolved with the passage of time but was engendered "as we know it today" and "as it will be at the end of the centuries," because—in the words of Msgr. Escrivá—"the Work of God comes to fulfill the Will of God," or whether you start from the postulate that it cannot change because, as Marx would say, "the poverty of Opus Dei is the expression of real poverty," the conclusion is always the same: denial of the historical dynamics and transformation of Opus Dei. This is surely one of the reasons why we still do not have any true history of Opus Dei.

Alberto Moncada, an author who belonged to Opus and after leaving it completed three books about it (Moncada, 1974, 1977, 1987) and wrote part of a fourth (Moncada, 1982), made some very interesting contributions, despite some clearly grave historical errors. He reaches the conclusion that "the real history of Opus will never be told. Not by those who know little, because they simplify everything, in one way or another. And not by those who know the truth, because of different sorts of fear or cowardice" (Moncada, 1982, 150). Moncada seems to imply that those who never belonged to Opus Dei can know little about it, and that in order to know "the truth" it is necessary to have been part of the Work (and, possibly, to have then left it). He is in any case doubtless correct in affirming that there are cases of fear and cowardice. (And not only fear and cowardice: we can be sure that there are also cases of laziness and lack of interest, or of conviction of having more interesting things to do, and perhaps even cases of notable ambivalence of feeling about Opus.) He is also undoubtedly right in warning of the danger of simplification by those who do not know everything. At the same time, however, it is not a matter of claiming, ambitiously and perhaps too naively, "*the* real history" of Opus Dei: "*a* real and reliable history" will suffice, a history which, while not the only possible and true one, at least avoids the temptation of becoming a mere exercise in *alternation*.

History and Alternation

Peter Berger introduces the concept of *alternation* to refer to the "process of the reconstitution of the past" through which the past is interpreted and reinterpreted in terms of the present. In our consciousness, "the past is malleable and flexible, constantly changing as our recollection reinterprets and reexplains what has happened" (Berger, 1963, 76; see also Berger and Luckmann, 1966, 157–61). The problem of alternation indisputably constitutes a constant danger to every historian. Thus it is not difficult to accuse the "official" versions of the history of Opus Dei (basically, the various biographies of Msgr. Escrivá) of having succumbed to it. But the concept Berger proposes is especially useful to the extent that it permits us to become aware that the phenomenon of alternation is general, that it affects us all, and that it is not at all at variance with what we usually call "sincerity," because, among other things, we frequently describe as sincere the "state of mind of a man who habitually believes his own propaganda" (Berger, 1963, 58, quoting David Riesman).

From this perspective, then, a "true history" of Opus Dei would be one that integrated the different possible interpretations, with an awareness of the inevitable limitations of each. It would also be aware of the problem of alternation, but at the same time it would be willing to interpret the past without forgetting to situate it in its own context, rather than reinterpreting it in terms of the present. In sum, it would understand the object of study—in this case, Opus Dei—as a socially constructed reality, thus avoiding reification.

A historian would probably have to say that what we are proposing here is less a "history" than a "sociology" of Opus Dei. Logically, I must admit this possibility, as the fruit of my own "professional deformation"; even further, I would almost venture to say that in the final analysis there is no boundary between the two.

Many years ago, José Luis Aranguren said that "a sociology of Opus Dei, written *sine ira et studio,* would be one of the most interesting topics for a social scientist in contemporary Spain" (Aranguren, 1965, 762). In fact, after more than a quarter of a century the statement is still valid, and the topic is still waiting to be addressed. In spite of several valuable books, such as José V. Casanova's thesis *The Opus Dei Ethic and the Modernization of Spain* (Casanova, 1982), on the whole there continues to be too much "wrath" and too little "study" in the analyses of Opus Dei.

This "sociology of Opus Dei" must necessarily have an essential historical dimension. "All sociology worthy of the name is historical sociology" (Mills, 1959, 146). Therefore, even though I basically share Daniel Artigues's view that "from the outset, Opus Dei has continued to be modified according to the times," I do not share what follows in his argument: "and for this reason we have planned our task as history and not as sociology; only an historical analysis can convey the complex itinerary of the Work" (Artigues, 2).

Let us be fair to Artigues. The ferocious criticism from some quarters (for example, Ynfante, 364–69), which almost treats him as a fellow traveler of Opus Dei, might make us forget the difficulties of the research the author accomplished for a book that was published by Ediciones Ruedo Ibérico in Paris under a pseudonym. (Later the same author published in Madrid, under his own name, translations of the two articles he wrote on Opus Dei for a book of essays; Bécarud, 1977.) Even with its defects, *El Opus Dei en España* is a valuable contribution and on some points is still highly relevant today.

In spite of that—and perhaps precisely because Artigues wants to do "history and not sociology"—I disagree with him, not so much with the things he says, and even less with the way in which he says them (which is what annoys Ynfante so much), but with the initial question he poses, which conditions the kind of answer he logically will attempt to find. Artigues writes, on page 1 of *El Opus Dei en España (1928–1962)*: "Is it an institution of purely religious character which places the rational use of technology and modern methods of action at the service of new forms of the apostolate? Or is it an almost secret society which aspires, in the first place, to capture the elites while at the same time pursuing ends that are little known in every case, retrograde, and more political than religious?"

Surely the question seems to us too derivative of those stereotypes and those dichotomies of which we spoke at the beginning. Probably one of the most interesting and innovative aspects of Opus Dei is its combination of a traditional religious countenance with rational and modern methods. And there is no doubt that during the initial phase of its history, Opus aspired primarily to "capture the elites." What does "institution of purely religious character" mean? No religious institution is ever "purely religious": from the moment it becomes an institution and has to begin to address the question of transmitting its content, every institution becomes a playing field for various forces and interests (social, economic, and political). What sense is there then in finding out "whether it pursues ends that are more political than religious"? Frequently the manner of formulating the question prejudices, or at least conditions, the response. Depending on the perspective one adopts, one could also say of the Vatican that it seeks political ends, and that it is not a "purely religious" institution. And in exactly the same manner one can show that a political party is not an institution of "purely political character" or that a soccer club does not have a "purely sporting character."

Artigues's initial question, then, just like the "official" literature of Opus Dei which affirms that it is indeed "strictly religious," brings us too close to certain stereotypes which lead one to dichotomous, black or white responses. The disjunctive black or white might be an excellent device for a television game show, but not for scientific research questions, because only in fiction or in fantasy does reality appear entirely black or entirely white. If what we seek is knowledge and understanding, things always turn out, for better or for worse, less pure and more textured, less simple and much richer.

So, without reifications and trying to avoid alternations to the greatest extent possible—with the consequent risk of not satisfying anyone completely—I believe that a history of Opus Dei has to make an effort to situate the object of its study in relation to the evolving dynamics of its context.

It would try, in other words, to see Opus Dei as a "child of its times," as the Spanish theologian José María González Ruiz wrote in a 1964 article that was "critical" in the best sense of the word: that is, it was based on the conviction that self-criticism is the only attitude that permits one to be truly critical. González Ruiz stands against "that species of conspiracy by the most diverse Catholic sectors against the only common target represented by Opus Dei," which, he believes, "can degenerate into a great hypocritical gesture that tries to hide its own spots by calling too much attention to the only expiatory victim." Later he adds: "Opus Dei was born in the Catholic Church, within the framework of the Spanish Catholicism of a specific epoch. . . . Unfortunately, the errors so insistently attributed to Opus Dei are an old collective patrimony of a great majority of the institutions of the Catholic Church in general, and of Spanish Catholicism in particular" (González Ruiz, 1964).

Five years later, Josep Dalmau, in a book that is frequently cited but whose prologue is systematically ignored, makes a similar statement: "I am convinced that the phenomenon of Opus Dei owes less to its founders or its responsible directors than to objective conditions or ruling mental patterns in the country. [I refer to] certain mental patterns surely favored by the victorious results of one band of the belligerents in our civil war, but never created by that same victory" (Josep Dalmau, 1969, 6).

Three Approach Routes

In spite of the great abundance of published material from Opus Dei, about Opus Dei, for Opus Dei, and against Opus Dei, this is the type of history that we lack today. As I understand it, "the present state of the question" might be synthesized by reference to an analogy, a parable, and a paradox.

The Iceberg Analogy

Using the iceberg analogy, what we know today about Opus Dei would be only the tip of its history, the tiny part that is visible; all the rest would be submerged and invisible. All the forays into "the secret world of Opus Dei" (Walsh, 1989) and all the descriptions from inside by those who belonged to it and left it (Moreno, 1978; Steigleder, 1983; etc.), along with all the more or less sensational journalistic reports (Hebblethwaite, 1983; Magister, 1986; Shaw, 1982; etc.), would be attempts to bring the hidden part, the secret underside of Opus Dei, to the surface.

This interpretive model is also implicitly present in those studies which see Opus Dei as a secret society or a "holy Mafia," those that are primarily concerned with Opus Dei's supposed political connections, or with its "financial empire," initially in Spain and later in Latin America, central Europe, the Philippines, and so on (see, for example, Walsh's bibliography, 1989, the last part of Hertel, 1990, or the book by Roth and Ender, 1984, which incidentally provoked a lawsuit).

A reading of the numerous texts that explicitly or implicitly turn to the iceberg analogy, in all their diversity, leads us to three kinds of observations:

(*a*) What stands out above all is the anguished and afflicted pathos of the personal testimonies of those who tell about their experience in Opus, and the trauma it represented for them. There are many examples of such testimonies, always respectable and often respectful (Moreno, 1976; Tapia, 1983; Felzmann, 1983; as well as the testimonies that form the material for Moncada's "oral history," 1987), concerning which Opus Dei has been accused of silence, at best, if not contempt, defamation, or persecution. In the interviews conducted during my research I was able to corroborate some of these elements: persons who left Opus many years ago were still traumatized and unable to talk about it calmly; doctors and priests confirmed the prolonged pathological consequences provoked by the experience.

(*b*) At the other extreme, passing over the most strictly sensationalist texts and concentrating on the most serious ones, a certain feeling of hypocrisy, perhaps involuntary, prevails. In fact, although they begin by rejecting the stereotype according to which Opus Dei would be "of purely religious character," they tend to judge it as if it were nevertheless obliged to be just that. In spite of the numerous and repeated protests by spokesmen for the Work to the effect that each member possesses complete freedom of action in fields such as politics, the professions, and economics, this class of texts seems determined to judge all the behavior of Opus members as if it were governed by obedience to certain slogans and subject to a kind of party discipline.

Probably the real freedom of an Opus member is less than its spokesmen claim, but certainly it is equal to or greater than that of a soldier in the army, a participant in a professional conference, or a representative in parliament. Every member of any society, simply by virtue of being a member, has his freedom restricted and conditioned (Berger, 1963). And at higher levels of instrumental rationality, there are more threats to individual freedom (Mills, 1959). Orwell's *1984* is an alarm warning not against the consequences of

Opus Dei, but against the consequences of the many *opera* we have fabricated and which could annihilate us.

There is also a certain amount of hypocrisy in the denunciation of Opus Dei's economic ruses: not because they do not exist, but because such ruses also exist—we do not even have to leave the religious sphere to find examples—in the Greek Orthodox Church, among Jehovah's Witnesses, in the German Evangelical Church, and in the Society of Jesus. As a monk pronounced with Benedictine wisdom, in the bosom of Catholicism "Opus Dei has inherited the place which used to be occupied by the Jesuits. And this business of having power means always making enemies" (Bernabé Dalmau, 1975, 12).

(*c*) Fragmentary as these considerations are, it would be unfair not to add one last observation. The leaders of Opus Dei had and still bear a good deal of responsibility for the appearance of these stereotypes and the persistence of these accusations. Intentionally or not, they have often fomented them with a strategy that seems to follow a literal interpretation of Escrivá's counsel, "Always keep silent when you feel indignation boiling inside you. And do so even when you are justifiably angry. Because, in spite of your discretion, at times like that you always say more than you want to" (*The Way,* no. 656).

Since this discretion, which goes to the extreme of renouncing any effort at "making oneself understood" (*The Way,* nos. 647, 650), is presented as simple "naturalness" (ibid., no. 641), it is not strange that it gives rise to suspicion when it translates, for example, into the obligation "to conceal the number of members from outsiders," "to always maintain prudent silence about the names of other members, and not to reveal to anyone that you belong to Opus Dei" (*Constitutions,* 1950, nos. 190, 191; this chapter of the *Constitutions* was in force until 1963, according to Fuenmayor et al., 1989, 349). Precisely one of the unresolved enigmas of the history of Opus Dei is how to know to what degree this insistence on the need for discretion constitutes an original feature of Opus, to what degree it is a consequence of the "persecution" the first members were initially subjected to, and to what degree it forms part of a tradition in many Catholic movements and organizations of that epoch.

But this question of the "historical enigmas" obliges us to go beyond the level of the iceberg analogy, and automatically leads us to our parable.

The Parable of the Puzzle

Given the abundance of extant historical materials, given their heterogeneity, and given the absence of any kind of systematic ordering, it is appropriate here to summarize the present state of the question by turning to what we will call "the parable of the jigsaw puzzle."

Imagine for a moment a situation in which you meet an individual who is about to put together one of those huge puzzles with twenty-five hundred or three thousand pieces. Besides a number of pieces that have very well-defined contours or else display some unequivocal detail that lets them be fitted in almost automatically, there are many other pieces that can only slowly be fitted together through a patient labor of reconstruction.

Our hypothetical historian of Opus Dei would be like this individual who takes on the assembly of the jigsaw puzzle. In the historian's case, however, there are three supplementary details, each of which constitutes an additional difficulty.

(*a*) In the first place, he finds that the puzzle has already been started. He is not the first to address the task; others have preceded him, leaving certain parts of the puzzle already assembled, or half assembled. At first this facilitates the job and permits him to save a lot of time. But soon he realizes that there are other areas where he finds himself unable to make any progress, because the pieces he has in his hands don't fit anywhere. When he gets tired, when his eyes can no longer make out the pattern of colors, he is as if paralyzed, and he begins to despair and to doubt not only his own ability to finish the puzzle, but even whether those who preceded him did their part correctly. Then suddenly he remembers the chapter on *alternation* in Berger's book, and he begins to think that perhaps the reason the pieces don't fit is someone else's fault. Perhaps in the parts previously assembled there are separated pieces that should go together, and vice versa, or even whole sets of pieces incorrectly positioned. This being so, the time he originally gained is now lost, and doubly so, first because he has to begin to take apart what has already been put together, but above all because the previously assembled picture—erroneous or not, but in any case coherent and plausible—influences him and obliges him to progress even more slowly.

Anyone who works with the biographies of Msgr. Escrivá, or in general with the "official" literature of Opus Dei, sooner or later must confront the temptation to start dismantling parts of the puzzle, even without being sure he is doing anything more than an unwilling Penelope.

(*b*) Furthermore, the situation gets worse as soon as our historian realizes that his puzzle is missing some pieces. And this time it is not mere doubt or suspicion but an undeniable certainty. The only thing he doesn't know is if the missing pieces are gone for good or if they are hiding in some corner, or if someone knows where the hiding place is. (It would be better to speak of hiding places, in the plural, in the form of different archives: those of Opus Dei first, of course, but not exclusively.) In any case, he knows perfectly well that access to them is forbidden to him: he is practically sure that, in "the present state of the question," no one has access to everything at one time. Moncada's positivist dream vanishes automatically, therefore, relative to the possibility of explaining "the true history of Opus" (Moncada, 1982, 150). It is a dream that all the partisans of the iceberg analogy share, and that now becomes unattainable.

In some cases the "official" literature of Opus Dei had already warned us about the absence of some of the pieces. "Excuse me for not offering more details about the beginning of the Work, because they are intimately united with the history of my soul, and belong to my interior life" (Escrivá, *Conversations,* no. 17). And then there was "a powerful organization, about which I prefer not to speak" (ibid., no. 30), which organized a campaign of calumnies, "although I do not like to talk about these things" (ibid., no. 64). In fact, the pieces of the puzzle that refer to the conflict with the Jesuits of the Marian congregations in the forties are not there; also missing are the pieces relating to the "crisis" of 1951 and 1952 (see the allusion to it in Sastre, 416). And these are not all the missing pieces, as we shall see.

(*c*) If this were not enough, on several other occasions our by now reeling puzzle-solver has a definite feeling—not quite a certainty, but more than just a suspicion—that those who started his puzzle had put extra pieces in certain areas. Some parts of the puzzle seem overpopulated, and appear excessively complete. On closer inspection, it appears that some of the pieces that seemed correctly joined at first glance have in fact been forced together. In other words, the situation seems just the opposite of what he thought a minute ago: the puzzle now has too many pieces. Some pieces, while appearing to be part of the puzzle, probably do not belong to it. Once more our protagonist has to remember something he read about the phenomenon of alternation, something that he found this time in Berger and Luckmann's treatise on the sociology of knowledge, when they speak of the "specific reinterpretations of certain events of the past" which an individual would like to be able to forget entirely;

> but to forget completely is notoriously difficult. . . . Since it is relatively easier to invent things that never happened than to forget those that actually did, the individual may fabricate and insert events wherever they are needed to harmonize the remembered with the reinterpreted past. And since it is the new reality rather than the old that now appears dominatingly plausible to him, he may be perfectly "sincere" in such a procedure—subjectively he is not telling lies about the past, but bringing it in line with *the truth* that, necessarily, embraces both present and past. [Berger and Luckmann, 1966, 160]

One only has to keep reading this text, where both authors add, "incidentally," that "this point is very important if one wishes to understand adequately the motives behind the historically recurrent falsifications and forgeries of religious documents" (ibid.), in order to see that here we find the key to the solution of this impossible puzzle. But to do this we must leave the parable and proceed to the paradox.

The Epistemological Paradox: Sherlock Holmes, Father Brown, and Opus Dei

The discussion in this section will be largely parallel to the foregoing parable. It is not, however, a simple repetition of it. It incorporates some new elements that are significant from the point of view of the present state of the question with respect to the history of Opus Dei. In addition, it makes use of a different language; I hope, incidentally, that the reader who has never thought of doing a three thousand-piece jigsaw puzzle will find himself in more familiar territory if it so happens that he is, instead, a fan of detective novels.

In a celebrated article, "Spie. Radici di un paradigma indiziario," the Italian historian Carlo Ginzburg presents an epistemological model for the social sciences based on a so-called paradigm of clues (Ginzburg, 1979; see also Ginzburg, 1981). Frequently the social scientist, particularly the historian, has only a few clues about the events of the era he wishes to study. He is obliged therefore to take them as a point of departure, to use them as pathways, and to find a lead that will take him from these clues to the reconstruction of events.

Ginzburg bases his epistemological model on the work of three authors, all doctors—and this is no coincidence—but all engaged in other activities as well. The first is Giovanni Morelli, an Italian whose method for the analysis of clues in pictorial techniques leads him to some surprising findings in the world of the falsification of works of art. The second is Sigmund Freud—who cites Morelli in his writings about art—whose psychoanalytic method consists of detecting the phenomena hidden behind certain clues or symptoms. And the third is Arthur Conan Doyle, the creator of Sherlock Holmes. Here we will concentrate above all on Doyle (relying especially on Terricabras, 1986, 32–37, and Estradé, 1986, 50–54).

For Holmes, the reality to be investigated is opaque, not transparent. The facts available to him are few. And behind this shortage of facts there is frequently premeditation on the part of the hidden reality, on the part of an individual being pursued, who does not wish to leave tracks or who even erases them. The investigator's problem is one of excessive distance from his object.

It might be said that all writers who have tried to study Opus Dei from the outside have availed themselves, without exception, of the clue paradigm. Every one of them has played Sherlock Holmes. It is possible that during the early years this was the only conceivable method; in any case it is clear that the very shortage of "official" literature, the literary genre of books like *The Way,* the secret character of the Constitutions, and so on, invited this approach.

But after a certain point in time, perhaps after the publication of the first anti–Opus Dei studies, and definitely after the death of Msgr. Escrivá and the definitive juridical configuration of the Opus Dei Prelature, the situation was substantially altered. "Official" literature began to proliferate. Often there was no shortage of data. Now the researcher found himself, sometimes, not with a few half-obliterated tracks, but with too many tracks.

Under such conditions, to attempt to read the most recent work, for example, the various biographies of Msgr. Escrivá or the study by Fuenmayor et al. entitled *El itinerario jurídico del Opus Dei,* using Holmes's epistemological model, is equivalent to resigning oneself to defeat. One can only end up in a state of confusion, of Durkheimian anomie, of pure desperation; one begins from the assumption of the absence of tracks and the need to find clues, and ends up lost, precisely because there are too many tracks.

Curiously, we discussed this years ago without having even dreamed of the phenomenon of Opus Dei as an object of study (Cardús and Estruch, 1986). On that occasion we said that, unlike the historian, the sociologist often has to abandon the "clue paradigm," because he is likely to find "a Babel of clues, and all of them, furthermore, want to become clues *par excellence,* to become collective evidence, to be the real reality itself." We described a situation in which the clue ceases to be a "precious and only track" and becomes a "camouflaging ideological artifact" (ibid., 94).

What has changed is not so much the object that resists being discovered but rather the situation of the investigator: a researcher who has to substitute Father Brown's *epistemology of paradoxes* for Holmes's epistemology of clues.

In the end, there is nothing surprising about this: it is not strange that in order to understand the "schemer" Msgr. Escrivá de Balaguer, the Catholic and priestly "scheming" of Father Brown proves much more suitable than the cold—and Protestant—analytical rationalism of Sherlock Holmes. And it is Chesterton, the master of the paradox and the great observer of the unintended consequences of social action (Estradé, 1986, 55ff.), who perhaps hands us the key to a correct sociological approach to the Opus Dei phenomenon.

Compared to the Holmesian epistemology of clues, the paradoxical epistemology of Father Brown is characterized by the following features (Cardús and Estruch, 1986, 96ff., based on the last episode in *The Innocence of Father Brown*).

(*a*) Irony and the use of paradox are the basic tools for approaching events, with the goal of staying receptive to their multiple and frequently masked forms of expression. This explicit recognition of the relativity of facts is not only the expression of a specific "intellectual" stance on the part of the observer, but also a recognition that "reality" itself is ironic and paradoxical.

We have already seen how the "official" literature of Opus Dei resists admitting the relativity of the events it narrates, and tends rather to reify them. But the truth is that all the Sherlock Holmeses who, with positivistic rationality, attempt to "unmask" Opus Dei by penetrating its "secrets" (the iceberg analogy) in fact share with the "official" literature the very same absolute conception of "reality" and "truth," denying precisely their ironic and paradoxical character.

(*b*) Father Brown intervenes in situations he is close to, becoming one more actor in the action. At the same time, however, he observes everything in great detail, with calculated—that is, constructed—distance and coolness. "There is no access without proximity, but no resolution is possible without strict distance" (Cardús and Estruch, 1986, 96).

It goes without saying that in the "official" literature of Opus Dei there is never any calculated distance, but rather a very strong affective involvement, while on the other hand the Sherlock Holmeses are cool and distant to such a degree that their access to understanding the phenomenon of Opus Dei is impeded. Sympathy and criticism, we said earlier, are not a mere juxtaposition of two apparently contradictory notions, but the paradoxical or dialectical synthesis of two complementary and equally indispensable ingredients needed to achieve understanding of the reality of what one is attempting to study.

(*c*) With the accumulation of evidence, one's immediate understanding frequently shapes a confusing reality. Reality appears to make no sense, to be absurd. What Father Brown does is to throw doubt on the evidence, until what had been absurd becomes comprehensible, until we experience "a strange clarity of surprise, the clarity with which we see for the first time things we have always known" (Chesterton, 1911, 181).

While the "official" literature of Opus Dei is devoted to accumulating evidence that has never been thrown into doubt, the Sherlock Holmeses follow the thread of other clues which should lead them to new evidence and to the "discovery of the truth" hidden beneath the iceberg, instead of seeking the "recognition" of a reality that perhaps only requires, in order to be "discovered," the elimination of an excess of evidence that is a "cover-up."

(*d*) The attempt to marry pieces of evidence without casting them into doubt implies a perception of reality as a puzzle in which all that is necessary is to fit the pieces together. Holmes's desire is always to find the missing piece that allows him to complete the puzzle. By contrast, Father Brown's doubts emerge rather from an excess of pieces. When he has too many pieces (or tracks), Father Brown says: "This isn't leading anywhere. . . . It doesn't fit. It doesn't pay" (Chesterton, 1911, 188).

The history of Opus Dei is, according to our parable, a puzzle in which there are *simultaneously* missing pieces and too many pieces. The clue method of Sherlock Holmes continues to be useful, then, to the extent that there are reasonable indications that the missing pieces have not disappeared for good. But even so—and I believe this is the first time this has been proposed—the paradoxical method of Father Brown must play a part in the investigation.

♦ ♦ ♦

The four characteristics on which we have based our attempt to synthesize Father Brown's modus operandi should serve, in any case, as a model for the elaboration of a true history of Opus Dei: a history that would be "true" not because it would be "the true history" that Moncada asked for and to which every Sherlock Holmes aspires, but rather because it would present itself with its "true" ironic and paradoxical character.

What follows is not by any means that history in its *totality;* it is, rather, a quick overview of some of the principal milestones that mark the historical trajectory of Opus Dei after its "founding" in 1928. The specific procedure we will adopt is as follows: each historical period we dwell on—without pretending to be exhaustive—begins with a synthesis (sufficiently accurate, I hope) of the "official version" of the facts, followed by any significant elements taken from the "nonofficial" literature. Second, I will explain the reasons for concluding whether this part of the puzzle is missing pieces or whether it has too many pieces. And after this I will suggest, third, some hypotheses derived from the application of the "clue paradigm" or the "paradoxical model." Only on a few occasions will we delve deeper into the theme, illustrating in some specific cases the application of the proposed method, and going further than the simple formulation of hypotheses.

2

José María Escrivá

> But if they ask me what his name is,
> how shall I answer?
>
> Exodus 3:13

"The Father": A Fragile Personal Identity?

Anyone wishing to abide by Umberto Eco's conclusion to *The Name of the Rose,* "nomina nuda tenemus," would be more than disconcerted by a confrontation with Msgr. Escrivá. When we consider this personage—an extraordinary man according to his followers, charismatic to many observers, as complex as anyone, enigmatic in spite of the thousands of pages that have been written about his life and work, and surrounded by an aura of mystery, surely as a deliberate consequence of this avalanche of texts—one of the first things that attracts our attention is precisely the variability of his name itself.

If the *nomina* of Umberto Eco are reflections, manifestations, or expressions of the reality they designate, we would have to conclude that the personality of Msgr. Escrivá was characterized basically by inconstancy and versatility. If we were to judge a person's disposition by his frequent changes of name, we would certainly find ourselves with a case worthy of psychological analysis, a person with a set of symptoms that would have to be interpreted as indicators of instability or fragility in the construction of his personal identity, and precariousness in maintaining that identity.

However, none of this appears among the facts presented by Escrivá's "official" biographers (basically, Berglar, Bernal, Gondrand, Helming, Sastre, and Vázquez de Prada), who tend rather to emphasize the coherence and internal strength of his personality, attributing little importance to the question of the multiplicity of names he uses, in some cases neglecting to mention it at all, and in others interpreting it in ways that are clearly open to dispute. It is worth our while to address this topic, attempting to avoid rash or intemperate conclusions, but keeping in mind that this is one possible approach to the complex and enigmatic figure of the founder of Opus Dei.

Escrivá's Names

José María Escrivá Albás. These, according to all appearances, are his original given name and surnames. For readers unfamiliar with Spanish customs pertaining to names, Berglar explains that it is customary to "add the mother's surname after the father's surname" (Berglar, 26). There would be nothing more to say about it were it not for the fact that in a volume which can also be considered a biography, albeit not an "official" one but rather a biography that has been an annoyance to Opus Dei members because of its ironic, almost sardonic tone (*Vida y milagros de Monseñor Escrivá de Balaguer, Fundador del Opus Dei*), Lluís Carandell informs us that in the registry book of the Barbastro cathedral, together with the baptismal certificate of little José María, an annotation dated June 20, 1943, indicates that the surname Escri*b*á was changed to Escri*v*á de Balaguer (Carandell, 80).

None of the "official" biographies makes any reference to this first possible name change in which the spelling was altered. It is clear, moreover, that he used the spelling Escri*v*á long before 1943: in the memento of his first mass, which he celebrated in 1925, both his name and that of his father are written with *v* rather than *b*.

Thus it is even more surprising that the bishop of Madrid, Leopoldo Eijo Garay, on May 24, 1941, sent a letter to the (assistant) abbot of Montserrat, Aureli Escarré (a letter, as we shall see, that is systematically cited in Opus Dei literature as the first document written by a member of the hierarchy in defense of Opus Dei and its founder), in which he mentions the name of the founder—whom he says he knows very well—three times, and all three times he calls him "Dr. Escri*b*á."

Of course there could simply be an error, both in the case of the bishop of Madrid and in the baptismal certificate. But it could also be the first of a series of manifestations of the Padre's fondness for changing his name.

José María Escrivá ***de Balaguer.*** In the memento of his first mass, alluded to previously, celebrated "in suffrage of the soul of his father D. José Escrivá Corzán," the newly ordained priest appears as José María Escrivá y Albás. The names of father and son differ in the use of the conjunction "y" (and), which appears between the son's patronymic and matronymic surnames, and whose presence in Castilian Spanish is usually associated with aristocratic families.

Even before the Spanish civil war Msgr. Escrivá regularly used this conjunction, and right after the war he applied for a change officially, together with his brother and sister. Again it is Carandell, who on this specific matter has good documentation, who provides the proof, reproducing the text published in the *Boletín Oficial del Estado* on June 16, 1940. The siblings request "that they be authorized to change their first surname so it will be written Escrivá de Balaguer," and they justify the petition "because the name Escrivá is common in the east coast and Catalonia, leading to harmful and annoying confusion" (Carandell, 78).

It takes a little effort to avoid the temptation to treat this matter with irony. Here we have the founder of an association which was tirelessly proclaiming to the four winds that it was made up of "common Christians" arguing that the confusion provoked by having a "common name" was annoying.

Unlike the question of the possible change in spelling, some of the "official" biographies pause to comment on this modification. Curiously, however, none of them refers to the official petition of 1940 or to the justification for it. The only one who says anything about it is the French writer Gondrand: "He took this name to distinguish his family from the family Escrivá de Romaní, originally from the Valencian region and very well known in Spain, something that could have caused confusion" (Gondrand, 167). It hardly seems that Escrivá, by itself, could be easily confused with Escrivá de Romaní, nor is the name Escrivá de Romaní exactly "common in the east coast and Catalonia"; furthermore, Gondrand places the change in 1945, not 1940. The documentary appendix in the volume by Fuenmayor, Gómez-Iglesias, and Illanes (to which we will refer frequently because it is so informative, in addition to being "official" Opus Dei literature) indirectly confirms the date of 1940. In fact, the documents from the prewar years are all for José María Escrivá y Albás, while all those reproduced after the beginning of 1941 are for José María Escrivá de Balaguer y Albás (Fuenmayor et al., 509ff.).

Gondrand goes on to say that on a trip Escrivá made to Barcelona in 1941 he was "already using this name, which refers to the city from which his paternal family came" (Gondrand, 167). The anecdote of the 1941 trip is also cited in the other biographies.

"When Don Josemaría decided to visit the *Palau* (the name given to the apartment where the first members of Opus in Barcelona lived), Monsignor Cicognani, the papal nuncio in Madrid, advised him to get his airplane ticket in the name of Josemaría E. de Balaguer. It was a cautious measure, since he was then known popularly as Escrivá. He would also have to stay at the house of a priest who was a friend of his, because in order to stay in a hotel he would have to present his papers" (Vázquez, 227).

The context in which this occurred was the open conflict between Escrivá and the Jesuit Father Vergés, of the Marian congregations in Barcelona, a topic to which we will return in a moment.

> Events have reached such extreme gravity that they compromise the safety of the Founder on his trips to Barcelona. He runs the risk of being arrested on account of false accusations of a political-religious nature. He has to restrict himself to coming and going on the same day so as not to have to stay in a hotel. The nuncio of His Holiness, Monsignor Gaetano Cicognani, advised him to make his plane reservations in another name so as not to activate the police, since he is known better at this time as P. Escrivá. [Sastre, 263f.]

The coincidences are as interesting as the discrepancies. All the "official" biographies quote the civil governor of Barcelona, Correa Veglison, as saying that if he had known Escrivá was coming to Barcelona, he would have sent police to the airport to arrest him. Every one of them stresses his extreme poverty, which obliged him to travel in a dilapidated old automobile, or by train although he scarcely had enough money for the ticket, and it seems to escape them all that in 1941 Escrivá took a trip by plane. They also claim that he did not sleep in a hotel, although they disagree about whether he spent the night in Barcelona.

All of them, finally, agree that it was the nuncio Cicognani's idea for him to change his name for safety's sake, and not the idea of Escrivá himself, as published in the *Boletín Oficial del Estado* in 1940. Cardinal Cicognani's interest in giving this paternal counsel seems even more touching in light of the fact that Cicognani himself, on July 3, 1941, addressed these words to Cardinal Segura, archbishop of Seville:

> You are not unaware, surely, Your Eminence, of the existence and functioning of the institution called Opus Dei. Since diverse assessments and opinions have arisen about it, I would sincerely appreciate it if Your Eminence would do me the favor of giving me your opinion on this organization and at the same time sharing whatever information and data you think appropriate and necessary so that I might duly inform the Holy See at the opportune time. [Rocca, 1985, 134]

Cardinal Segura answered this letter on July 29, 1941, saying that he hardly had any information at all, that this same lack of information "indicates the secret, strictly secret character with which it operates," but that even so he knew that "Sr. Escrivá" had dealings with two Spanish bishops, and that he did not know if the organization was "political, social, or apostolic." He concluded: "I have very little confidence in these ways of proceeding, so alien to the tradition of the Church's Apostolate" (Rocca, 1985, 134f.).

Josemaría Escrivá de Balaguer. It was not only Msgr. Escrivá's surname that underwent modifications. His first name also evolved over time: thus from the "common" José María it came to be a very original Josemaría.

"Years later—around 1935—he joined his two first names—Josemaría—because his single love for the Virgin Mary and Saint Joseph were equally inseparable" (Sastre, 22; identical explanation in Vázquez de Prada, 34). Gondrand is somewhat more precise, informing us that at this time he joined the two names "only in his signature" (Gondrand, 106). After twenty years, in fact, on official documents the two names still appear separated, and only in the seventies did the use of "Josemaría" become systematic.

The name change has been incorporated into modern editions of *The Way*, in Spanish as well as in other languages, and is used in the "official" literature to refer to *any* period in the life of Msgr. Escrivá, to the point where the use of one or another spelling constitutes, at the present time, a good indication of the position a particular author adopts toward Opus Dei.

Mariano. During the Spanish civil war, Escrivá signed some documents and letters with the name Mariano, "because of his devotion to the Virgin and so as not to attract attention" (Helming, 47). Helming reproduces a document bearing this signature dated Daimiel, April 20, 1939, "The year of the Victory," addressed to Dolores Fisac, one of the first members of the female branch of the Work, sister of the architect Miguel Fisac, also a member of Opus during the forties.

But the use of this new name was not confined to the war years. In 1949 he sent a letter from Milan to "his children in Portugal," which closed with "A big embrace for all. The blessing of your Father. Mariano" (Berglar, 284). The custom was still in use in 1960 and 1969 judging by letters reproduced in other biographies (for example, Vázquez, 322f., 349). Apparently, then, this was the name he used to sign his correspondence with members of Opus.

There were other occasional modifications as well: the Latin form appeared in many official Vatican documents, alternating with the Italian Giuseppe; Josep María in the first Catalan edition of *The Way* (1955); an amusing Joe Mary on a postcard drafted in English and sent from Francoist Pamplona to Republican Madrid by way of France on December 29, 1937. Aside from this, what we have is an initial José María Escrivá (or Escribá) converted to Josemaría Escrivá de Balaguer.

Perhaps it is far-fetched to use this evidence, as Carandell does throughout his book, to support the thesis that Opus Dei is a pure reflection of the personality of the Padre. Nevertheless, the truth is that Opus Dei also frequently changed its name—its "robes," in official terminology—during its history. And there is nothing absurd about Carandell's supposition that behind all of it lay a desire on Escrivá's part to erase his origins. Thus, while on the one hand we are told of the desire to preserve the Padre's room in the Madrid house on Calle Diego de León exactly as it was when he lived there in the forties, on the other hand the house where he was born in Barbastro has disappeared. It was demolished in the seventies "and replaced by a new building which currently houses an Opus Dei center" (Berglar, 25): that is, the demolition was deliberate and tends to corroborate the hypothesis that he wanted to deny his origins.

In any case, it is certainly no less far-fetched to claim that Escrivá's life demonstrates "an inner coherence and an unalterable continuity," and that "there were no ruptures, sudden changes, or unexpected transformations" (Berglar, 327). One does not need to be a psychologist or a private detective or a novelist, as Berglar goes on to say, in order to find notable surprises in the life of Msgr. Escrivá, beginning with surprises about his name and continuing with the attributes and titles that accompany it, as we shall now see.

Escrivá's Titles

***Doctor** Escrivá.* If we look for suspense in the life of Father Escrivá (Berglar, 25), even his doctoral degree title from the university seems surrounded with mystery.

According to the biographical profile appearing in the most recently published editions of his works, Msgr. Escrivá de Balaguer was a doctor of laws from the University of Madrid and a doctor of theology from the Lateran University of Rome, as well as holding an honorary degree from the University of Zaragoza and the title of grand chancellor of the universities of Navarre and Piura (Peru), both Opus Dei universities.

Of the Roman doctorate in theology, obtained when he was fifty-three years old, nothing is known, not even the topic of the thesis, which was never published. In his bibliographical study of the works of Msgr. Escrivá, Lucas F. Mateo Seco (in Rodríguez et al., 469–572) does not mention it. Although he devotes four pages (495–98) to Escrivá's research on *La Abadesa de Las Huelgas,* he neglects to mention that this was the topic of his doctorate in law. But we know this from many other sources.

To sum up, the version offered by the "official" biographies is as follows. He received his law degree from the University of Zaragoza, and in 1927 he received permission from the bishop of the diocese to go to Madrid for two years to earn his doctorate. The topic of his thesis was "Priestly Ordination of Mestizos and Quadroons in the Sixteenth and Seventeenth Centuries" (Bernal, 118). Nine years later, when the war broke out, Escrivá was still in Madrid but had not yet finished his thesis. In 1938, now settled in Burgos, a strategic city "in which the government of the national zone resided" (Vázquez, 188), he had to begin a new thesis, because "the notes he had been making had been abandoned in Madrid" (Helming, 43). In Burgos, in fact, the monastery of Santa María de las Huelgas, where the ceremony of the "exaltation of Caudillo Franco as Head of State" took place, gave Escrivá "a new research topic on the canonical-theological problems presented by the jurisdiction of the abbesses of said monastery during the middle ages" (Gondrand, 136).

This time the thesis went much more quickly, in spite of its being a very scholarly topic, since it involved studying "all the documents of the specialized literature" and "quite a few publications in the German language" (Berglar, 398). Some writers have expressed surprise at this, since Padre Escrivá did not know German, and have attributed a substantial role in the preparation of the thesis to some disciple—specifically, Amadeo de Fuenmayor. Whatever the case, according to the official version, in Burgos the Padre spent many hours in the archives of the monastery (and there is no written evidence, furthermore, that Fuenmayor was in Burgos, or even that he knew Escrivá before their 1939 meeting in Valencia). When the war was over and he was back in Madrid he mentioned to his friend the Augustinian priest José López Ortiz, future bishop of Tuy-Vigo and future military vicar general, that he was still "working on his doctoral thesis on the Abbess of Las Huelgas" (Berglar, 59).

Before the end of the year, on December 18, he read the thesis to the Faculty of Law in Madrid, and passed with highest marks (Vázquez, 506). It is surprising how little attention the biographers pay to this important academic event. Some of them do not even allude to it, and those who do mention it pass over it quickly: in no case, for example, are we told the names of the members of the doctoral committee, or of the friends and disciples who surely must have accompanied the Padre and been present at the presentation. Equally surprising is the absence of any kind of concrete detail, in a group of books that are so thoroughly documented and that, in the great majority of cases, are characterized precisely by their attention to every detail.

All this makes the reaction of Antonio Pérez somewhat comprehensible.

One of Escrivá's principal collaborators, ordained in 1948 and a former Opus Dei general manager (*Anuario Pontificio,* 1956, 880), Pérez states that "Father Escrivá was not a great jurist, as we were later led to believe. I even have serious doubts about whether he studied law at all. I never saw his bachelor's degree, and the way things were in the Work, if he had it, he would have put it in an impressive gold frame. But he might have lost this document, like so many others, during the war" (Moncada, 1987, 19). Of course, his undergraduate diploma could have been lost during the war, but not his doctorate, because it came later.

Another fairly surprising piece of information is that even when Escrivá stated in 1939 that he was working on his thesis, which he submitted in December, a copy of Escrivá's curriculum vitae sent to Rome in 1943 by the assistant bishop of Madrid, Casimiro Morcillo, includes among his publications a "historical-canonical study of the ecclesiastical jurisdiction 'nullius diocesis' of the Most Illustrious Abbess of the Monastery of Santa María La Real de Las Huelgas," published in Burgos in 1938 (Fuenmayor et al., 523). When keeping in mind that Escrivá did not go to Burgos until January 1938, it is difficult to imagine how he could have completed the work by the end of the year. Furthermore, no study from this place and year is cited in any other publication.

In fact, according to all sources, the study on *La Abadesa de Las Huelgas* was published in 1944 (Madrid, Editorial Luz). The text is said to be "an expanded revised version of his doctoral thesis in law" (Sastre, 272). I have consulted this first edition, and again my attention is drawn to the fact that there is no mention of its being a doctoral thesis. The prologue, signed by José María Escrivá and dated in Burgos, March 31, 1944, says only that "here I present you, gentle reader, with the famous Abbess of las Huelgas" (Escrivá, 1944, 5). He does add a very significant sentence—having nothing to do with the question that concerns us here—when he writes: "You will see her govern as would a queen. . . . You will see her rule as *Mother* and *Prelate.*" These two attributes, out of the mouth of the *Father* and founder of what would become, in the eighties, the *Personal Prelature* of Opus Dei, seem like an authentic premonition. . . . But of the doctoral thesis, not a word.

Nor is there any explicit reference to its origin as a thesis in the review published in *Arbor* (November–December 1944, p. 395) by M. Giménez Fernández. This magazine, published by the Consejo Superior de Investigaciones Científicas, was founded and directed by a group that included many members of Opus Dei. The author of the review comments that "Don José María Escrivá has confirmed the opinion of extraordinary worth he well deserves in his other activities," and "everywhere, through the calmly objective exposition, one senses the beating of the supernatural fervor that overflows from the truly priestly soul of the author."

Independent of what might have happened in December 1939, then, it seems indisputable that five years later a choice was made to stress the "supernatural fervor" of the priest over the "scientific virtues" of the doctor of law. And in all this there is a little bit of "suspense" (Berglar, 25), Berglar notwithstanding.

***Monsignor** Escrivá de Balaguer.* On April 22, 1947, the Padre was named "Prelato Domestico di Sua Santità," a title that conferred him the right to be addressed as monsignor. Gondrand assures us that before accepting he hesitated, "since he wanted nothing for himself. If, in the end, he accepted, it was so as not to anger those who had nominated him" (Gondrand, 181). Thanks to Vázquez de Prada we know that his name had been proposed by Álvaro del Portillo, in the name of the Opus Council, "without his knowing about it" (Vázquez, 249). The note of humility is sounded when he states that "he rarely wore the showy prelate's robes, or wore the buckled shoes. He felt the weight of the purple vestments as a hair shirt; but on notable occasions, knowing *how much the color entertained his children,* he followed the path of good humor" (ibid., 249; emphasis mine).

It is interesting to observe to what extent these opinions contrast with, not to say radically contradict, the testimony of those who are not, or have ceased to be, members of Opus Dei (Hertel, Moreno, Steigleder, Walsh, etc.), and with the statements of the persons interviewed during this research project who personally knew Msgr. Escrivá. They all agree about his fondness, not only for luxury, but for what he fancied as aristocratic refinements, as well as honors, titles, and the symbols of prestige.

Anyone caring to do research focusing on this specific period might work on the hypothesis that he did in fact accept his appointment as *domestic prelate* in 1947 with little enthusiasm, but that this was because it came as compensation for his having been passed over for a higher position. What Opus Dei's general procurator, Álvaro del Portillo, had tried to secure for him was the title of monsignor, not by way of appointment as domestic prelate of the Holy Father, but by way of the *episcopate.* The goal was for Padre Escrivá to become bishop (or archbishop). And the hypothesis states that this was on the point of happening, but at the last minute the project was blocked (through intervention by members of the Society of Jesus), and that the appointment of April 22, 1947, was a kind of consolation prize.

Monsignor Escrivá died in 1975 having seen several Opus priests named bishop in Latin American dioceses, but the Founder was never awarded this rank.

*"The **Father.**"* But presbyter José María Escrivá, who became Monsignor Josemaría Escrivá de Balaguer, or simply Mariano in personal letters to his friends, is above all known as and habitually referred to as "el Padre," "the Father." "More than sixty thousand persons called him Father" was the title of an article printed in a Milan newspaper on the occasion of his death (*Il Giorno,* July 26, 1975).

Even in the early years the custom of using this form of address was widespread among the members of Opus Dei. He, in turn, always addressed the people of the Work as "sons and daughters." He even requested this epitaph for his tomb: "Pray for this sinner, who engendered sons and daughters" (Vázquez, 309). The idea is already present in *The Way* (no. 28), where he states: "We shall leave children, many children, and a lasting trail of light, if we sacrifice the selfishness of the flesh." But for once they paid no attention to him, and on his tomb is engraved the truly lapidary epithet: "El Padre."

The examples and references relating to the custom of calling the founder of Opus Dei "the father" are endless. We will limit ourselves, consequently, to two or three observations.

First we should state that the custom was converted into law. In the Constitutions in force in Opus Dei before its conversion into a Prelature, references to the figure of the Father are very frequent, and it is expressly stipulated that the president general is called Father by the members ("a sociis vocatur Pater"; *Constitutions,* 1950, no. 327) in article 2 (*De Patre*) of Part III, Chapter 1.

During Msgr. Escrivá's life, the polarization of Opus around the figure of the Father helped make possible the image of the organization as a big family. The family analogy pleased the Father, and he mentioned it frequently. Thus at the beginning, Escrivá's mother was called "grandmother," while his sister was "Aunt Carmen." "We are very glad," Escrivá is said to have affirmed, "that my mother and my sister were ready to take charge of the administration of our first centers. Especially Carmen, who was the one who got the most involved in everything. If not, we would not have had a real home; it would have been a kind of barracks." Helming adds that "the members of the Work began to call her 'Aunt Carmen' spontaneously, not only because she was the Founder's sister, but because they were so fond of her. She knew the names of all the first members of the Work, she knew what their favorite dishes were and what they liked for dessert. She gave them candy and if a button was about to fall off, she sewed it back on" (Helming, 58). The text may be considered a good illustration of the difference, for Escrivá, between a barracks and a home, as well as of the role assigned to women, a topic we shall have occasion to touch on elsewhere.

Even later, when the group of students and young college graduates who followed the Father had become an international organization of considerable size, an effort was made to preserve this image of Opus Dei as a big family. "Father of a large, poor family," is the title of the last chapter in Bernal's biography. "Family reunions" are what Gondrand (p. 284) calls the large assemblies that gathered to hear the Father speak during his long journeys around the world in the seventies ("a supernatural marathon," according to Sastre, Chapter 9). And at the moment of his death Álvaro del Portillo said that "it does not matter if we shed tears, because we are a family and we love each other" (in Vázquez, 486).

It is true that sometimes this designation of Father might lead to some ambiguity. Thus when the "Father" writes to his "children," "what trust, what rest and what optimism it will give you, in the midst of trouble, to feel you are children of a Father, who knows all and can do all" (quoted in Gondrand, 67), one must cast a sharp eye to see which "Father" is being referred to in each case. And this ambiguity may have become even more accentuated following Escrivá's death because, as Álvaro del Portillo said at the first mass *de corpore insepulto,* celebrated that same day (June 26, 1975), "as well as having God the Father, who is in Heaven, we have our Father in Heaven, who looks down on all his daughters and all his sons" (quoted in Vázquez, 486). And from the thirties on Don Josemaría counseled his children to recite the prayer Our Father often, "meditating above all on the first two words, Our Father . . ." (Gondrand, 82).

Marqués de Peralta. Of all the honors, titles, and decorations received by Escrivá during his lifetime, this one is perhaps the most surprising, and he requested it himself in 1968. When the matter became public knowledge, it provoked quite a stir and started a rash of rumors about the possible hidden motives that might lie behind such a decision (see, for example, Carandell, 65, and Ynfante, 31f.). The news was grist for many ironic commentaries, it caused some consternation in certain ecclesiastical circles, and it was not easy for many members of Opus itself to digest. The matter is not even mentioned in the official biographical profile that appears at the beginning of recent editions of his books. Some biographers tiptoe around it, and others present it in such a forced manner that it seems frankly to be an embarrassment.

Thus Bernal refers to the topic in a subchapter in which Escrivá is paradoxically described as "the definition of poor" (Bernal, 327–43) and presents it as a "heroic decision" made with full awareness "of the criticism his petition would provoke," a decision based on the notion that "the casting off of human possessions or symbols of honor can never be an excuse for shirking one's own duty" (ibid., 342). Posing the question in terms of "duty" might be interpreted as a transposition of the beginning of one of the maxims in *The Way* (no. 332), whereby: "There is no excuse for those who could be scholars and are not" becomes "There is no excuse for those who could be noble and are not." Of course this is not the interpretation of the official biographies, which is that Escrivá acted as he did in order to compensate his family for the many sacrifices they had made for the Work (Gondrand, 250ff.; Vázquez, 348ff.).

Since the only member of his family still living was his brother Santiago, the title would be for him—after one year had passed, according to Gondrand; after four years, according to Bernal. Neither of them mentions that along with Escrivá's petition for the title of marquis, another title, that of baron, had simultaneously been requested for his brother. Neither of them feels it necessary to reproduce the decree (published in the *Boletín Oficial del Estado,* August 3, 1969) acceding to the petition to revive the title of marqués de Peralta. Vázquez de Prada, on the other hand, does reproduce the 1972 document showing that the title passed to Don Santiago Escrivá de Balaguer y Albás (Vázquez, 350). The same author, in the biographical index at the end of his volume, lists all the titles, appointments, decorations, and honors received by Msgr. Escrivá (ibid., 550); the title of marquis does not appear anywhere in it.

All the biographers do concur in describing in detail earlier consultations with "some of the Cardinals who enjoyed reputations for great discretion, and with the Secretary of State of the Holy Father" (Bernal, 342). The episode oddly recalls another episode in 1941, mentioned earlier, in which the same authors attribute the initiative for the Father's trip to Barcelona under the name E. de Balaguer to the nuncio Cicognani, as if both manifestations of Escrivá's aristocratic pretensions ought to be legitimated by the blessing of the ecclesiastical authorities. In fact, if legitimation were to be sought in his own literary production, it could be easily found in this passage: "Honors, distinctions, titles . . ., things of air, puffs of pride, lies, nothing" (*The Way,* no. 677).

Saint Josemaría?

To crown this set of "mutations" by Msgr. Escrivá's names, his "children" would like to be able to present him posthumously with the greatest honor of all: canonization.

Immediately after his death, thousands of letters began to arrive at the Holy See asking for the process to be opened. Letters from "heads of state and governments, from ministers, senators and deputies, from families and individuals of every class and from the world over; also from 69 cardinals and 1,300 bishops, more than a third of the world episcopate, which is a unique event in the history of the Catholic Church" (Le Tourneau, 19). The reputation for saintliness of the founder of Opus Dei spread progressively, "with meaningful spontaneity," according to the decree of introduction of the cause of beatification and canonization (the whole text is reproduced in Seco, 196–205).

Monsignor Escrivá's successor, Álvaro del Portillo, quickly named a Postulator of the Cause. The Reverend Flavio Capucci has officially held this position since 1978. In 1980 the introduction of the Cause was solicited. In 1981 the process was begun. Monsignor Escrivá became "the Servant of God Josemaría Escrivá de Balaguer."

In different countries the respective "vicepostulations of Opus Dei" periodically publish information bulletins "with ecclesiastical censure of the Congregation for the Causes of the Saints," which keep track of the progress of the process, reproduce fragments of those letters from bishops from all over the world, and publish new letters from persons thankful for the "favors" obtained through the "intercession" of Msgr. Escrivá, to whom they direct their prayers.

With the same "meaningful spontaneity," the vicar general of Opus Dei, Javier Echevarría, joined the Congregation for the Causes of the Saints as a consultant (*Anuario Pontificio,* 1982), followed by the prelate himself, Msgr. Álvaro del Portillo (*Anuario Pontificio,* 1983). They were not to be the only ones. Later other Opus Dei priests joined the same congregation: Joaquín Alonso, also as consultant, and José Luis Gutiérrez Gómez, as reporter, for example.

In 1990 a papal decree was published "on the heroic exercise of the virtues of the Servant of God Josemaría Escrivá de Balaguer" (*Hoja Informativa,* of the Vicepostulation of Opus Dei, no. 12). Monsignor Escrivá became "the Venerable Josemaría Escrivá de Balaguer."

But the process does not necessarily end here. "Venerable" can become "Beatified,"* and "Beatified" can become "Saint." In general these processes are exceedingly slow, usually taking many years. Until now, the Cause of the "Venerable Josemaría Escrivá" has progressed at a speed which many ob-

* And did indeed become so in May 1992.

servers in Rome regard as unprecedented. In 1991, also in Rome, it was being said that the rush was over and that the process would move forward at the usual speed (that is, the usual slowness).

Only time will tell whether Opus Dei will be able one day to celebrate, along with that ambiguous "our Father in Heaven," a no less ambiguous "Holy Father," or whether the words pronounced many years before by Escrivá (quoted in Vázquez, 54) will be converted into prophecy: "Don't have any illusions! I'm not of saintly timber."

3

The Founder of Opus Dei and His God

The Biographies of Monsignor Escrivá

In a book published originally in 1948, *El valor divino de lo humano,* which during the fifties and sixties was almost as widely read in Spain as *The Way,* Jesús Urteaga, an Opus Dei priest, devoted his first chapter to the saints and the "lives of the saints."

It is shameful, says the author, to see what many Catholics' conception of a saint is—a kind of fetish they turn to "in order to ask favors" (Urteaga, 1948, 30). Forty years later, when we read the kind of documentation sent to the Archive of the Postulation of the Cause of Msgr. Escrivá, we have the inevitable sensation that Opus Dei runs a risk of seeing its founder converted into one of these "fetishes."

The biographies of Msgr. Escrivá, quoted repeatedly in the previous chapter, become comparable to those "lives of the saints" criticized by Urteaga in 1948: "It is so difficult to find human weaknesses in the subjects of these biographies! How well they are concealed! The authors are really afraid to tell us that they were men. How inspiring it would be for us to contemplate the natural shortcomings of the saints and to see what they did to overcome them!" (Urteaga, 1948, 32).

We must say that the Benedictine Paulus Gordan is right when he writes in his review of Peter Berglar's book that it is nothing more than the fabrication of an "apologetic hagiography and a golden legend," and that a figure as undeniably significant as Escrivá "deserves something more: he deserves an authentic biography" (quoted in Rocca, 1985, 128). The same could be said of all the biographies of the Founder published to the present day.

In other words, no one has written a genuine biography of Msgr. Escrivá. Writers from outside Opus Dei who have dared to address the topic, regardless of their greater or lesser accuracy and their greater or lesser polemical intent, have encountered the same difficulties of access to the material that have hindered those who have tried to write a history of Opus. The most logical thing would be, consequently, to wait for a "true biography" to be written by a historian of Opus Dei who had access to the same archival documentation available to all the "official" biographers, from Bernal (first edition, 1976) to Sastre (first edition, 1989), but who would not write, as they did, one of those "lives of the saints" so harshly criticized by Urteaga.

There is no doubt that this true biography cannot be written by someone who is not both a historian and a member of Opus Dei. No one should expect to find it, then, in these pages. However, insofar as the figure of Msgr. Escrivá is inseparable from the history of the first fifty years of Opus Dei's existence, and insofar as his life is indispensable to an understanding of the meaning of Opus Dei as a movement and as an organization, we will try to avail ourselves of an attentive comparative reading of the "official" biographies in order to emphasize some of their lacunae and to point out some complementary features, in the manner of musical counterpoint, to use the felicitous phrase from a suggestive essay by Josep Dalmau (*Contrapunts al Camí de l'Opus Dei*, 1969).

In exploring the issue of the different names used successively by the Padre, the previous chapter highlights one theme that requires a more contextual analysis than is offered in the literature written by members of Opus Dei: the topic of the complexity and, to some extent, the enigmatic character of the figure of Father Escrivá, in contrast to the stereotype of his transparency and seamless consistency.

Underlying this question is a point which the official biographers seem to avoid deliberately, a point which, in any real biographical study, would have to be taken as fundamental: the topic of the evolution of Escrivá's thought, of the intellectual influences that shaped it and the social context in which his thinking emerged and developed.

In lieu of this, the "official" biographers choose to present Msgr. Escrivá's thinking as something static, fixed, and immutable. It is perfect, and therefore untouchable and unsurpassable. Let us look at a couple of examples. Urteaga's book, mentioned above, opens with the author's dedication to Msgr. Escrivá, of course:

> I wrote this book, above all, with the pure and simple hope of reproducing as faithfully as possible your doctrine and your words, and it is natural for it to be written in your words, many times literally; now, even I cannot tell which words I put in to connect or paraphrase your own, and which ones I repeated. Fortunately, there is a way to tell them apart. Perhaps there are some that you do not recognize as your own, or that perhaps you reject: those are the ones I wrote. [Urteaga, 1948, 13]

We find the same thing, stated in a slightly different way, in the prologue to a book by Salvador Canals, like Urteaga an Opus priest: "When you read *Ascética meditada* and your eye stops on a burning, stirring phrase, have no doubt: thanks is due to Msgr. Escrivá de Balaguer, because he is the principal author of these thoughts now put on paper" (Canals, 1962, 12).

This attitude of the disciple with nothing to teach toward the master who has nothing to learn is the first step toward reification. He aspires to nothing except to reproduce, in the best case, what is already perfect; any modification would necessarily make it worse. Yet a third priest of Opus Dei, José María Casciaro, expresses it this way:

> The clarity, expressive force and simplicity with which the first High Chancellor of the University of Navarre explains the most profound truths of Christian existence are impressive. So much so that in trying to comment on the texts, one has the impression that any interpretation one might make either adds nothing that is not already clearly expressed, or even that one runs the risk of damaging, with clumsy hands, something that is a work of art produced by an exalted intelligence, a totally extraordinary pastoral religious experience. [Casciaro, 1982, 125]

This reification has become total by the time Escrivá's successor states that his biography "can only be explained and understood in the context of a divine design which, in traversing his whole life, shapes him as an instrument of God, chosen precisely to remind Humanity what God had engraved unequivocally on his own soul" (Portillo et al., 1976, 19). In this sentence, one has only to replace "remind" with "reveal to" to turn the corner not simply into the beatification, but into the divinization of the Founder.

It seems obvious that starting from these coordinates any biography would have to become a hagiography. Or we might say what is almost the same thing: any true biography of Msgr. Escrivá would have to abandon these coordinates, which are an attempt to remove any human mediation and to place the accent exclusively on the direct and immediate relation between God and the individual who has been "chosen as his instrument." Thus the "official biographies" of the "Father" are not "true biographies," and the stance adopted by the representatives of Opus Dei is—as we have just seen—so extreme that any attempt to focus on the issue from a different perspective is automatically denied as synonymous with misunderstanding, of "lack of supernatural sense," as they say, and definitely as an evil intent not very far from blasphemy. Instead of a model of complementarity and "counterpoint," the leaders of Opus Dei tend to prefer a black and white model without nuances.

This is why the "official" biographies specify neither the intellectual influences that shaped Escrivá's thought nor the social context in which it emerged. This will be illustrated a little later, when we discuss the founding of Opus Dei and the form in which it is officially presented. But it is also patent in many other events of the "Father's" life. Under the pretext of needing a "supernatural" key to interpretation, the strategy usually consists of linking his thought and his actions to an *anecdote,* and not to the concrete circumstances of his life.

For example, what are the reasons that led the young José María, son of a relatively modest family which underwent serious economic reverses that obliged his father to change jobs and move to Logroño with the whole family, to enter the seminary? Instead of an analysis of the objective access offered by the seminary to an education superior to what was generally available to bright, ambitious boys in the semirural and unmodernized Spain of the twenties, and an analysis of the repercussions of this on priestly vocations, the "official" biographies prefer to use anecdotes, elevated of course to symbolic levels. Thus was born the legend of "the footprints in the snow," described down to the last detail in all the biographies, and even used as the title of one of them (Helming).

The plot is always the same. A novelistic description is always the point of departure: "The countryside was white. The snow completely blanketed all the streets. An intense cold had frozen the lakes and icicles had formed on the trees and the eaves of the houses" (Vázquez, 72). It is Logroño, winter of 1917. Escrivá is about to celebrate his sixteenth birthday. "During these days of intense cold, with the ground all covered with snow, Josemaría Escrivá de Balaguer, still an adolescent [remember that the adolescent was called neither Josemaría nor de Balaguer], was walking the streets of Logroño" (Illanes, 1982, 71). Now we arrive at the symbolic event: "Suddenly something powerfully drew his attention: the tracks left in the snow by the bare feet of a Carmelite brother. He felt something like a loud knocking in his soul: is there someone who confronts the cold for the love of God, and I, what am I doing for Christ? That thought pierced his soul and it urged him on. It was not something fleeting, but a profound disturbance that demanded a definite response" (ibid., 71). It was the birth of Escrivá's priestly vocation.

In order to convey the message that this was literally a vocation, that is to say a call from God, the emphasis is placed not on those factors that might make it perfectly comprehensible but, on the contrary, on those aspects that make it seem unforeseen, providential, and, finally, supernatural. "Would he have to become a priest? Only a few months earlier, just the idea of it would have made him smile" (Gondrand, 31). In fact, he had never thought about the possibility of entering the priesthood: "furthermore, the idea of becoming a priest someday bothered me so much that I felt I was anticlerical" (in Bernal, 61). When he had to study Latin in school, he thought, "Latin is for priests and monks" (Gondrand, 22); "for monks and for priests" (Vázquez, 56). In anecdote after anecdote, we meet a boy who does not dream of the priesthood or want anything to do with Latin, but who will be a priest and who will pray "in labored but pious Latin" (Escrivá, 1975, 26), one who would found an association made up of "ordinary Christians," an organization destined to seek "the sanctification of those of simple faith who live in the midst of the world, just like everyone else," but which would have its constitutions drafted in Latin, with an express prohibition against translating them into "vulgar languages" (*Constitutions,* 1950, art. 193).

Escrivá's initial resistance (Gondrand's Castilian translator goes so far as to use the word "repugnance") serves not only to emphasize the heroic character of his final submission to the calling, but also to introduce in passing a certain degree of *negotiation* of the conditions under which he would become a priest. The footprints in the snow signaled the beginning of a long period of "guessing" (Vázquez, Chapter 2). What did the boy "guess"? *"That God, one fine day (who knows when?) was going to demand something of him (who knows what?)"* (Berglar, 37; emphasis mine). But when the Carmelite brother who had left his footprints in the snow, under whose spiritual tutelage Escrivá placed himself (Vázquez, 73), proposed after a few months that he enter the Carmelite order, "although he could not see clearly what it was God wanted of him, he saw that this was not the divine will" (Illanes, 1982, 72). On the other hand, he did not see himself, either, as "following a career in church affairs"; he was not enthused by the idea of becoming a parish priest, a canon, a member of the diocesan curia, or the director of the seminary (Gondrand, 33). "The prospect of an ecclesiastical career did not attract him. His vocation was something else, although he still saw it only vaguely" (Sastre, 50). He himself says that "I did not want to be a priest for the sake of being a priest, to be the *cura,* as they say in Spain. I had veneration for the priesthood, but I did not want that kind of priesthood for myself" (Vázquez, 76). As Berglar says, "if he did not know what God wanted of him, he clearly knew what God did *not* want" (Berglar, 56). The explanation advanced by Msgr. Escrivá himself many years later, when reminiscing about this period of his life, was that *"God came into my life without asking my permission,"* and he repeated this on many occasions. So we have a situation where anyone wanting to know more about his priestly vocation would have to ask God directly. This in fact is the argument he used in an interview during the sixties when a reporter asked him for the reasons behind the success of Opus Dei. Monsignor Escrivá replied: "Can you explain it? I can't. Humanly, it has no explanation. It is a work of God, and only He could satisfy your curiosity" (quoted in Berglar, 346).

After a time, José María Escrivá entered the seminary in Zaragoza and studied law at the university. In this case also there is no explanation of the reasons for this unusual double course of study except, of course, the "supernatural" one: "God had chosen him for a mission with profound and broad legal implications" (Berglar, 56). Above all, we are told very little about the ideas and influences he was exposed to. We do have lists of the courses he took, and even the grades he received, but we do not know what kind of theology he studied, what kind of Bible training was taught in Zaragoza, what kind of spirituality was imparted to the seminarians.

The obligatory explanatory anecdote in all the biographies is, in this case, the story of his promotion to the rank of inspector or superior in the seminary. The nature of this job is not clearly specified, but it had something to do with vigilance over his fellow seminarians. We will pursue this point by following Vázquez de Prada's exposition, which, although not substantially different from that of his colleagues, is perhaps the most complete.

Escrivá was not like the other seminarians, above all because he was outstanding "for his external distinction, his piety and talent" (Vázquez, 75), and second because the others *only* aspired to be parish priests, while Escrivá "transcended these limitations." This allows the author to justify the idea that the others did not understand his way of being and doing, which explains

Escrivá's poor relations with his classmates. "The prying eyes of some seminarians . . . soon found reasons to criticize him" (ibid., 78). They teased him and called him "mystic rose." There was even a fight, doubtless the "other's" fault, but even so Escrivá "earned a punishment" (ibid., 87). We are not told what the punishment was. Similarly, it is not very clear if the envy of his classmates was due to "his humanistic culture, his elegant eloquence, his care in his toilet and his urbane manners at table" (ibid., 78), or rather to the fact that his position as superior gave him the right to "a separate table, special food, a separate bedroom and a servant to make his bed and serve him" (ibid., 82).

These are nothing more than anecdotes, but they are revelatory of a climate which, as it was the predominant climate in seminaries of the twenties, helps one to understand the Spanish clergy in the first decades of Francoism both from a human and an intellectual point of view. It also probably helps us understand the future of Msgr. Escrivá de Balaguer, who from this closed provincial ambience "discerned other spiritual horizons" (Vázquez, 75).

Meanwhile, the future monsignor was ordained a priest on March 28, 1925, and two days later celebrated his first mass. Again we have the inevitable anecdote: while the new priest's highest hope was to be able to give communion first to his mother, another woman came forward and Escrivá, "so as not to commit an affront to the Sacred Host in his hand, had to administer first to the stranger" (Vázquez, 93).

The next day Escrivá, who did not want to be a parish priest because God did not want him to (Berglar, 56), was sent by the bishop to an obscure village in the diocese. Again the presently Beatified Josemaría had an opportunity to manifest his "heroic virtues," but our authors cannot refrain from showing their respectful objections to his treatment, no doubt a reflection of Father Escrivá's *contrariness*. "He obeyed promptly, but it was clear that there was something unusual in this measure, which was so unfavorable to him" (Gondrand, 43). The diocesan curia had more than enough time "to have made a different decision earlier" (Vázquez, 93).

Escrivá spent only a month and a half in the village of Perdiguera. He returned to Zaragoza, finished his legal studies, and taught in a private school that prepared boys for entrance into the General Military Academy. In January 1927 he received his bachelor's degree in law. In the spring he was sent to another hamlet in the diocese, but almost simultaneously he obtained permission from the bishop to spend two years in Madrid to study for his doctorate. By April he was already in "the capital of Spain," always following "the inspirations that God dictated to him in his prayers" (Sastre, 78).

We could go on like this indefinitely, from anecdote to anecdote and legend to legend. But it does not make much sense to continue along this road. By way of example this path is illustrative of the initial period of Escrivá's life, prior to the founding of Opus Dei. But after 1928, it is undoubtedly more useful for us to take up the thread of the history and evolution of the movement, especially since—as we shall see—the official sources tend to continue to nourish the evolution of anecdotes rather than make any attempts at explanation.

However, before abandoning our central focus on the figure of the young priest destined to become the "Father," we must ask ourselves how far we can

Later, Escrivá personally oversaw and closely controlled the construction of the cental headquarters of Opus Dei in Rome. Afterwards, he contemplated with emotion the creation of an entire university campus, the work of Opus Dei, in Pamplona, and the erection of many other "magnificent buildings" all over the world, which would culminate with the construction, a little before his death, of the monumental Marian sanctuary at Torreciudad, a short distance from the place of his birth.

The young priest who had wanted to be an architect would become, paradoxically, the builder of one of the most singular structures in the Roman Church in modern times.

Professor?

The ambitious twenty-five-year-old priest fled to Madrid from a world he found too small. He was running away from dioceses like that of Zaragoza, where the career of a priest like him meant obligatory service in the rural parishes of tiny hamlets that were not even on the map: Perdiguera, Fombuena. . . . Consequently his journey to Madrid was not to be a round trip. Although he had only a two-year leave, after a few months he moved his entire family to Madrid. And although the leave he was granted was to study for his doctorate in law, he immediately took up a frenetic round of activities: teaching school to earn a living, visiting the sick, teaching catechism in far corners of the city, plus the classic priestly activities.

But the old, the infirm, and the inhabitants of the city's outskirts were not the only people with whom he associated. He also had contact with university academics (all men, at first), with the aristocratic establishment in Madrid (see the list of marchionesses and countesses, for example, in Sastre, 85, 87), and also with the "vital forces" of Catholicism in the Spanish capital, among whom the Society of Jesus occupied a preeminent place. They were so preeminent, in fact, that while on the one hand Escrivá could not but have been inspired quite directly by the Jesuits when he was planning his Opus Dei, on the other hand the Society was to become the fiercest and most persistent enemy of his work, from the moment the Jesuits perceived the Opus of the 1940s as a potential threat to their monopoly.

Before the civil war, however, Escrivá's ambition had not yet crystallized into a precise objective: he wanted to command, he wanted to have power, he wanted to be important. But he did not know how or where. In politics or the Church? Did he want to be a bishop or a nobleman? (See the previous chapter; he would become monsignor as well as marquis.) Or did he want to be a university professor? The "official" biographers of Msgr. Escrivá expressly state that his mother had made overtures to the bishop of Cuenca, a distant relative, who offered him a canonship, which Escrivá refused (Vázquez, 139; Sastre, 168). But the same biographers are much less explicit with respect to a possible plan for a university chair, in spite of the fact that there are numerous oral testimonies from witnesses who overheard discussion on the matter. Sastre attributes the initiative in this case also to Escrivá's mother: "Why don't you apply for a chair?" (Sastre, 167); others attribute it to "some friends," without being specific (Vázquez, 353). There are those who feel, at any rate, that an academic career was his initial plan when he moved to Madrid and started work on his doctorate, and that Father Pedro Poveda, founder of the Institución Teresiana (whose goals and trajectory were quite similar to those of the Opus Dei of the early years), played a decisive role in his change of direction toward dedicating himself to the guidance of young university graduates and away from devoting himself directly to teaching. (A study of the relationship between Poveda and Escrivá, which is poorly treated in the "official" biographies, would be another possible monograph of indisputable interest. One hindrance would be that all the documentation that used to be in the archives of the Institución Teresiana was requested by the headquarters of Opus in connection with a digest of the material for the introduction of the cause of Msgr. Escrivá's beatification, a request to which the authorities at the institute acceded—"ingenuously," they now say, "without even making photocopies." But the incomparable benefit would be that, if the hypothesis is even narrowly confirmed, we would have via this route an explanation of the origins of Opus that would be a little more concrete than pure and simple divine revelation to Msgr. Escrivá without any intermediary whatever.)

In any case, if the young man who wanted to be an architect turned out to be the "Father" of architects, the hypothetical candidate for a university chair turned out, after the civil war, to be the "Father" of a host of Spanish university professors.

The Leader

If Father Escrivá was an ambitious man, he was also at the same time an exuberant, cordial, and very likable person. One need not turn to the "official" biographies to confirm this. Many who knew him agree on this point, from those who never belonged to Opus Dei to others who were members and later left. Among the latter there must be some who grew to hate the Padre, but this is not the dominant feeling. Artigues states that even his adversaries recognized his "magnetism"; he speaks of the atmosphere of "a feeling of trust that Father Escrivá often managed to create around himself," and he emphasizes the personal appreciation that many former members of Opus Dei still feel for the Founder, "not for his intellectual merits, but for the undeniable human warmth that emanated from his person" (Artigues, 93f.).

Other authors stress his extraordinary qualities of "mise en scène" (Carandell) or his capacity to "dominate the masses" (B. Dalmau). Of María Angustias Moreno's book *El Opus Dei. Anexo a una historia,* members of the Work have said that it was the work of a resentful person, but what most attracts the attention is her obsession with the figure of Escrivá, with his omnipresence and his absolute power in the bosom of Opus Dei. In different ways, they all recognize in Msgr. Escrivá the characteristic features of what Max Weber called a "charismatic leader." This is clearly a typical case of what Richard Hutch has studied and analyzed under the generic category "religious leaders" (Hutch, 1991). The disagreements stem, in every case, from the attempt to classify Escrivá within a specific typology of such religious leaders.

Thus, for Casanova, he was "a typical religious prophet who devoted his life to proclaiming the religious truth of a salvation obtained, according to him, thanks to a personal revelation" (Casanova, 1982, 140). The notion of prophet used here is the one that Weber uses in his *Sociology of Religion;* Weber contrasts this notion to that of the priest and states that only rarely do prophets emerge from the priesthood, so Casanova is obliged to admit that "in this sense, Escrivá was an atypical prophet, since he was a priest," and he tries to solve this problem by adding that "he was not by any means a typical Catholic priest" (ibid., 141), falling into the trap of identifying whatever fits the prior definition as typical, and whatever does not fit as atypical.

When Vladimir Felzmann, in turn, comments that he still has great admiration for the founder of Opus Dei (in an article entitled "Why I Left Opus Dei," Felzmann, 288), Michael Walsh interprets this as equivalent to the "Stockholm syndrome" in the case of "authoritarian religious movements, or sects" (Walsh, 18f.). Although we will return to this, we can say in advance that according to our way of looking at things, the desire to cast Msgr. Escrivá as a specific type of religious leader, or to argue that a movement like Opus Dei possibly has the character of a "sect," is the wrong road to an understanding of the person and his work.

The Impulsive Man

At the same time, this likable and cordial person was also a very impulsive individual who was easily annoyed and irascible. "Restrain yourself, Father," was an expression often heard on the lips of his colleagues, who would intervene to calm him. Escrivá was a man perfectly capable, at any given moment, of uttering insults and obscenities (and then apologizing). Although the context is not exactly the same, in *The Way* he speaks on one occasion of the "strong-language apostolate" and then specifies, "when I see you I'll tell you—privately—some useful expressions" (no. 850).

"From boyhood he was distinguished by his strong and friendly character," it says in the text of the decree of introduction of his cause of beatification. The biographies illustrate this "strong character" with anecdotes extracted from childhood episodes. When he was still a little boy, once in a while he would fly into a rage (Vázquez, 50). For example, one day "he threw the eraser at the blackboard because the mathematics teacher asked him something that had not been explained in class" (Gondrand, 22). By way of justification, it is explained that Escrivá thought that "the professor had been unfair to him" (Sastre, 36). Earlier we mentioned his fight with another seminarian, which "earned him a punishment" (Vázquez, 87). Many later anecdotes could be added, some of which have been written down (to someone who arrived at a meeting two minutes late he said, "I will not and cannot stand for my children to arrive late"; Sastre, 346); most are explained with a genuine mixture of fascination and terror by those who were victims of the Father's outbursts. If collected, they could be made into a regular anthology, perhaps amusing but probably not very useful.

More useful would be a study of the profound irritation, based on misunderstanding, with which Msgr. Escrivá experienced the events of the Second Vatican Council. (So as not to be unfair, I must add that such a study ought to be undertaken, not only on Escrivá and Opus Dei, but on a whole significant sector of Spanish Catholicism, led by a large part of the episcopate.) But there is no room for doubt that this is one of the most impenetrable chapters in the history of Opus Dei.

(*a*) This is true first because, as we might expect, the "official" literature never mentions it. Just reading between the lines of a book like Berglar's, for example, one can sense their opposition (Berglar, 293–305). One can say the same of Escrivá himself, who, speaking about the council and the *aggiornamento,* stated that Opus Dei "will never have any problems of adaptation in the world: it will never find it necessary to bring itself up to date. Our Lord God brought the Work up to date once and for all" (*Conversations,* no. 72). Nor is it certain that a detailed analysis of writings by Opus members during the first half of the seventies would provide much specific proof; we do know that during the council there were systematic and repeated attempts to sabotage its work in Rome among certain groups of bishops (remember the famous *Coetus Internationalis Patrum*), among certain groups of advisory experts, and in certain circles of the Vatican Curia. But this type of behavior, almost by definition, is not usually documented in writing.

(*b*) In the second place this is an impenetrable chapter because Opus Dei has systematically covered up every manifestation of Msgr. Escrivá's irritation on the topic. Relations between the "Father" and John XXIII and, above all, with Paul VI were frankly poor. All the biographies assert and defend exactly the opposite. Escrivá's opinion of the role played by these two popes was negative. Not a word of it! It is false, they say, that Msgr. Escrivá said, after the election of John XXIII and the convening of the council, that the devil was wriggling into the Church (Moncada, 1987, 26); and it is false that in a temper tantrum over the election of Cardinal Montini—who as archbishop of Milan had serious conflicts with the Franco regime at a time when several of Franco's ministers were members of Opus Dei—he had asserted "that everyone who had contributed to his election would be condemned to hell" (ibid., 27). It is also false that he asked his children, again apropos of Paul VI, to pray that God in his mercy would take him (Walsh, 78). In a long footnote, Hertel relates the intriguing episode of the disappearance and replacement of a document in the original version of which Msgr. Escrivá had written, "it seems as if the Mystical Body of Christ is a decomposing corpse that stinks" (Hertel, 15–18).

(*c*) In the third place, after 1965 the "official" discourse of Msgr. Escrivá and of Opus Dei not only ceased to be the discourse of irritation, but began a process of evolution that goes from serene and thoughtful opinions about the council and the situation of the Catholic Church after the council (for example, Escrivá, *Conversations,* nos. 1, 2, 23) to the affirmation, which appears more and more frequently, that in fact the council did nothing more than corroborate what Msgr. Escrivá had been saying all along. This transmutation is a perfect illustration of something that I myself wrote in those years—without any reference to Opus Dei—namely, that in the Catholic Church the introduction of any innovation passes through three distinct stages: first it is said that the new thing is false; second it is said that even if it is not totally false, it is extremely dangerous; and finally it is said that the Church has always held it to be true (Estruch, 1972, 10).

In this way Opus Dei began to construct an image of Msgr. Escrivá as a precursor of the Second Vatican Council. This is a kind of leitmotif in the set of interviews collected in the volume *Conversations:* one review of the book says that in the Founder's doctrine "the adoption of the Second Vatican Ecumenical Council was nothing more than a confirmation of what had sounded to many like heresy forty years before" (Benito, 1968, 634). This image, propagated with great persistence, finally jelled to the point where in the decree of introduction of his cause of beatification we find the following sentence, which neatly synthesizes the current official version: "In having proclaimed the universal vocation of holiness, since he founded Opus Dei in 1928, Msgr. Josemaría Escrivá de Balaguer has been universally recognized as a precursor of the council precisely in that which constitutes the fundamental core of his teaching, so fecund for the life of the Church" (Seco, 1986, 196).

When we know that in 1928 Msgr. Josemaría Escrivá de Balaguer was not a monsignor, was not called Josemaría, and did not have the last name Balaguer; when we harbor serious doubts about whether, strictly speaking, 1928 was the date of the founding of Opus Dei; and when we hope to be able to show that the core of Escrivá's teaching was not, in 1928, the proclamation of the universal vocation of holiness, it is hard not to interpret this whole process of evolution simply in terms of strategy. It is the same strategy, at root, adopted by Laureano López Rodó, one of the best known of the Opus members who collaborated closely with Franco, who on December 15, 1965, said in an audience with the general: "We must ensure that we are perceived as the carriers of the banner of reform demanded by the council documents" (López Rodó, 1990, 591).

The Man of Scruples

This ambitious, impulsive individual was at the same time a man of very traditional Catholic background, defender of a no less traditional spirituality. Religious scruples obliged him to rein in his ambitions, to sublimate, to repress his impulses to the point of denying them. Texts like *The Way* or the former *Constitutions* of Opus Dei naturally lend themselves to a reading from this perspective. It is a reading that has often been rehearsed by the "critical," so it is not necessary to insist on it. The so-called acts of indemnification and reparation, the mortifications of every sort, the hair shirts and the *ad corpus castigandum* disciplines, "to punish the body and reduce it to servitude" (*Constitutions,* 1950, art. 260) were habitual practices, "pious customs" that Escrivá imposed upon himself, as we are reminded in all the "official" biographies (sparing no detail—drops of blood as a consequence of flagellations, and so forth). Escrivá did all this "to tame the colt" (quoted in Vázquez, 278), exactly the same expression used by Freud when he spoke of the domination and repression of drives and basically of sexuality.

But in speaking of the scruples which obliged him to rein in his ambitions we are not referring exclusively to the topic of sexuality (at least, in its manifest dimensions). One of the expressions consecrated in all the hagiographic literature of the Padre to the point of making it into a kind of slogan, practically a motto for his entire life, is the statement according to which "to hide myself and disappear is mine, so that Jesus alone may shine." All mention it, Vázquez de Prada eight times (p. 560); Gondrand (p. 284) puts it in Msgr. Escrivá's mouth just before he dies, as if it were a recapitulation of his life; Helming (p. 79) ends his book by saying that in life he managed in fact to "hide himself," but that after his death he failed in his will to "disappear"; Bernal (p. 14) uses the phrase in his introduction.

"To hide myself and disappear is mine." This is the supreme argument used by the biographers to emphasize the humility of Msgr. Escrivá. It is the denial of any ambition for power or prestige, a denial in the strict sense of the word, when an author like Berglar explains that though Msgr. Escrivá permitted himself to be filmed during his great "catechetic journeys," he did it for only one reason: "so Christ would shine also under the lights and the cameras" (Berglar, 260f.).

"So that Jesus alone may shine." This "shining" is, in addition, an expression typical of the bullfight. There has been, at the very least, one "torero" in Opus Dei, and it seems that the language of the bullring was not distasteful to Escrivá (Sastre, 607ff.): "the Father performed a *verónica,* and continued, joking: . . . Enjoy yourself in the fight, like an artist, with love! This is also what we must do with God Our Lord." On another occasion he said, according to Álvaro del Portillo: "When I die, pray a lot for me, so I can leap like a bullfighter over purgatory" (Bernal, 167). And in Mexico, before the image of the Virgin of Guadalupe: "Make us see that you are our Mother, shine!" (Gondrand, 257). On another occasion, according to the same author, he spoke of the need to "carry the cross with style" (ibid., 276).

A study of the language and style of Msgr. Escrivá is undoubtedly the topic for another monograph that ought to be done someday, but this is neither the place nor the time.

There are many examples illustrating this mechanism of denial of impulses (see the entries under "humility" in Vázquez's analytical index; 259f.). We will pick only one more, which is no less profusely illustrated in all the biographies: Msgr. Escrivá's habit of comparing himself to an ass, a donkey, or, more exactly, a "scruffy little *burro.*"

Patience, docility, and humility are the great virtues of this animal, who works without stopping until the hour of his death. "That is the way he liked to see the members of the Work, that is the way he wanted them to be" (Berglar, 306). Even as a young man, he used to sign letters to his spiritual director with the name, "followed by *b.s.* (*burrito sarnoso*)" (Vázquez, 318). Later, in Rome, when someone would ask him for a photograph of himself, Msgr. Escrivá would offer a picture of a donkey, saying "This is I." (In my interviews with members of Opus I often came across "photographs" of this kind.)

Vázquez also tells how "one day when he was praying to the Lord, saying to him in his prayer, 'Here you have your little scruffy donkey,' he heard these words in reply: 'A donkey was my throne in Jerusalem'" (Vázquez, 123; see also Gondrand, 73f.). The humble ass represents the denial of all personal ambition, but at the same time the sublimation of the drive that made Msgr. Escrivá into one of God's chosen.

The Chosen Man

On certain occasions one might also speak of "messianism" in the case of the founder of Opus Dei, and not only in the terms the "official literature" uses to refer to his illumination or vision in 1928, presenting him as "the instrument chosen by God to realize on earth the divine enterprise of Opus Dei" (Bernal, 109). It is Escrivá's life as a whole, "consecrated now as one of the classics of spirituality of all time" (Orlandis, 1974, 131), he who appears replete with divine interventions, which the biographers call the "divine locutions" (see again Vázquez's useful analytical index under "divine locutions," p. 562, and "supernatural facts," p. 559).

In his early childhood he was saved by a miraculous cure when he was two years old (a legitimation of sorts for the construction of the Marian sanctuary of Torreciudad), and even the birth of his younger brother, in 1919, partook of a miraculous character, according to the biographers. These miraculous interventions were repeated, in fact, throughout Escrivá's life, although not all of them are known to us since, in humility, he generally preferred not to speak of them. But really, proclaimed Escrivá himself, "there were extraordinary interventions, *when it was necessary*" (Vázquez, 124; emphasis mine).

The hagiographic literature emphasizes and magnifies the notion of Msgr. Escrivá as "chosen," as "sent," as "anointed." But it is the founder himself who, with his manifestations, frequently contributes to this notion. In the last part of his life he did so by means of two kinds of comparisons: first, he presented Opus Dei as "a small relic of Israel," as a group of those who for their fidelity and orthodoxy had been chosen by God to carry out the mission of preserving the faith of the Church (a choice and a mission of which Escrivá was, historically, the instrument par excellence). Second, he stressed that Opus Dei was a new reality in the life of the Church, comparable only to the early Christian communities. Within the Catholic Church, in fact, this claim of a direct connection with the primitive Christian communities, through a no less direct divine inspiration, has been one of the characteristics of all the movements of the messianic type.

Even as a young man Escrivá, using different and more explicit language, would make this same kind of comparison, ultimately identifying himself not merely with some prophetic figure, but with the figure of Christ himself. ("To be another Christ" was not an unusual expression for him.) For example, in recalling the beginnings of the Work, Escrivá says,

> I began by not speaking of the Work to those who joined me: I put them to work for God, and that was that. This is the same thing the Lord did with the Apostles: if you open the Gospels, you will see that at first he did not tell them what he wanted them to do. He called them, they followed him, and he had private conversations with them, and others with small or large groups . . . ; this is the way I acted with the first ones. I said to them: come with me. [quoted in Fuenmayor et al., 35]

During the forties, young people who joined Opus were told about the early years, when the Work was limited to the group made up of the Padre and the "first twelve," among whom there was also a Judas. Sastre writes that after 1937 Escrivá called Álvaro del Portillo "by the affectionate name of *Saxum:* rock" (Sastre, 158). This new Peter, in fact, became Escrivá's successor and the first prelate of Opus Dei.

Money

An instrument chosen by God, an object of special favors, Msgr. Escrivá was totally convinced that everything he did was by divine will, that the Work of God was the will of God ("God is working to make it happen"), and that anyone acting against him and against Opus was acting against God. This was why throughout his life supernatural interventions were produced—as we saw earlier—"when it was necessary."

This "when it was necessary" denotes the peculiar conception of the kind of relation Escrivá had with God. Without getting into a theological discussion of the theme (although it would be a good idea for the theologians to address it someday), we can say that insofar as Escrivá was an instrument of God, God had to be an accomplice of Escrivá. Given that "God did not ask my permission to come into my life" (Bernal, 63), now God was going to have to take the consequences.

Here is one of Escrivá's biographers' favorite anecdotes. When he was twelve or thirteen years old, his sister and some friends were making "a precariously balanced house of cards. Josemaría suddenly came into the room and knocked it down with a sweep of his hand. The girls could not believe it: it was not like him. Why did you do that? they asked, annoyed." The answer was: "This is just what God does to people: you build a castle and, when it's nearly finished, God knocks it down" (Sastre, 40).

This arbitrary and incomprehensible God—who had allowed the death of three little sisters and a serious economic crisis in the family—was a God who had to be domesticated, who could be manipulated. Escrivá used expressions like the following throughout his life: "If you do this, God will be happy"; or prayers of this type: "Lord, don't do this to me." The same can be said in the case of the Virgin, and those whom Escrivá called "Guardian Angels": "Win over the Guardian Angel of whomever you want to draw to your apostolate. He is always a great 'accomplice'" (*The Way,* no. 563); "Get into the habit of giving advance thanks to the Guardian Angels . . . , to obligate them more" (*Forge,* no. 93).

The relationship with God thus becomes, quite often, a bargaining relationship. Perhaps Sastre's biography has the largest number of examples. In 1943, on the eve of the death of Isidoro Zorzano, a young engineer who had been the first disciple, "the Father gave him an important job. At that moment he had an important bargaining chip with which to ask favors of God: he was going to give him life, at the peak of his prime. Isidoro took note in his mind and in his heart of the needs of the Work: the first thing was the real estate on which to build, finally, a University Residence in Madrid" (Sastre, 273).

Two years earlier, Escrivá had to go to Lleida to lead some spiritual exercises. He left his mother in Madrid, very ill. When he reached the town "he went to the tabernacle of the chapel:—Lord, look after my mother, since I am busy with your priests" (Sastre, 280).

In 1947 when Opus wanted to buy the building that would be the headquarters of Opus Dei and they did not have enough money, Escrivá felt that "if God needs this place for his work" he would take care of getting it (Sastre, 341). And in fact, said Escrivá, in Opus "we strike a good bargain: we exchange felicity for fidelity" (Sastre, 388).

It is not surprising that in this language through which the bargaining relationship is expressed, money and striking a good bargain play an important role. As is well known, the theme of Opus Dei and money is the topic that has elicited the greatest number of comments in relation to the work of Msgr. Escrivá. The topic—perhaps by way of defense—crops up in all the biographies of the founder. Bernal, for example, dedicates the entire long last chapter of his book to it, with the title "Father of a Large and Poor Family" (Bernal, 313–51). But this is not the time for us to address this whole question; for now the question that interests us is Msgr. Escrivá's personal relationship to money. Did his ambition for power and prestige include ambition for money? Can we speak, in the case of the Padre, of the *sacra auri fames* (the hunger for money, or zeal to enrich oneself) which Max Weber discusses in *The Protestant Ethic and the Spirit of Capitalism?*

The question permits an affirmative answer only if we preserve the adjective used by Weber, and accentuate it: the zeal for money is in every case, for Escrivá, a "sacred" zeal. In other words, the recourse to sublimation, as a defense mechanism against impulses that religious scruples would oblige him to reject, is never so obvious as in this case. Reading the biographies of the Padre one draws the conclusion that he spent his whole life without money, and that he went begging his whole life, sponging off people and bargaining. In exchange for the money they gave him, those he sponged off would "receive their money in heaven, multiplied a hundredfold . . . and eternal life" (Sastre, 172).

So Escrivá never has money (he is utterly poor, says Bernal, 327); or he never has enough. If he never has enough, it is because the money he wants is not for himself but for God. Speaking of the things of God—the liturgy, for example—"all the richness, majesty, and beauty possible would still seem too little to me" (*The Way,* no. 527). In the Old Testament, God had ordered that "everything had to be gold. All gold!" (in Bernal, 349). Yet in the Temple of Solomon "Jesus Christ was not really and truly present, as he is on our altars and our tabernacles" (ibid., 347). With costly religious objects, this author says, Msgr. Escrivá "wanted to show his love"; Escrivá acted the way lovers do when they give each other expensive gifts, "the best they have." "When that changes, we will change our minds as well" (ibid.).

With respect to the things of God, then, there is no need to scrimp. On the contrary, Escrivá's motto in such cases was "spend what you owe, even if you owe what you spend." And if there was no other way, the Padre "asked for financial help, with great pain in his soul, but without shame": it was time to "sponge off friends, relatives and strangers" (Vázquez, 255).

It is clear that Opus Dei itself belongs to this set of things that are things of God. If Opus Dei is the will of God and is the Work of God, which "happened in spite of me" (Vázquez, 471), because it is God who "is working to make it happen," in accordance with the logic of the bargain it is natural that God also has to contribute something. Thus, "when especially serious financial problems arose in 1944," with his proverbial good humor the Padre "would say: Ask the Lord to give us money, which we need very much, but ask for millions, because if everything is his, it's the same to ask for five as for five thousand million, while you're asking" (Berglar, 334; apparently the quote is taken from the text of the postulator of the cause of beatification of Msgr. Escrivá).

A few years later, when the building for the central Roman headquarters was being bought, the sellers demanded payment in Swiss francs, and Escrivá exclaimed: "Since we don't have anything, what difference does it make to the Lord whether he gives us Swiss francs or Italian lire?" (Vázquez, 250). All the biographers relate this anecdote. None of them considers it necessary to remind us, however, of the Spanish law in that period regarding the possibility (or rather, the impossibility) of taking money out of the country. And since the money had to come fundamentally from Spain, some of the Padre's children, in addition to praying, must have had to work on "opening roads to the Lord" (via Portugal, via Andorra) so it would reach Rome, transformed into Swiss francs. "Are we working for ourselves? No! Well, then, let us say without fear: Jesus of my soul, we are working for Thee, and . . . are you going to deny us the material means?" (*Forge,* no. 218).

Dennis Helming, who in his illustrated biography of Msgr. Escrivá also reminds us of the existence of "extraordinary interventions, when it was necessary" (Helming, 69), locates one of them in 1958 in the "City," London's "banking and big business district." Escrivá was visiting, and "he was overcome with a feeling of impotence on seeing such a great display of wealth and power." Then he heard the voice, which said, "You cannot! I can!" (Helming, 66).

Sexuality

We have said that Msgr. Escrivá was a man who restrained his ambitions, and we have tried to indicate—although in a fragmentary way—how he denied and sublimated his impulses. But although we have avoided the topic until now, it is obvious that the repression of drives has a basic sexual component, and that when Msgr. Escrivá speaks of the need to "tame the colt" this dimension is very present.

The topic is, however, sufficiently delicate to make us prudent and cautious. This is so not only because Opus Dei has stood up as a stalwart defender of very conservative positions in this field (and believe me, in the bosom of the Roman Church it is not easy to stand out as exceptionally orthodox in matters of sexual morals!) but also because, speaking from a perhaps slightly less orthodox position, one might feel that a research project has no right at all to interfere in such an intimate sphere, an area that belongs strictly to the conscience of the individual. From this angle, then, I will ignore all elements of information from other sources, oral or written (and it hardly needs to be said that there is plenty of gossip), and limit myself exclusively to the "official" information that various members of Opus Dei consider themselves authorized to divulge in their publications, as well, of course, as the texts of Msgr. Escrivá himself.

Perhaps the most famous of all the Padre's maxims is the one that appears in number 28 of *The Way:* "Marriage is for the rank and file, not for the officers of Christ's army." To understand the meaning of this statement, we must put it into context. Many years after writing this sentence, Escrivá himself saw to it that he was asked, in an interview later published in the volume *Conversations,* about the compatibility between this statement and his "doctrine on marriage as a holy path" and vocation (*Conversations,* no. 92). In his reply, Escrivá explains: (1) that "the greater excellence of celibacy is not my theological opinion, but a doctrine of faith in the Church"; (2) that when he wrote the statement the Church tended to value celibacy exclusively, to the detriment of marriage, while Opus Dei always valued both vocations; and (3) that when he wrote that marriage was for the rank and file "he was only describing what had always been the case in the Church," and that this did not contradict the idea that "in an army the troops are as necessary as the officers, and can be more heroic and deserve more glory."

We may assume that what is indeed his "theological opinion" and not "doctrine of faith" is the comparison of the Church to an army, where the bishops constitute the high command and the laity the troops. For a "precursor" of the council which would speak of the Church as "People of God," the reference to this pyramidal model of the army might seem surprising. It is not so surprising, however, if we take into account the historical context in which *The Way* was published, as well as the kind of persons to whom it is addressed. In fact, *The Way* was published in 1939, immediately after the end of the Spanish civil war ("The Year of Victory," it says in the first edition), and Opus Dei would be structured from that date on in accordance with a pyramidal model, strictly hierarchical, with distinct categories of members (as spelled out in the Constitutions of 1950), to the point where Escrivá himself defined it on more than one occasion as an "ordained army" (*acies ordinata*).

No matter how much it is insisted today that *The Way* is addressed to men and women, single and married persons, people of every social class and every profession, the truth is that it was published fundamentally for men—young men of good family who were university-educated and ready to commit themselves to a life of celibacy. It was addressed to men called precisely not to be of the rank and file, but, on the contrary, to be "caudillos." "You were born to be a caudillo!" (*The Way,* no. 16); "Make your will manly so that God can make a caudillo of you" (no. 833); "You told me you wanted to be a caudillo" (no. 931).

It is true that right before he speaks of marriage as being "for the rank and file," the Padre uses the expression "vocation to marriage" (no. 27). But it is also true that later (no. 360) he repeats this entire point again in order to add a joke that the woman ought to be "good, pretty and rich," along with the advice that one ought to first ask oneself if God might not ask for more, that is, the renunciation of marriage.

It is also true that today Opus Dei has both single and married members, both men and women. But during the first twenty years of its existence (starting from the date of its "official" founding), it admitted only unmarried persons. As for women, initially they had no place at all in Escrivá's plans. "There will never be women—not even in jest—in Opus Dei" (quoted in Gondrand, 63). A few days after writing this sentence, one of those supernatural interventions to which we have alluded made him see the necessity of including women in the framework of Opus Dei. "By the will of God, then, Opus Dei will have two sections from this day forward, one for men and the other for women" (Sastre, 102). Thus, according to Escrivá, while "the foundation of Opus Dei happened without me, the women's section happened against my personal opinion" (Berglar, 83), "against my inclination and against my will" (Sastre, 101). The female section of Opus Dei, in any case, would remain "radically separate" from the male section (*Constitutions,* 1950, art. 437), because, as Escrivá said, recalling an old Castilian proverb: "between female saint and male saint, a wall of stone and mortar" (*Forge,* no. 414; Vázquez, 264).

One wonders, of course, what reasons necessitated the existence of the women's section, an arrival received with the hope and surprise of "an unplanned child" (Vázquez, 264). We might even be tempted to explain it in terms of the functions assigned to women in the Opus Dei Constitutions: "the management of the houses" in the case of the "regular members" (art. 444.7), and "the manual tasks and domestic service" in the houses of Opus Dei in the case of the "auxiliary members" (in Latin, *inservientes,* art. 440.2). But such an explanation would oblige us to refer to the "critical" literature, and in this section we have promised to use only "official" sources. These were, in any case, the basic tasks of Escrivá's mother and sister, who "accepted to take charge of the management of our first centers" (Vázquez, 212). At the same time, however, neither of them ever belonged to Opus. Why? Because they did not have the "vocation for Opus Dei" (ibid., 211), in spite of the fact that (according to the same author, five pages earlier) in Burgos in 1938 "the Padre told Álvaro del Portillo about a thought that kept recurring insistently: whether or not to *put* his mother and sister into the Work" (ibid., 206).

Be that as it may, although the "official" founding of the female section of Opus dates from 1930, women are practically absent in the book *The Way.* One could say that the whole volume was conceived in the form of reflections and advice directed at a certain "thou," and this "thou" is always a man. "Let your prayer be manly. To be a child does not mean to be effeminate" (no. 888). Woman is scarcely mentioned, and when she is it is not—except on rare occasions—in the second but in the third person, as in another famous phrase from *The Way:* "Women need not be scholars: it is enough for them to be prudent" (no. 946).

Even though "we cannot disdain the cooperation of women in the apostolate" (*The Way,* no. 980), in fact, the role of women appears in the writing and thinking of Msgr. Escrivá as not only different from, but subordinate to, the role of men. While in the case of men "there is no excuse for those who could be scholars and are not" (no. 332), in the case of women, they "do not have to be scholars: it is enough for them to be prudent" (no. 946). To the women who run the Opus centers, the Padre gives advice about how to prepare meals, recommends "food magazines," and, on returning from his trips, gives them "new recipes" (Helming, 55). As for the married woman, the counsels of the Padre are "those of an experienced confessor, full of common sense and supernatural sense." They should be well groomed and ought to be concerned "above all with age, fixing up the facade a little, using the available cosmetics. It is an act of charity for all and a way of looking youthful for their husbands" (Gondrand, 278). The home thus becomes "an oasis of peace, happiness, love, and beauty" (Bernal, 57).

We are not very far from the cultural stereotype, so very Spanish, of woman as "the warrior's repose," are we? Víctor García Hoz, represented by Sastre (p. 597) as one of the first "matrimonial vocations" in Opus, wrote in a 1945 review of Escrivá's book *Santo Rosario:* "It is another act of good judgment on the part of Don José María Escrivá to revive the Rosary, not as an idle pastime for old women, but as a weapon, as something men should use when they are occupied with matters of war" (García Hoz, 1945, 594). Urteaga, commenting on item 22 of *The Way* ("Be strong. Be virile. Be a man."), defends "strength" and "virility" (Urteaga, 1948, 63–70), while at the same time asking "why we have allowed so many souls to make that marvelous intimacy of the soul with God effeminate" (ibid., 66).

In spite of this Ana Sastre, who is a woman, affirms that "Don Josemaría has, by this time, a broad and accurate view of the role women must play in the world and in the Church" (Sastre, 103).

Monsignor Escrivá uses the words sex and sexuality very little. Generally he approaches this topic indirectly, avoiding the explicit use of the term, a fact possibly meaningful in itself. "Do not be alarmed when you notice the rubble of the wretched body and the human passions: it would be foolish and naively childish for you to find out now that 'that' exists" (*Surco,* no. 134). "That" is "the prickling of the wretched flesh, which sometimes attacks with violence" (*Forge,* no. 317; the remedy is to "kiss the Crucifix"); it is "the temptations of the flesh" (*Surco,* no. 847; the remedy is to "closely embrace Our Mother in Heaven"); it is "base tendencies" (*Surco,* no. 849; one must "pray slowly to the Immaculate Virgin"); it is "the dry and dirty concupiscence" of "the emotions of the earth" (*Forge,* no. 477); the "vile slime of the passions" (*Surco,* no. 414); it is, finally, the "instincts of males and females" (*Surco,* no. 835).

On one of the rare occasions in which Escrivá expressly uses the word "sexuality," it is to assert: "In these times of violence, of brutal, savage sexuality, we must be rebels. You and I are rebels: we have no desire to let ourselves be carried with the current, to be animals" (*Forge,* no. 15). But "they are not more manly, nor more womanly, because they lead that disorderly life. It is obvious that those who reason thus find their ideal in whores, in homosexuals, in degenerates . . . , in those who have rotten hearts and will not be able to enter the Kingdom of Heaven" (*Surco,* no. 848). But when it comes to the question of entering the Kingdom of Heaven, rich men find that difficult too, according to the Evangelist (Matthew 19:23), and it might even be that some "whores" will go ahead of the kind reader, or this writer, or Msgr. Escrivá himself (Matthew 21:31).

From all this the Padre deduced that it is appropriate to treat the body "with charity, but with no more charity than you would show toward a treacherous enemy" (*The Way,* no. 226).

In his own personal case, since boyhood "his good looks must have made quite an impression on the girls; and because of this he began to be careful about looking, not letting what he saw arouse his curiosity" (Vázquez, 71). Once his priestly vocation was established, it was clear that even though "I bless that love (between married couples) with both hands . . . the Lord has asked more of me" (*Friends of God,* homilies, no. 184). And in spite of the fact that "one feels the bad inclination of nature as much at fifty as at twenty years old," chastity, Msgr. Escrivá goes on, is "a triumph that gives us a marvelous paternity, far superior to that of the flesh" (Berglar, 52). He had already written the same thing in *The Way,* precisely in the reflection on "marriage for the rank and file" with which we began this set of observations, and which ends: "A desire to have children? Behind us we shall leave children, many children . . . and a lasting trail of light, if we sacrifice the selfishness of the flesh" (*The Way,* no. 28).

Quite understandably, the biographies are sparing in their comments on sexuality in the personal life of Msgr. Escrivá. According to Berglar, "we know that there are women for whom nothing is more attractive than men who want to live in celibacy, who want to be priests or who already are. We are not going to delve into this topic now. We will only say that even the young Josemaría was not exempt from such ambushes. Like any young person, he had to overcome temptation" (Berglar, 52).

But although this author concludes that "in this field there is nothing sensational worth mentioning," the truth is that there are a few episodes which, while not "sensational," are nevertheless enlightening. There is, for example, the fact, mentioned by Vázquez, that when Escrivá used to walk down the street with his mother in Madrid in the thirties, he made her walk apart from him: "We cannot walk together in the street," he told her, "becuase it is not written on my forehead that I am your son and I do not want to run the risk of shocking anyone" (Vázquez, 136).

In 1936, a little after the outbreak of the Spanish civil war, the Padre had to go into hiding and move his domicile frequently; he ran the risk of being shot merely for being a priest. At one especially precarious moment when he had no place to take refuge, the following episode took place (Gondrand, 118; Sastre, 198; Vázquez, 170). A friend offered him the key to a home in Madrid; the family was out of town and the doorman could be trusted. There was only a maid in the apartment, who could also be trusted, and who could look after him. Escrivá asked how old she was. Twenty-two or twenty-three, was the reply. "The Founder looked at this man who had probably walked all over the city to find him a hiding place where his life would be safe, and he said: 'My son, don't you realize I am a priest and that, with the war and the persecution, everyone's nerves are shot? I cannot and do not want to stay shut in with a young woman, day and night. I made a promise to God which comes before everything else. I would die rather than offend God, rather than break this promise of Love'" (Sastre, 198). "With tact and firmness he made him understand that since he had made a promise of Love to God, it was reckless to stay closed in day and night with a woman, even more so in light of the circumstances of physical and moral erosion in a time of persecution" (Vázquez, 170). He rejected the offer. But he did not return the key to his friend. "He said: 'You see this key you gave me?' Then he walked over and stopped at a sewer drain. And standing over the grating, he dropped it in" (Sastre, 198). "'You see this key?' he said. He walked over to a sewer and dropped it into the darkness. With the key thrown away, the remotest possibility of temptation was gone" (Vázquez, 170).

"To defend his purity, Saint Francis of Assisi rolled in the snow. Saint Benedict threw himself into a thornbush, Saint Bernard plunged into an icy pond. . . . You, what have you done?" asked Escrivá (*The Way,* no. 143). Throw a key into the sewer!

In this same chapter of *The Way,* dedicated to "holy purity," there are two consecutive items (121 and 122) that seem inspired by this *Pensée* of Pascal, *"L'Homme n'est ni ange ni bête"* (man is neither angel nor beast). Escrivá writes: "There is need for a crusade of virility and purity to counteract and nullify the savage work of those who think that man is a beast" (no. 121). He continues: "many live like angels in the middle of the world. Why not you?" (no. 122). Because—Pascal goes on—*"et qui veut faire l'ange fait la bête"* (He who would act like an angel acts like a beast).

The Originality of Monsignor Escrivá

Having reached this point, one might well ask: and what is so original in all this? Better said, having reached this point, I wish someone would ask: and what is so original in all this?

Well, frankly, not much. Of course, a reader geographically far removed from the Iberian world, culturally unfamiliar with the history of Spain in the twentieth century, or religiously disconnected from traditional Roman Catholicism, might find very original the personality, disposition, and behavior patterns of a person who doubtless appears to him as exceedingly exotic. On the other hand, to a reader who is Catholic, Spanish, and sixty years old, the figure of Msgr. Escrivá would seem much less original, precisely because the context is not so exotic to him.

More than "a classic of the spirituality of all time," Escrivá de Balaguer is at bottom *a child of his time:* he is the product of a specific country, a specific epoch, and a specific church. These are the Spain of General Franco and the church of Pope Pius X. If Opus Dei had "never seen the need to bring itself up to date," as Escrivá maintained, Opus would today be a paramilitary, profascist, antimodernist, and integrist (reactionary) organization. If it is not, it is because it has evolved over time, just as the Catholic Church, the Franco regime, and Msgr. Escrivá himself evolved.

But evolution never erases the past all at once, as if it had never existed. The past leaves traces and "blazes trails" (as Escrivá used to say; *The Way,* no. 1). The present-day Catholic Church is heir to the Church that existed before the Second Vatican Council, and even in the most progressive circles one can detect "tics" of authoritarianism and a kind of clericalism more suited to a monolithic culture and monopolistic regime than to true pluralism. The present Spanish system of democratic monarchy retains numerous and notable "tics" inherited from the long period of Franco's dictatorship, in the bosom of which—with several notable exceptions—its current political leaders were socialized. And in the *"chapelle ardente"* of Msgr. Escrivá de Balaguer, converted into a supposed "precursor" of the council, he held in his hands the crucifix Pius X—pope of the most belligerent Roman Catholic fundamentalism—had when he died (Sastre, 632; Vázquez, 486).

Having stated that there is nothing very original about the figure of Escrivá, that he is fundamentally the product of a specific country, epoch, and church, we must nevertheless not fall into the trap of presenting him as a "typical product" of that church, that epoch, and that country. The ideas, behavior, spirituality, and devotions of Father Escrivá are the fruit of a set of social and intellectual influences that have been made to disappear by his biographers' systematic sleight of hand. But this does not mean that Escrivá was a "typical" religious reactionary (integrist) or a "typical" Francoist, or even that he was "typically Spanish." And this is so because, however much he was a child of his time, he was not *just any* child of his time. He had a strong personality; like it or not, he was an authentic religious leader, and the work he left as his legacy is not simply one more of the many foundations of the Catholic Church in the twentieth century.

Whether through divine inspiration, personal skill, or through a whole set of those jokes, unforeseen and unforeseeable, that history likes to visit on us, Msgr. Escrivá was a man who generally knew how to be in the right place at the right time. At the beginning of the century a career in the Church provided a platform for social advancement for ambitious young men from certain social classes and from certain parts of Spain. But Escrivá went to seminary in Zaragoza and not in his native diocese. After he was ordained he found a way to get to Madrid, where he created a whole network of relationships. In 1938 he went to Burgos, where those who would win the war had established their headquarters. He returned to Madrid on the same day that Franco's troops entered the capital city. His incipient organization began to spread all over the peninsula. It had its own university residences, it filled university chairs with its people, and it exercised nearly monopolistic control over research resources. In 1944 Escrivá was in charge of Franco's spiritual exercises.

In 1946 he moved to Rome. Shortly thereafter, the Vatican created a new category of ecclesiastical organization, the secular institutes, and Opus Dei became the first institute approved by the Holy See. In a few years the organization founded by Msgr. Escrivá had passed from a diocesan status to a pontifical one. And Opus Dei had expanded to the international level. Parallel to this seeding of Opus members (at that time still nearly all Spanish) throughout the entire world, members of the organization assumed key positions in universities and, gradually, in world finance and banking, Spanish public administration, and the mass media. Its base for recruitment broadened with the inclusion of married members. Simultaneously, a few members of the Work penetrated the world of the Vatican Curia and its diverse "Sacred Congregations."

Why continue? Its subsequent development is fairly well documented in the "official" literature. Opus established its own training centers, schools, clubs, residences, even its own centers for educating priests and its own universities. The juridical cloak of the secular institutes became too small, or ceased to be adequate. Patiently, Opus Dei followed a path of conversion that brought about its establishment as the first—and until now the only—Personal Prelature in the Catholic Church. At the moment of its formal application for this juridical transformation Opus Dei had—according to a report signed by the president general, Álvaro del Portillo—72,375 members of 87 different nationalities who worked in 479 universities and high schools, 604 newspapers and magazines, 52 radio and television stations, 38 information and publicity agencies, and 12 movie production studios and distributors, among other things (reported in Fuenmayor et al., 601, 609; see also the long—and encomiastic—account of its activities and accomplishments in William West's little book *Opus Dei: Exploding a Myth*).

Was all this the work of one man? Of course it was, say some, overwhelmed by his charisma, his power of attraction, the sense of security he inspired, the fascination he elicited. Of course not, say others: Escrivá was an impulsive, simple, primitive man, a man of rosaries and hair shirts, of sacraria, confessionals, mortifications, and penances, of the guardian angels and the Holy Virgin of Beautiful Love, progressively manipulated by cold and calculating individuals, men who were distant and controlling, clever, good strategists, who wrote down what they wanted him to say and even instructed him on how to part his hair. A third group counters that neither of these views is true: when he was manipulated, it was because he let himself be manipulated, because it suited him and was in his interests, and because in the end it was to his benefit. All of this is false from top to bottom, says the "official" version of Opus Dei. The question is wrongly put. Who wants to say whether it is "the work of one man"? One need only look at the name: it is *The Work of God.*

Conclusion

Work of Escrivá, Work of God through Escrivá, or rather Work of Escrivá's entourage—the truth is that the history of Opus Dei, on which we shall concentrate in the chapters to come, is no less paradoxical than the personal history of Msgr. Escrivá de Balaguer.

Having come to the close of the first section, then, can we say that we have reached some kind of conclusion?

(*a*) The preceding pages are not, nor did they ever pretend to be, a biography of Padre Escrivá. At best they constitute a fragmentary, partial, incomplete, and provisional approximation of his life and some of the most interesting aspects of his personality. They do demonstrate, however, that the books emanating from Opus Dei and officially purporting to be biographical studies of the Founder are not real biographies either. Jokingly, the Padre once wrote that since he had often had to sleep "under the dining room table with a theology book for a pillow, perforce I am a good theologian" (Vázquez, 256). The "official" biographers have likely had the privilege of sleeping with their heads on top of the documents of the "Historical Register of the Founder," but they do not realize that this is not enough. They wanted to mythologize the Padre, and they have converted him into a fetish (Urteaga, 1948, 30), an "abnormal being," a "case to be studied by a modernistic doctor" (*The Way,* no. 133). No good study of Msgr. Escrivá exists, and there deserves to be one.

(*b*) Again jokingly, to a bishop friend of his who showed him a paper in which he was harshly attacked, Escrivá said that if the writers had known him better, "perhaps they could have said worse things" (Vázquez, 223). The good study that Escrivá deserves should not be written to say "worse things" about him, but to understand him better, to undo the fiction of his transparency, of his "inalterable coherence and continuity" (Berglar, 327), and to penetrate his complexity. Escrivá's personality is a rich, and complicated one, and this is what these pages have tried to bring to light: *the paradoxes* of Msgr. Escrivá. If we had to risk formulating an opinion as bold as those of Berglar, we would dare to say that if anything about Escrivá was unalterable, it was his sincerity. At least until 1946 (after this date we are not so sure and would hesitate a little more; in the chapters to come we shall see why), Escrivá was a man of extraordinary sincerity. "He is not putting on an act. It is impossible for it to be an act" (Calvo Serer, in Martí Gómez and Ramoneda, 28). He was sincere in the sense that he was the first to believe whatever he said and whatever he did; he was the first to believe his own propaganda.

(*c*) Given the complex, interesting, and paradoxical character of this personage, to summarize his thinking in a few sentences is impossible, because the choice of the aspects to be emphasized would necessarily be largely arbitrary. Even so, and recognizing the simplification in which we are engaging, but also realizing that our choice closely parallels that of Raimon Panikkar (from a distance, but from a very direct knowledge; see Moncada, 1987, 135f.), we propose a synthesis based on two highly complementary texts by Escrivá, taken respectively from *The Way* and *Forge.*

On this occasion, clearly, the Padre was not joking:

> Caudillos! . . . make your will manly so God will make you a caudillo. Don't you see how those infamous secret societies operate? They have never won over the masses. They get a few demon-men together in their dens who agitate and stir up the multitudes, making them go wild, so that they will follow them over the precipice, into every excess . . . and into hell. They spread an accursed seed.
>
> If you wish . . . , you will spread the word of God, blessed a thousand thousand times, which can never fail. If you are generous . . . , if you correspond, with your personal sanctification you bring about the sanctification of others, the kingdom of Christ: May all go with Peter to Jesus through Mary. [*The Way,* no. 833, quoted in its entirety; the last words are in Latin in the original: "omnes cum Petro ad Jesum per Mariam"]
>
> The enemies of God and of His Church, driven by the indestructible hatred of Satan, are ceaselessly stirring and organizing. With "exemplary" constancy, they ready their regiments, maintain schools, directives, and agitators, and, acting deceitfully—but effectively—they propagate their ideas and spread their seed to home and workplace, destructive of all religious ideology.

What must Christians do to serve our God, always with the truth? [*Forge*, no. 466]

If we had not already said that in a selection like this simplification is inevitable, we might well be tempted to exclaim: "Here it all is!" (1) We have the division of the world into good and evil, into Christians and enemies of God and the Church: on one side, the saintly, the generous, the virile, the apostolic, the devout; on the other, the demoniacal, the destructive, marked by hatred, agitation, excess. (2) We have the language, loaded with negative connotations: "infamous secret societies," "den, precipice, hell," "accursed seed," "indestructible hatred," "deceitfully," "destructive seed," and the "manly will" of the "caudillo" as the only instrument capable of standing up to the situation. (3) We have the "supernatural" means which this "caudillo" must employ: the "personal sanctification," the apostolate ("spread the word of God"), fidelity to the Roman Church ("cum Petro"), devotion to the Virgin. (4) We have the "material" means: "what must Christians do. . . ."

Let us reconstruct Escrivá's sentence, changing only the qualification: the members of Opus Dei, "with exemplary constancy (no quotation marks), prepare their regiments, maintain schools, directives, and apostles (not agitators), and, acting discreetly (not deceitfully) but effectively, propagate their ideas, and spread to home and workplace the blessed (instead of destructive) seed of religion." This is not at all a bad definition of the pattern of conduct of Opus members: the "good guys" have to oppose the "bad guys" with the same methods the latter use themselves, the only guarantee of successful confrontation.

(5) We have Escrivá's final objective: "the kingdom of Christ," closely linked to the whole tradition of devotion to "the Sacred Heart of Jesus," promulgated above all by the Jesuitical spirituality of the epoch: "I will rule in Spain," the symbol and slogan of Roman Catholic integrism, of antimodernism, of the desperate effort to put the brakes on what has since come to be called the "process of secularization." They had to put a stop to it "with the lash of the whip!" if necessary (Urteaga, 1948, 97–108): "If in your fatherland there is a cowardly hand that brings its lighted torch near the Church of the Powerful, do not be afraid to raise your weapon. We cannot let them kill us! Let no one laugh at a Christian," because "the Christians of today are not cut out to be martyrs, but warriors," and all the world had better know that "the Christians of this generation are ready to die killing" (ibid., 97, 98, 99).

Is this Opus Dei? If we agreed with the "official" literature that Opus Dei has not changed since the day it was founded, then this is Opus Dei. Our thesis is that it has changed, and that today it is not this. But it was in the Spain of the forties.

(*d*) While no less arbitrary in choice of passage, let us be much briefer in the presentation of a text that will serve as a synthesis of some of the principal features of Escrivá's personality.

At a meeting with the first French members of Opus Dei, Msgr. Escrivá said in 1959: "I want you to be pious, cheerful, optimistic, hardworking, and schemers" (Sastre, 381). This is not bad as an approximate portrait of the

Padre. All these are qualities of this paradoxical personality, and the last one is perhaps a good key to the understanding precisely of the paradox. In *The Way* Escrivá states that "the standard of holiness the Lord asks of us is determined by these three points: holy intransigence, holy coercion and holy shamelessness" (no. 387). We might add a new virtue: that of "holy scheming," which is closely linked to the question of methods mentioned above. "When intelligence does not suffice to you, ask for holy scheming, better and more to serve all" (*Surco,* no. 942).

In spite of the fact that in traditional Catholic worship there are already patron saints for practically everything, we propose, with the contagious sense of humor of the Padre, that if the Church decides in fact to canonize him someday, and if the seat is vacant, that the new Saint Josemaría be made the patron saint of schemers.

♦ ♦ ♦

Perhaps some readers think this was meant to be an amusing ending. I hasten to add that to me it does not seem particularly amusing, nor do I believe it could be the ending. The Padre's sense of humor was not the only contagious thing about him. The paradoxes are too. As in those pieces of music that reach an apparent end only to return again to the initial phrases of the movement, I must also return to something I have already said. The whole thing is more exotic than really original; Msgr. Escrivá was a child of his country, of his epoch, and of his Church.

"Scheming" is not the exclusive patrimony of Msgr. Escrivá; whether "holy" or "secular," it is cultivated by Spaniards as a national sport. Influence peddling, tax fraud, using the same methods as your enemy as an effective method of combat, are all virtues practiced "discreetly" but "brazenly," with "holy shamelessness," and in the same way "holy intransigence" is used to defend those realities defined as untouchable: monarchy, constitution, army, and "holy unity of the fatherland."

The whole "model" Spanish political transition that took place after the death of Franco was built on the base of a tacit consensus to camouflage the past, to forget it as if it had never existed. Alexander and Margarethe Mitscherlich say that the "inability to mourn" was a typical feature of postwar Germany. What I say is that the same is true of post-Franco Spain.

Finally, with respect to the "caudillos" and the "lashes of the whip," the "generous Christians" and the Christians "ready to die killing," Msgr. Escrivá and Opus Dei are not especially original either. A large part of the Spanish clergy of the forties thought and said the same thing. The Spanish bishops applauded Escrivá, and some of those who did not applaud, as well as those who turned against him, led by the Jesuits, did not attack Escrivá to express radical disagreement with his ideas: they attacked him because he was giving them competition, and because they wanted to be the ones to have absolute priority when push came to shove.

Thus, if any readers feel like interpreting these pages as a "devastating" critique of Msgr. Escrivá's character, let them at least know that this has not been my intent. I want to try to understand; and in the process of seeking that understanding, simple discrediting turns out to be just as misleading as pure mythologization. Escrivá was the child of a specific church and a specific society. The Spaniards of today are also, if not the children, then the grandchildren of those institutions, and that includes those who, while not Catholics today, were socialized in that culture. Spanish society of today, on the left and the right, and today's Spanish church, both the "restorationist" and the "progressive," are replete with little imitations of the Padre. If anyone, after having tried to understand him, wants to pass judgment, to pass out merits and demerits, prizes and penalties, that person must make sure the distribution is equitable.

4

From the Official Founding of Opus Dei to the Beginning of the Spanish Civil War (1928–1936)

Synopsis of the Official Version: The Founding of Opus Dei

(*a*) In April 1927 José María Escrivá arrived in Madrid. In June he was named chaplain of the Patronato de Enfermos, and he began to "develop a tireless apostolic labor" (Berglar, 387). He earned his living by teaching at the Academia Cicuéndez. It was not yet exactly the period of "guessing" (that "one fine day, who knows when, God was going to demand something of him, who knows what"; Berglar, 37). It was, rather, a period in which he hoped and prayed that God would reveal his will: *"ut videam"* (let me see, Lord).

(*b*) On October 2, 1928, "he saw," "he saw Opus Dei," "God made him see Opus Dei." He saw it "as God wanted it, as it was going to be at the end of the centuries" (Vázquez, 113). And he saw it, furthermore, with total clarity: "It was not a generic inspiration, destined to become concrete with the historical task, but a precise and specific illumination" (Illanes, 1982, 87). He saw, in final form, "the specific vocation that divine providence had reserved for him since eternity" (Berglar, 68).

The Opus Dei authors comment on this date and its significance ad infinitum. Vázquez de Prada explains what the weather was like that day in Madrid (p. 16), what was in the papers (p. 20), what number was drawn in the lottery (p. 21), and what films were playing in the theaters in Madrid (p. 24), finishing up with a description of the voyage of the Zeppelin on that same October 2, the day that Hindenburg celebrated his eighty-first birthday (p. 25).

On a different level, Fuenmayor, Gómez-Iglesias, and Illanes define "the physiognomy of the Work just as the Founder saw it," enumerating no less than twelve fundamental features: the call to the valuation of professional work; the call to live the faith radically, in response to the "intervention of God in history," which the birth of Opus Dei presumes; the call to personal sanctification, because "it is not a matter of carrying out a human enterprise, but of participating in the divine adventure of Redemption"; the call to the apostolate; the call to unity of life; the call to men and to women (although on this point the authors admit that it was not until 1930 that "God made him understand that the light received a year and a half before had to be communicated to women also"); the call to single and married persons (reserving "specific functions of direction or training" for the first); the call to priests and laymen; the call to the valuation of intelligence (hence "the appreciation don Josemaría manifested toward the intellectual professions, aware of their social transcendence"); the call to the recognition of the "complete freedom of the members in all professional, social and political questions"; and, finally, the call to the universality or international nature of Opus Dei (Fuenmayor et al., 39–47).

(*c*) After this day on which "God deigned to enlighten him" (Berglar, 67), Padre Escrivá did not rest. He intensified his life of prayer and mortification and began to seek out "persons who could understand and live the ideal that God had shown to him" (Berglar, 387). From that moment on, Escrivá said, "I never had a moment's peace, and I began to work, reluctantly, because I resisted making an effort to found anything; but I began to work, to move, to do: to lay the foundation" (quoted in Illanes, 1982, 66).

Other testimonies corroborate this resistance to undertaking a new foundation: "I was not interested in being the founder of anything" (quoted in Vázquez, 115); "I did not want, I never thought about or wanted to make a foundation" (quoted in Sastre, 96). The Padre began to study several Catholic organizations that had arisen recently in Italy and central Europe, "thinking that if any of them resembled what God had shown to him, he would join it to be the last one of all" (Helming, 23). But the information he found convinced him that this was not the case; therefore, there was nothing left for him to do but "enter the breach alone" (ibid., 23).

Allegedly, Opus Dei had no immediate precedents. As Álvaro del Portillo writes, "If one keeps in mind the length of time—many centuries—between the holy life of the first followers of Christ and the spirituality of the Work, one can see why I cannot point to any immediate precedent for Opus Dei" (Portillo, 1978, 40). What is this long parenthesis which closes with Padre Escrivá's "vision"? "At the end of the centuries, he again remembered the whole humanity that man had been created to work" (Sastre, 93). That, according to the "official" version, is what made Msgr. Escrivá into "a pioneer of lay holiness" whose mission "has become one of the fundamental elements of the ecclesiastical renewal provoked by the Lord during the last few decades" (Alonso, 1982, 229).

(*d*) In February 1930, shortly after Escrivá wrote that there would never be women, "not even in jest," in Opus Dei, God made him see that he indeed wanted women in Opus. Another thirteen years would pass before God, "interfering in his life again," showed him that he also wanted there to be "a priestly body or nucleus," since "that is the structure of the Church, and the one which must also be reproduced in Opus Dei" (Fuenmayor et al., 118f.).

This is how the triple foundation of Opus Dei was consolidated: the original Opus Dei, the women's section, and the priestly society. Of the first, Escrivá said, "And I have to say that I did not found Opus Dei. Opus Dei was founded in spite of me. It was the will of God that has come true and there it is. I am a just poor man who was simply in the way, so do not call me founder of anything" (quoted in Vázquez, 472). On the women's section he said: "I assure you with physical assurance—yes, physical—that you are daughters of God. You did not have a founder: your Founder was the Holy Virgin" (ibid., 116). Of the priestly society he said: "On February 14, 1943, after seeking and not finding the juridical solution, the Lord wanted to give it to me, precise and clear" (ibid., 233). In sum, again according to the Padre: "The founding of Opus Dei happened without me; the women's section went against my personal opinion, and as for the Priestly Society of the Holy Cross, I had been looking for it and not finding it" (ibid., 234).

A month after founding the women's section, on March 24, 1930, the Padre wrote the first letter to those whom, from then on, he would call "my children." Berglar dwells at length on the significance of this document, reproducing it and commenting on several passages (Berglar, 94–101). The letter, which has "great importance *for the history of the Church,*" has been translated into Latin, "the language of the Church," and is known as *Singuli dies,* since "as is normal *in these cases,*" such letters are customarily cited using their initial words (Berglar, 94; emphasis mine). We should point out that this is, in fact, the procedure customarily used in the case of papal encyclicals. We should also call attention to the fact that the Padre wrote the letter *Singuli dies* at a time when there was no one, *not a single member,* in Opus Dei, which certainly is an innovation with respect to pontifical customs.

(*e*) Toward the end of 1930 the first members of Opus Dei began to appear, "as the fruit of the enormous labor of the Founder with persons of every condition: men, women, priests, students, workers, ill people" (Berglar, 387). In the chronology established by this author at the end of his volume, for the period 1931–1932 he verifies only that Escrivá ceased to be chaplain of the Patronato de Enfermos and became chaplain of the Recollet Augustine nuns of the convent of Santa Isabel, and that "on Sundays he went with a group of students to visit the patients in the General Hospital" (Berglar, 387).

(*f*) In December 1933, with the republic now in full swing, and six months after a law was passed prohibiting religious congregations to create or run private schools (Vázquez, 135), Escrivá inaugurated his first apostolic work: "the first corporate labor was the academy we called DYA—Derecho y Arquitectura [Law and Architecture]—because classes were given in those two fields; but for us it meant Dios y Audacia [God and Audacity]" (Escrivá, 1975, 27).

The academy "functioned like a cultural and teaching center"; in addition to classes "in professional subjects, there were some courses in religious and apologetic training," taught not by Escrivá but by a priest friend (Sastre, 173;

see also Bernal, 197). At the start of the next academic year, the academy moved to more spacious quarters, which also housed a student residence, "in a distinguished neighborhood" of Madrid (Pérez Embid, 1963). Escrivá asked the bishop for permission to install a chapel in the building. This petition is the first document in the documentary appendix of Fuenmayor et al. (p. 509). In it, Escrivá presents himself as spiritual director of the academy-residence, whose technical director is the architect Ricardo Fernández Vallespín, one of the Padre's first "disciples." According to the petition, classes in religion were given at the academy, "as well as the cultural ends that go with it." In addition, "we practice works of devotion with the students and residents of the House and with other students of all the faculties and special schools, explaining the Holy Gospel to them, going on monthly retreats, teaching the catechism in outlying neighborhoods, etc."; and because of all this they wanted to be able to have a "Chapel and Sacrarium reserved for His Divine Majesty" in the house (Fuenmayor et al., 509).

After the academic year 1935–1936 the residence moved again, "to a noble palace on the same street, Calle de Ferraz" (Pérez Embid, 1963); but one week after authorization was requested to move the chapel as well (Fuenmayor et al., 510), the military uprising of July 1936 broke out, and during an attack on the barracks next door, the academy's building was destroyed.

(*g*) Finally, it must be pointed out that in 1934 Escrivá published, in addition to *Santo Rosario,* a "profoundly poetic and intimate work" (Berglar, 168), a little book entitled *Consideraciones espirituales,* published by Imprenta Moderna, the old press of the Seminary of Cuenca, a city whose bishop, a distant relation of the Padre, had offered him a canonship some time back.

Consideraciones espirituales, a work of which it is practically impossible to get a copy today, is presented in the "official" literature of Opus Dei as a first version of *Camino,* with the latter being simply an amplified second edition. For example, it is said that *Consideraciones espirituales,* "in its second edition published in 1939 without modifications and with a few additional chapters, was given the name *Camino*" (Gómez Pérez, 1976, 253). Escrivá himself writes: "In 1934 I wrote a large part of this book (*Camino*), summarizing my priestly experience for all the souls—whether of Opus Dei or not" (*Conversations,* no. 36). In the preliminary announcement of *Consideraciones,* Escrivá had said that it was "notes, written without literary or publicity pretensions, in response to the needs of secular university youths taught by the author." In the same year, 1934, in a letter addressed to the vicar general of the diocese of Madrid, he writes of the imminent publication of a "pamphlet," explaining that it is "notes that I use to help me in teaching the young students," and adding that "they have no pretensions nor importance," and that "they are only useful for certain souls, who really want to have the interior life and excel in their profession" (Bernal, 199).

In any case, the little 1934 book "has 438 points of meditation" (Gondrand, 94), while *Camino* has 999, more than twice as many. No one has ever done a comparative analysis of the two texts; it would no doubt be a very interesting project.

Some Unanswered Questions

It is logical that this initial and relatively distant period would raise many unresolved questions, or at least questions not resolved in a satisfactory manner. It is convenient to subdivide them—even if just for the sake of greater expository clarity, since in fact all of them are closely related—into questions which basically pertain to the personal history of the Padre, questions relating to Opus Dei as a movement and nascent organization, and questions relating to the appearance of Escrivá's first followers and, hence, the earliest members of Opus.

Padre Escrivá in Madrid

Although apparently these are minor questions, in the case of the history of Padre Escrivá during these years one must take into consideration, on the one hand, the topic of his incardination, or assignment, in the diocese of Zaragoza and of the permission granted to him to reside in Madrid and, on the other hand, the reasons and objectives for this move from Zaragoza to Madrid.

The first of these questions has been posed rather insistently by Giancarlo Rocca (1985, 9–14), according to whom there might have been some irregularity produced in the juridical situation of Padre Escrivá. After the two-year permission that had been granted in April 1927 expired, without his having returned to Zaragoza or having any intent to do so, it is not known exactly what Escrivá's official position was. Rocca explains that in spite of several attempts he made to clarify this, he could not find out from the bishop of Madrid if Escrivá had become part of the clergy of the Madrid diocese, and in the diocesan curia of Zaragoza there is no proof at all that he ceased to belong to that diocese (Rocca, 1985, 13). The author comes to the conclusion that Escrivá was not incardinated in the diocese of Madrid until 1942: that is, thirteen years after his permission expired, with the Spanish civil war intervening. Fuenmayor's book, without ever mentioning the questions raised by Rocca, confirms this last detail (Fuenmayor et al., 26 n. 2). Before the publication of Rocca's book, the "official" literature never alluded to the question (Gondrand, 47; Berglar, 59); afterwards, however, it is mentioned, but without any precision whatever: the initial period of two years "would keep being extended" (Sastre, 82); the "opportune canonical authorizations" were renewed for Escrivá in 1929, 1930, and 1931, "the last time for a period of five years" (Fuenmayor et al., 26).

None of this would be of any particular interest for us here except that the question might be closely related to the reasons for Escrivá's desire to leave Zaragoza. Apparently Escrivá the seminarian had maintained excellent relations with the then archbishop, Cardinal Soldevila. But Soldevila died in 1924 (the victim of an assassination), and his successor, Msgr. Rigoberto Doménech, did not treat Escrivá with so much deference. Thus Escrivá, who wanted to be a priest but didn't want, as we have seen, "to be the typical parish priest" (Bernal, 65), on the day after he sang his first mass was sent to a tiny rural parish. After a month and a half he was already on his way back to Zaragoza, possibly under the pretext that he had to finish his law studies. As soon as he received his degree, he was sent to another village, arriving on April 1, where he stayed until the 17th of the same month; on the 19th he left for Madrid, with the famous permission to study for his doctorate.

Leaving aside the fact that once in Madrid he devoted himself to many other activities, but without getting his doctorate, the question remains of why he wanted the doctoral degree. And it looks as though either it was merely a pretext to flee Zaragoza, or else he wanted it because at that moment he planned to devote himself to teaching. It might be pure coincidence, but it is not certain that it was mere coincidence: on the epoch of the seminary and the University of Zaragoza, Ana Sastre mentions only the names of three of Escrivá's friends: Don Félix Lasheras, Professor Legaz Lacambra, and Msgr. José López Ortiz (Sastre, 68). The first, a priest and military chaplain, taught Latin for many years (bibliographical references are not necessary in this case: he was my teacher at a school in Barcelona); the second, professor of law (see López Rodó, 1990, 27) and—another coincidence?—the Spanish translator of Weber's *The Protestant Ethic and the Spirit of Capitalism;* and the third, also a professor before becoming bishop and "military vicar general" (ibid., 29).

The Date of the Founding

These are relatively minor questions, however. For this period (1928–1936), it is clear that the basic theme is the foundation of Opus Dei, officially dated 1928. There are also unresolved problems about this matter, or rather problems resolved in an unsatisfactory manner.

From an analysis of the official literature it can be deduced that before October 2, 1928, Escrivá had no intentions of founding anything. At no time does he speak of the existence of any project prior to the divine "illumination." Without denying the possibility that it really might have happened that way, there is no doubt that the argument serves, in passing, to avoid any recognition of the existence of any type of influence on the origin of Escrivá's idea. By positing things in this way, in fact, the atmosphere that Escrivá discovered on his arrival in Madrid would have absolutely nothing to do with the founding of Opus, since (*quod erat demonstrandum*) Opus is the direct manifestation of the will of God.

Still, what was this atmosphere like? Remaining faithful to the principle of relying as far as possible on authors belonging to Opus Dei, let us look at how Antonio Fontán describes the ambience of the Madrid university where Escrivá registered for his doctoral studies: "The dynamic sectors of the Spanish university faculty were predominantly pro-revolutionaries, a-Catholic or neutral in the religious sphere, and in some notable cases, openly anti-Catholic" (Fontán, 1961, 22). "Professors who were members of religious organizations were few, and in the most important universities, as for example in Madrid, the cultural and scientific movement was led by men of the left or agnostics and secularists, even Marxists" (ibid., 23). There was a "decline in intellectual creativity among Spanish Catholics . . . [who forgot] that their mission was to maintain an active presence in the affairs of the world, and especially in the most noble and fecund areas: science, education, culture" (ibid., 27).

We find ourselves, Fontán continues (basing his argument on a book by Suñer, *Los intelectuales y la tragedia española,* 1937), with an authentic revolution, the work of "the political ideologues who conquered the university in the first third of the twentieth century," and who postulated "the implantation of a new Spain, secularized, running counter to the Catholic tradition, which had been the backbone of the whole Spanish national tradition" (Fontán, 1961, 31). So, "the point of departure for this profound revolution was Giner and the *Institución Libre de Enseñanza,"* an initially pedagogical movement that became political, fought against General Primo de Rivera and against the monarchy, and contributed to the establishment of the republican regime, allying itself "with the extremists of socialism and Spanish anarchism," and finally becoming "a revolutionary clique of dubious intellectual sincerity, which proposed to take over all the leadership positions in Spanish cultural life, without worrying about what sort of means they used to accomplish this" (ibid.).

When we read this last sentence we can understand how Aranguren could maintain, many years ago, that Opus Dei was precisely "a parody of the Institución Libre de Enseñanza" (Aranguren, 1962, 12), and that Escrivá's project consisted, in the last analysis, of copying the model but inverting its objectives. If the Institución founded by Giner, with its Committee for the Amplification of Studies, had contributed to the de-Christianization of Spain, the Opus Dei of the forties, with its control over the Higher Council of Scientific Research (which even occupied the same offices!) had to "re-Christianize Spain from top to bottom, working from the university" (ibid., 4).

The "official" studies of Msgr. Escrivá and the founding of Opus Dei, however, say not a single word about any of this. Whenever any member of Opus Dei writes about the Institución Libre de Enseñanza, it is done in a context in which no comparison with Opus could fit (in addition to Fontán, 1961, see, in a much less critical tone, Orlandis, 1967, and Cacho, 1962), whereas when they write about the origins of Opus Dei, they do not even mention the Institución (on the possible relations between the Institución and Opus, see, in addition to the text by Aranguren already quoted, Casanova, 1982, 152ff., and especially Artigues, 1971, 20ff.).

Likewise, in reference to the founding of Opus Dei the "official" literature does not point out the multiple initiatives of the Jesuits with which Escrivá came into contact when he arrived in Madrid. We will discuss this later, when we talk specifically about the role played by the Jesuits in the history of Opus Dei. For now, we will simply say that a person like Ángel Ayala, founder of the National Catholic Association of Propagandists—a movement with many similarities to Opus Dei, and to which many of Escrivá's first disciples belonged—is never once mentioned in any of the biographies of the Padre. In the year Opus was founded, the president of the Propagandists was another key person, Ángel Herrera Oria. Bernal and Sastre don't mention him either; Vázquez mentions him once, to report that in 1933 Herrera offered a job to Escrivá, and that Escrivá humbly refused it (Vázquez, 139). Berglar has the effrontery to pretend that Herrera proposed an alliance between his association and Opus Dei, because "the youths of the Work whom he had met had impressed him and he wanted to get them for his movement" (Berglar, 231), when it is a fact that some of these youths had joined Opus Dei after having belonged to the Association of Propagandists. In 1932 (one year before the creation of Escrivá's law and architecture academy), the members of the association founded another academy in Madrid (Centro de Estudios Universitarios) for the study of law. This is not mentioned in any of the biographies.

An exceptional case is that of another Jesuit, Padre Valentín Sánchez Ruiz. He is important for two reasons, first because, according to the legend, it was he who gave Opus Dei its name when he asked Escrivá one day, "so how is this work of God going?" (Gondrand, 62; he goes on to say, "It was like a revelation. If it had to have a name, let this be it: the Work of God, in Latin *Opus Dei.*") The second reason is that Padre Sánchez was Escrivá's confessor, his spiritual director since 1927, that is, before the official date of the founding (Vázquez, 106). One might naively suppose, then, that Padre Sánchez played some role in the origin of Opus Dei.

But no. We immediately see from the way Vázquez de Prada speaks of him from the beginning that the answer is no. When the young Escrivá used to go visit him, he had to "walk a long way." When he arrived, they made him wait, and sometimes "it was a long wait." There were even days when "the wait was interminable. No one came. No one made excuses." And finally, it might happen that after he had waited such a long time they would tell him "that Father Sánchez Ruiz could not see him." Vázquez exclaims, "Anyone else would have taken this as a serious rebuff" (Vázquez, 106). But not Don Josemaría, of course.

Given the situation, Vázquez de Prada chooses to strip the Jesuit of the honor of having found—even unintentionally—the name for Opus Dei. When he tells the anecdote, he does not specify who asked that question, "How is that work of God going?" (Vázquez, 117). Bernal had already written, "*Someone* asked him, how's that work of God going?" (Bernal, 116), and according to Le Tourneau, it was "a friend of Josemaría" (Le Tourneau, 8). In Berglar's and Sastre's indexes of names, the poor Jesuit does not appear.

Even so, knowing to what extent Opus Dei uses the sacrament of confession, and knowing what Msgr. Escrivá says in *The Way* about the figure of the spiritual director ("one owes great obedience to the director," no. 56; "Director.—You need him.—To offer yourself, to surrender yourself . . . , by obedience. And a Direcor, who knows your apostolate . . ." no. 62), we can assume that if Sánchez Ruiz played no direct part in the origin of Opus, he at least had to know about the apostolate begun by the Padre. But this is not admitted either: "Of course, the confessor did not involve himself in matters of Opus Dei; and anyway Don Josemaría would not have permitted any interference" (Vázquez, 116).

By now we begin to suspect that the matter is going to come to a bad end. In fact, that is the case. Since Gondrand is the only one to explain that it was Padre Sánchez who suggested the name of Opus Dei, let us hear the end of the story from him. When the war was over, Don Josemaría "once more went to confess to him regularly," but one day the Jesuit told him that "the Holy See will never approve of Opus Dei." "Don Josemaría was upset by this sudden change, since his confessor had always seemed convinced of the divine origin of what had happened in his soul on that second of October, 1928." Escrivá could not explain this change of attitude unless it was because the Jesuit "had been pressured to dissuade him from founding Opus Dei." (This expression raises an interesting question: how could Sánchez, in fact, want to "dissuade him from founding" a Work that had been "officially" founded twelve years earlier?) Padre Escrivá changed confessors (Gondrand, 115).

To sum up, then, in 1928 Escrivá had no intention of founding anything, and the atmosphere and the people he met in Madrid exerted no influence on an event that was produced by the direct intervention of God. Furthermore, even after this intervention the Padre continued to resist founding a new organization. He did not decide to do it until, after looking in vain for the existence of some similar initiative, he arrived at the conclusion—as Msgr. del Portillo put it—that "there were no precedents." We will not address here the debate over the existence or nonexistence of precedents: the specialists would have to say if, without even looking outside Madrid, an initiative like Father Poveda's Institución Teresiana has many similarities to Opus Dei; outside Spain, many preexisting foundations more or less resemble it.

To make a long story short (and run the risk of simplifying a little too much), if the specificity of Opus Dei really resided in what its highest authorities claim today ("[its goal] to promote among persons of all classes of society the desire for Christian perfection in the midst of the world" [Escrivá, *Conversations,* no. 24]; and "from the first moment the only objective of Opus Dei has been what I have just described: to contribute to there being, in the midst of the world, men and women of every race and social condition who succeed in loving and serving God and other men and do it in the course of their ordinary work" [ibid., no. 26]), perhaps it would be best to invert the terms of the question. Perhaps, instead of asking if any precedent exists or does not exist for Opus Dei, we should ask if there exists in Opus Dei any truly original element.

The unresolved issues surrounding the founding of Opus Dei can be summarized in a threefold question. What was new about what Escrivá founded in 1928? Does the 1928 foundation match the description that Opus members give of it today? In what sense can it be said that 1928 was the year Opus Dei was founded?

The First Disciples

Not even the question of who Escrivá's first disciples were is easy to answer. The list has to be patiently reconstructed from the labyrinth of names and facts presented in the different biographies, a disorderly mixture which seems deliberately designed to create confusion rather than clarity, either with the aim of avoiding saying exactly who and how many were the members of Opus during these first years, or else through application of the principle—unwritten but easily verifiable—that the names of those who have left the Work must not be mentioned.

We saw earlier that, according to Berglar, after 1930 Opus Dei began to include "persons of every condition: men, women, priests, students, workers, ill people" (Berglar, 387). In accordance with the clearer scheme of Fuenmayor et al. (pp. 36f.), we will concentrate on the first three categories, and later we will see if there are any students or workers (any of them could be ill). According to these authors, in fact, "the sum total of those who have heard his message and followed him" (Fuenmayor et al., 36) are:

- Seven or eight priests, to whom he spoke of the Work "and on whose cooperation he counted, to a greater or lesser degree." There is only one name, that of José María Somoano, who died in 1932, "probably poisoned" (Vázquez, 135). In the biographies a couple of other names are mentioned: Lino Vea, "another priest who wanted to join the Work" (Berglar, 118), and who died at the outbreak of the civil war; and Norberto Rodríguez (Sastre, 121). The rest of the names may not be mentioned because, as Vázquez says, "some of those good priests turned out to be his 'crown of thorns' when they did not adjust to the same spirit as the Founder" (Vázquez, 232). In any case, the truth is that on the day before the priestly ordination of Álvaro del Portillo and two other members in 1944, the only priest in Opus Dei was José María Escrivá.
- "A few women" (Fuenmayor et al., 36), without giving names or specifying the number. The very first one seems to have been María Ignacia García Escobar (Sastre, 114), who died in 1933. But none of them "absorbed the specific spirit of Opus Dei." In 1939 Escrivá "decided to begin this work over again almost from scratch" (Fuenmayor et al., 37). As can be verified, then, by the end of this initial period Opus Dei had among priests and women its first "martyrs" (Berglar, 118f.), but not a single member.
- "A small group of men, members of the Work" (Fuenmayor et al., 36). Of those who joined before 1933, one died in 1932—Luis Gordon—and "the rest did not last," with the single exception of Isidoro Zorzano, who died in 1943. But in 1933 "the panorama changed," and when the war broke out Escrivá "now had ten or twelve men" (ibid.; it is so unlikely that the authors do not know whether there were ten or twelve that the only possible conclusion is that they do not want to specify).

Who were these ten or twelve?

Isidoro Zorzano. "Outside the Work, his name will be known when his Cause of Beatification, initiated in 1948, is concluded" (Berglar, 131). Since this could be a long wait, as it is obvious that Zorzano's "cause" is not progressing at the same pace as that of the Padre, we will jump ahead a little to say that he had been a colleague of Escrivá in Logroño and that he was from Argentina, a fact that allowed Josemaría, ever attuned to the "universal nature" of the Work, to remark in 1930, "We already have persons from both hemispheres in Opus Dei" (Sastre, 144). Zorzano was an engineer.

Juan Jiménez Vargas. When he met the Padre he was a medical student, and he joined Opus Dei in January 1933 (Gondrand, 87). When the war broke out he lived with the Padre, in hiding; he was arrested and imprisoned; he accompanied the Padre in his flight to Barcelona and the Pyrenees; when they reached the Francoist zone, he was mobilized (Sastre, 220). Made a professor at the age of twenty-seven or twenty-eight (Barcelona, 1941), he was one of the promoters of the Faculty of Medicine of the University of Opus Dei in Navarre beginning in 1954 (ibid., 422).

José María González Barredo. He was a chemistry student who joined Opus Dei in 1933. He put up the money to rent the first apartment where the DYA academy started out (Sastre, 150). It was he who, in the early months of the war, gave Escrivá the key that ended up in the sewer drain (ibid., 198), and later was one of those who took refuge with him in the Honduran legation in 1937 (Berglar, 172). As a professor of physics and chemistry, he taught at two American universities (Sastre, 148). He also translated *Camino* into German and English (ibid., 150f.).

Ricardo Fernández Vallespín. An architecture student, he met the Padre in May 1933. A year and a half later, Escrivá spoke to him about the Work, "of which he still has no news" (Sastre, 154), and he decided "to be of that" (a "that" to which "he did not succeed in putting a specific name") (ibid., 155). He was director of the DYA academy. The day before the military revolt, he went to Valencia, where he had to organize another residence (in Madrid Zorzano took his place). Son of a soldier, he had two brothers in jail for having participated in the military "pronouncement" of 1932 (ibid., 152). He was a university professor ordained as a priest in 1949. He pioneered the expansion of Opus Dei in Argentina (ibid., 398).

Álvaro del Portillo. He met Escrivá through a relation: Escrivá used to ask all the women who volunteered at the Patronato de Enfermos "if they had any young relatives, students" (Sastre, 120). He was a highway engineer who became a member of Opus Dei in 1935 and hid with the Padre in Madrid during the war. In 1938 Portillo's mother and Escrivá both found themselves in Burgos: "The Founder, by divine illumination, told her categorically: On the 12th your son will cross over" (Vázquez, 196). On the "intuited" day, in fact, he crossed the dividing line of republican Spain; he was mobilized and sent to a village near Valladolid. When he was ordained in 1944, he was one of the first three Opus priests. An inseparable companion of the founder, he has lived in Rome since the forties, and he became Escrivá's successor in 1975.

José María Hernández de Garnica. He found out about Opus Dei in 1935. A student of mining engineering, he was "a wise guy, and *madrileño* to the core" (Vázquez, 155). Imprisoned in Madrid in 1936, he was about to be shot but was moved to a prison in Valencia (Sastre, 159) and joined the republican army (Vázquez, 191). Ordained priest at the same time as Portillo in 1944, he was initially in charge of the women's section of Opus. Later he went to France and England (ibid., 298), and during the sixties he was one of the Opus pioneers in Germany (Berglar, 168).

Francisco Botella. A student of architecture when he met the Padre in 1935 (Vázquez, 161; Sastre, 162, says he was a student of exact sciences and architecture), he was one of the residents at the DYA academy. From Valencia he went to Barcelona "with false papers" (ibid., 210; he was a "deserter from the republican army," according to Berglar, 178) to join the Padre's little group on its journey across the Pyrenees to Franco's Spain. Mobilized to be sent to Pamplona, a month later he was sent to Burgos, where he stayed in the Hotel Sabadell with Escrivá. A professor of analytic geometry (Barcelona, 1941), in 1946 he was ordained a priest.

Pedro Casciaro. A comrade of Botella, he followed the same course of studies, and their paths immediately before and during the war were identical. They are the "clowns" of the Opus Dei of these years (Berglar, 189). Casciaro was the son of a professor with republican convictions (Bernal, 49), and his family background made him suspect among the military at Burgos (Vázquez, 190f.). A younger brother, José María, would become dean of the Faculty of Theology in Navarre (Bernal, 294). After 1940 he ran a residence in Valencia, a source of future distinguished members of Opus (Sastre, 255f.). Ordained a priest in 1946, three years later he was sent to America, where he became the first to promote the expansion of Opus Dei in Mexico (Berglar, 92).

Up to this point all the biographies of Escrivá concur. They give different information, but once put in chronological order, it is consistent. There were eight young men, all bachelors (and all committed to celibacy). In spite of Opus's insistence on the "lay" character of their spirituality, by 1944 most of these first disciples were ordained as priests. In spite of the fact that, according to Berglar, the first members of Opus already included "persons of all conditions," they are all university graduates or university students. And in spite of the ideological pluralism which Opus claims it has always had, it is clear toward which side its preferences leaned in the context of the Spanish situation of the times: some of them were caught in republican Madrid; they hid, possibly to escape induction into the army; in some cases they are expressly recognized as "deserters"; and once they cross over from "red" Spain to "Francoist" Spain, all their resistance to the military evaporates immediately.

But for the period leading up to the beginning of the war we should have had "ten or twelve" (according to Fuenmayor et al.), and the fact is that up to this point the biographies of Escrivá mention only eight names.

For example, José María Albareda does not appear on the list, because he did not meet the Padre until 1937 (some sources say 1936; Álvaro d'Ors, 1974). Dr. Albareda was not only one of the members of the expedition that crossed the Pyrenees into the "national" zone, but he also played a prominent role in making it possible. He was already by that time a distinguished scientist who had been to Germany, Switzerland, and England. After 1939 he became the key member of the Consejo Superior de Investigaciones Científicas. He was not ordained until 1959. In 1962, four years before his death, he became rector of the University of Navarre. (On Albareda, see Gutiérrez Ríos, 1970).

Tomás Alvira was in the same initial situation. A friend of Albareda, he also joined the expedition that fled republican Spain in 1937. In Alvira's case, however, there is a decisive factor that prevents him from being on the list of Opus's first members: he had a "vocation to marriage" (*The Way,* no. 27).

Another person who might be on the list (and in fact some authors seem to include him; for example, Vázquez, 155) is José Luis Múzquiz. His trajectory is very similar to that of those we have already discussed: he met Escrivá in 1934 or 1935 (Bernal, 303); during the war he appears once in a while at the hotel in Burgos where the Padre was staying (ibid., 133). A highway engineer, he became a priest in 1944 (one of the first three, moreover). He had a lot to do with Opus Dei's first contacts with people in Barcelona (Sastre, 261) and Valladolid (ibid., 259). In 1946 he was sent to Portugal (ibid., 366), and in 1949 to the United States (New York and Chicago), where he became a pioneer for Opus Dei's work ("Father Joe"; ibid., 388f.), before beginning the work in Japan nine years later (ibid., 464). But Berglar expressly states that he became a member of Opus Dei only after the Spanish civil war was over (Berglar, 219).

In order to complete the list of "ten or twelve" first members, we shall have to look for other names, although the information offered by the biographies of Msgr. Escrivá is scanty.

Pepe Isasa, for example, who met the Padre in 1932 and died on the front during the war, is not mentioned in most of the biographies. In fact, the only one who mentions him is Sastre, but she seems to consider him a member of Opus (Sastre, 146).

This author is also the only one who seems to date the relationship between Escrivá and Fernando Maycas before the war (Sastre, 120); Maycas was the man who started the work of Opus Dei in Paris (ibid., 376).

Of Jenaro Lázaro, artist and sculptor (Bernal, 344), it is said that he was one of those who accompanied Escrivá on his first visits to patients in the hospital (Vázquez, 133). For once we would have someone who was not an engineer, an architect, or a doctor; but there is no evidence that he was ever a member of Opus Dei.

We could also mention Pepe Romeo, another of "the boys who accompanied him on his hospital visits" (Sastre, 143). We do not know his profession, but we know that Escrivá was a friend of his family (Bernal, 162). And, above all, we know that the Padre had gone to his house on the same day he later met Isidoro Zorzano, and that Romeo was also directly involved in the first meeting between Escrivá and Ricardo Fernández Vallespín. No one states, however, that Romeo was ever a member of Opus.

What about Manuel Sainz de los Terreros? An engineer, he had accompanied the Padre on several occasions before the war (Sastre, 184; Bernal, 288) as well as on the 1937 expedition. Berglar, for example, mentions his name repeatedly throughout his account of this episode (Berglar, 176–83), but after this event he disappears definitively from all the biographies.

Was another student "who frequented the DYA residence" (Sastre, 201), named Eduardo Alastrué, a member of Opus? He was one of the men with whom the Padre took refuge in the Honduran legation in March 1937. Furthermore, during this period it was he who wrote down "the conversations and meditations of Don Josemaría" (Sastre, 202). In 1938, he accompanied Álvaro del Portillo on his flight to Burgos (ibid., 236). Why is his name not even mentioned in the rest of the "official" literature? If, according to the legend of the forties, there was a "Judas" among the "twelve apostles" of the Padre, it was probably one of the names just mentioned.

Finally, two more people remain who became members of Opus, although the biographies hardly mention them. Vicente Rodríguez Casado, a soldier's son, was another of those who went into hiding in Madrid at the outbreak of the war. He stayed in the Norwegian embassy for nearly two years (Sastre, 236), until he fled with Portillo and Alastrué (Vázquez, 196). Bernal tells us he was a professor (Bernal, 164), and Sastre says he already belonged to Opus Dei in 1936 (Sastre, 199; see also Gondrand, 111).

As for Miguel Fisac, an architecture student, the "official" literature tells us only that he hid in his house in the region of La Mancha at the beginning of the war (Berglar, 178), and that he also joined the 1937 expedition across the Pyrenees. If we had to depend on Vázquez de Prada, we would not even know his name, since all he says is that the expedition was joined by Pedro (Casciaro), Francisco (Botella), "and another student" (Vázquez, 179). Miguel Fisac tells us a little about his sojourn with Opus Dei and his subsequent separation from it in one of Alberto Moncada's books (1987).

An Alternative Hypothesis

In these last few pages I have progressively, almost imperceptibly, but consciously and deliberately, begun to adopt the approach introduced earlier as "the epistemological model of Sherlock Holmes." Now I would like to close the chapter on this early period in the history of Opus Dei by proposing an alternative hypothesis based on the other model, "the Father Brown model."

In fact, using the scattered and unsystematic information provided in the "official" literature, we can attempt to reconstruct a few facts—for example, the list of the first members of Opus Dei and the date of their first commitment to Msgr. Escrivá's Work. But in so doing we are indeed taking for granted the decisive element: the very existence of Opus Dei, from the moment of its official founding in 1928.

If we were to contemplate all this from a different perspective on the other hand, perhaps we would not reach the conclusion that the "official" literature is offering us, stingily, only few scattered and insufficient paths, but rather the contrary, that this literature is unnecessarily leading us down too many paths, perhaps with the objective of making us detour around the principal subject.

Consequently, instead of asking ourselves who the "ten or twelve" first members of Opus were, and whether there were ten or twelve, and when they joined Opus, and if they were all academics or rather came from "all social conditions," why don't we simply ask *what* Opus Dei was in this period, if it was anything?

When we turn the question on its head this way, the first surprising item we find is that while on the one hand "after October 2, 1928, the Founder of Opus Dei preached, with inexorable clairvoyance and power, the holiness of the laity in the world, in professional work, in the family, in all the affairs of men" (Sastre, 93), on the other hand we find that "for many long years" he never spoke of October 2, 1928, out of "humility" and "prudence" (Berglar, 69).

Thus, with the young men he initially met, "the Padre still did not speak of this Work of God whose foundations he was laying" (Gondrand, 58). During the first years, the meeting place was Escrivá's mother's house, where they had "long talks and drank coffee" (Vázquez, 140); but Opus Dei is not mentioned. Just before being appointed director of the DYA academy, Ricardo Fernández Vallespín "has not heard any news about Opus," and responds by saying that "he wants to be of *that,*" without giving it "a specific name" (Sastre, 154). Not even Escrivá's own family knows anything: "he said nothing to his family" (Gondrand, 56); in 1933, when his mother questioned him, Escrivá still "answered in an evasive manner," because "he had not told her what happened in his soul on that October 2" (ibid., 87). His mother and sister remained in the dark until 1934; only then did he "speak to them openly about this divine desire" (ibid., 99). And what is "the reason for his telling them all this?" at that time? An uncle of Escrivá's—Mosén Teodoro, a priest—had just died in Aragon. He had left some property, some land (Vázquez, 141). Now it was a matter of selling this property and of asking the family "to renounce their inheritance" since with the proceeds of the sale "he could underwrite the cost of the Residence he was thinking of opening in October" (Gondrand, 99).

The first conclusion, then, is that for nearly six years Padre Escrivá said nothing about Opus Dei; and one cannot even be certain that when he did mention it for the first time to his relatives, he told them about anything more than his project to open a student residence.

Second, we must keep in mind the fact that during this entire time Opus Dei had no name. "Nor did the Founder at the beginning want his apostolic work to even have a name" (Vázquez, 117). It is true that according to the legend the Jesuit Sánchez Ruiz had unintentionally "baptized" Opus Dei, when he asked Escrivá, "how is that Work of God going?" And it is no less true that simultaneously he had begun to write those letters, like the *Singuli dies* to which Berglar refers, and others repeatedly cited by Fuenmayor, Gómez-Iglesias, and Illanes, in their study. But as long as there is no public and complete edition of all this documentation, no matter how much Opus Dei might be mentioned in it, it is difficult to know exactly what was in the letters and when they were written. In the rest of the documents which are available, the name Opus Dei never appears. In Escrivá's correspondence with the vicar general of the Madrid diocese, for example, there are allusions to the book *Consideraciones espirituales* and to the DYA academy, to "our priestly apostolate among the intellectuals" (Vázquez, 143), and to the "works of devotion with students" (Fuenmayor et al., 509), but the name of Opus Dei is not mentioned.

Various authors (Artigues, Hermet, Walsh, Ynfante) make reference to a so-called *Socoin* (Society of Intellectual Cooperation—or Collaboration) as a name that Escrivá might have initially given to his project. Berglar is the only one of the "official" authors who takes note of this—albeit without quoting any specific author—and he comments that it was an association that "had been founded by several members of Opus Dei in order to give a juridical character to their cultural and apostolic activities" (Berglar, 228).

In any case, the fact that in the entire book *The Way* (whose first edition, you will recall, was published in 1939) *there is not a single reference to Opus Dei,* the fact that the expression "Opus Dei" does not appear even a single time, would seem to lend plausibility to our alternative hypothesis: if during all these years Escrivá never spoke of Opus Dei, and if Opus Dei did not have a name, it is because Opus Dei did not exist.

To pose the question in these terms is the equivalent of taking the jigsaw puzzle of our initial parable, verifying that too many pieces don't fit or are in the wrong place, and taking a good portion of it apart. The possibility exists that if we proceed in this manner we will be taking away more pieces from the full set. We would be ready to admit that probably that is the case, and even that we may have removed some pieces that perhaps did fit correctly. On the other hand, we must recognize that as the puzzle is presented in the "official" literature, there are too many pieces. In other words, to attempt to *totally* eliminate the date October 2, 1928, from the history of Opus Dei is probably going too far; but it would be necessary to have an explanation of the real significance of this date, without exaggerating it, and without making a myth out of it in the worst sense of the word myth.

It would be enough if we were told that the stories about the founding of Opus Dei ought to be read the way one reads the first chapter of the book of Genesis. Then we would know that they do not have to be judged according to the *usual* standards of what is "true" and what is "false." But as long as one pretends that the literary genre of these stories is in the Western historiographical tradition, one must say that what remains of our (incomplete) puzzle is the following:

Before 1936, Opus Dei did not exist. Padre Escrivá, a young priest who did not want to be just a priest and who wanted to devote himself to teaching, went to Madrid in 1927. In Zaragoza he had gotten some experience teaching classes at an academy; in Madrid he resumed this activity, teaching at an academy founded by a priest, José Cicuéndez, which was devoted "exclusively to preparatory courses for the law degree," and which functioned at the same time as a residence for about eight students (Sastre, 81). Escrivá, who did this work "to earn the money he needed to live on and support his family" (ibid., 103), conceived the notion of copying the model, creating a similar academy of his own. The objective would be to see that, as in the case of the Cicuéndez academy, "many students from this academy would go on to occupy important positions in professional life" (Berglar, 81).

In addition, with the proclamation of the republic (1931) and the passing of a series of laws that dismantled the Catholic educational system (the expulsion of the Jesuits, the prohibition of schools run by religious orders), this objective acquired a character of great urgency even for someone like Escrivá, who dreamed—like the Catholic forces with whom he was in contact—of the possibility of a "re-Christianization of Spain." Hence the birth of the Academia-Residencia DYA ("Derecho y Arquitectura" = "Dios y Audacia"). Convinced of the need to "pay great attention to apostolic work with university students" (Fuenmayor et al., 85), Padre Escrivá adopted the arguments of Ramiro de Maeztu as his own:

> We found that what we needed most at that moment were not reasons, but swords, but to have the swords we needed reasons; we had cultivated swords for decades and at the same time we had allowed the men who wielded them to be educated in

> schools where they were not taught what the Spanish monarchy was, what Catholicism was in the life of the nation, what they meant to national unity and to the defense and conservation of the religious spirit in Spain. [quoted in Fontán, 1961, 37]

The program of people like Maeztu, combined with the limitations imposed by republican legislation, prepared the way for Escrivá's project. In 1933 he commented to Juan Jiménez Vargas that "we have to create, among other things, unofficial education centers, imbued with Christian sentiment from top to bottom, but never calling themselves Catholic" (Bernal, 196).

The outbreak of the Spanish civil war meant that for three years, "swords" became more important than "reasons." Escrivá and the students who were his followers saw very clearly on which side they had better put both their "swords" and their "reasons." And the outcome of the war would change everything: it would no longer be necessary to worry about "unofficial education centers." Opus Dei, nonexistent as such before 1936, was born during the final months of the war, or immediately after it, with a sweeping gesture.

We repeat that in the state in which we have just left the puzzle there are missing pieces, and that surely we have eliminated some that actually belong. But at least those that are left all fit together. And, as Father Brown used to say, perhaps we are just beginning "to see for the first time things that we have always known." Perhaps, as we covered the story, we have *discovered* something in the most literal sense of the word, by removing what had been *covering it up*.

5

Three Years of War (1936–1939)

Synopsis of the Official Version: Escrivá and the War

In the preceding chapter, especially in reference to the Padre's first followers, we have already introduced many elements relating to the war years, so we will be able to proceed more succinctly here.

(*a*) When the so-called civil war broke out in July 1936, the persecution to which many priests and monks were subjected forced Escrivá to go into hiding. Furthermore, because there were frequent house searches, he had to repeatedly change his hiding place. As usual, the biographies are brimming with details and anecdotes: various situations of imminent danger, the episode of the key thrown into the sewer because the Padre did not want to live in the same apartment with a young maid, the time he took refuge in an insane asylum, literally passing himself off as crazy. In March 1937 he found shelter in a building that enjoyed diplomatic immunity, the Honduran legation. Meanwhile, as we have already seen, the boys who were with him at the DYA academy found themselves in various situations: some had been mobilized into the republican army, others had been arrested and put in jail, and others were living in hiding with the Padre.

(*b*) In October 1937 the chain of events that would lead Escrivá and a small group into Francoist Spain began. They met in Barcelona, where the clandestine trip to the border was organized, and after traversing Andorra and southern France they reached the Basque Country. In addition to Escrivá, the group comprised Albareda, Jiménez Vargas, Casciaro, Botella, Fisac, Sainz de los Terreros, and Alvira. They had been preceded a few months before by Fernández Vallespín (Berglar, 193). A few others remained in refuge in various diplomatic locations in Madrid: Portillo, Rodríguez Casado, González Barredo. The first two, along with Alastrué, crossed the military front in 1938. Except for Escrivá (and, apparently, Albareda), they all joined different units of Franco's army.

(*c*) In January 1938, Escrivá was living in Burgos, the provisional capital of the "new" Spain. He stayed in the Hotel Sabadell, in the company of Botella and Casciaro (who, probably thanks to the Padre's influence, had been allowed to stay in the city) and Albareda, who performed "consulting services for the Dirección General de Enseñanza Media" (Gondrand, 134). Escrivá, as always, was tirelessly busy: he wrote letters, prepared the edition of *The Way*, and worked on his doctoral thesis in the archives of the Monasterio de las Huelgas. He also traveled a great deal; there are references to his trips to Cordoba, Zaragoza, and Santiago (Gondrand, 135f.), and to Vitoria, Bilbao, Palencia, Valladolid, Salamanca, Avila, Leon, and Astorga (Vázquez, 189). According to Vázquez de Prada, he was "at the front lines as often as at the rear guard," and sometimes even "on the front in the midst of combat" (ibid., 194).

(*d*) The year 1939 began with clear indications that the war was ending. And from the point of view of Escrivá and his friends, it was ending well: "War has been for us," says number 311 of *The Way*. On March 28 Escrivá arrived in Madrid, the same day that the first troops entered the city. "His holy impatience drove him to join the first provisions column to enter Madrid" (Vázquez, 197).

At that moment, Helming explains, Opus Dei "had about the same number of members as the number of years that had passed since 1928" (Helming, 44).

Some Questions: Escrivá and the Franco Regime

Generally the biographies of Msgr. Escrivá devote a lot of space to this period, and with a little patience it is not too difficult to reconstruct the chain of events that affected the little group clustered around the Padre, whom the "official" literature now calls "the members of Opus." Details and anecdotes abound, as always, to appeal to all manner of tastes: from "supernatural interventions" on various occasions to the fact that in one of the apartments were he lived in hiding with Jiménez Vargas, Portillo, and one of Portillo's brothers, Escrivá taught them "to play cards, cheating shamelessly to amuse them" (Vázquez, 169).

Notwithstanding all these details, there remain a number of unclear or insufficiently explained issues. Let us look at some of these, starting from the "clues" offered by the "official" literature itself.

(*a*) Even admitting as logical the need for Escrivá to hide in republican Madrid in 1936, simply because he was a priest, it is obvious that the reasons for which the others went into hiding were not the same. In the case of Portillo, for instance, why could any soldier "ask for his papers and immediately put him in jail" (Sastre, 196)? As for Jiménez Vargas, "they put him in jail and were on the point of executing him" (Gondrand, 119). Hernández Garnica, the same author reports, "was saved from death after being sentenced by a peoples' court." Is it credible that the victim "did not know the reasons" for his sentencing (ibid., 119)? Fisac lived in hiding "to escape certain death" (Sastre, 274). Except for Zorzano, who was an Argentinian national, all the rest should have been mobilized, and if the Opus boys were not in the army, it was not because they did not want to take up arms, but because they wanted to fight with Franco against the republic, as is demonstrated by the fact that as soon as they reached the other side "they presented themselves to the military authorities" (Gondrand, 133). In fact, as Berglar states about three of them, they were "deserters from the republican army" (Berglar, 178).

(*b*) Nevertheless, the biographies of Msgr. Escrivá tend to disguise the clearly Francoist sympathies of all of them. At times they even try to project an image of the Padre as a person who did not take sides in the conflict. However, they do not insist too much, presumably because they realize that it would be perfectly ridiculous to try to deny something so obvious. Quoting this passage from *The Way:* "War has a supernatural end that the world is unaware of: war has been for us.—War is the greatest obstacle to the easy path. But in the end, we will have to love it, as the religious must love his disciplines" (no. 311), Vázquez de Prada tries to explicate it as if "war has been for us" means "to struggle against ourselves" (Vázquez, 196), but Escrivá's language is unequivocal.

Without concrete proof, we would not be so bold as to agree that Escrivá "led a violent anti-Marxist, anti-Masonic, and anti-Jewish campaign" (Ortiz, 8); but neither would we say that Escrivá considered "Franco's victory as the lesser evil" (Berglar, 159), no matter how much the author insists that "these words, lesser evil, are the correct formula." Berglar wrote his book more than forty years later for a German-speaking public, and perhaps he thought that his readers would not notice it; but in 1939 Escrivá—like the immense majority of the Spanish clergy—would have thought that the formula "lesser evil" was not only not correct but a grave insult to "Caudillo Franco" and his "Crusade." (The literature on this topic is vast; for many years Alfonso Álvarez Bolado has been preparing a detailed analysis of the documentation of those years; see, pending the publication of the final version, Álvarez Bolado, 1986–1990).

(*c*) In his volume of memoirs, *Descargo de conciencia,* Pedro Laín Entralgo relates a curious episode that took place in the city of Burgos. The author stayed at the Hotel Sabadell, and he observed that seated at the table next to him in the dining room were a regular group of three or four, one of whom was "a priest, not fat, but roundfaced." After several months of silent daily coexistence, Laín received a letter from a priest who was a friend of his, saying: "I know that you are in Burgos and that you eat every day next to the priest José María Escrivá, whom I know; he is the one who told me. Well, Escrivá says he would like to have a conversation with you. Will you agree to do so?" After noting his surprise at such timidity on the part of someone who would later owe his fame to "holy shamelessness," Laín narrates his conversation with Escrivá, in which the latter did not tell him "a single word about the I don't know whether already born or projected Work" (Laín Entralgo, 238–41).

This episode contrasts noticeably with the impression produced in the biographies of Msgr. Escrivá, which rather emphasize the energy and initiative of the Padre and his many conversations with all kinds of people, "thus preparing the future vocations or spiritual orientations of many of them" (Vázquez, 194), as well as his frequent absences from Burgos. During this entire period, in fact, Escrivá traveled a lot. But beyond the anecdotes about his exhausting himself, about the inconvenience of the journeys, about his lack of money, which at times was not sufficient to buy a train ticket, what remains unexplained is the reason for his trips, and the type of permission or safe conduct he must have had in order to make them.

We do not know, in other words, who he was going to see on these trips, or his reasons for seeing them. If he was going to visit friends or acquaintances on his own initiative, it hardly seems likely that he would be able to travel freely from one city to another, especially if we recall that on several occasions he reached the front itself "in the midst of combat" (Vázquez, 194), without some kind of special authorization. And if he was going on some official mission or assignment, we do not know what it was or who had ordered it. Did he perhaps have, at this time, some kind of connection to the army? Lacking any kind of regular relationship, how can we explain how he was able to "enter Madrid with the soldiers of the army" (Sastre, 239), "at the same time as the first provisions column" (Bernal, 251)? When the bishop of Avila wrote him in 1938 that he wished to work "like you, not as Captain, but at least as a good soldier of Christ" (Vázquez, 192), should we assign any special significance to this phrase?

Hypothesis

These are some of the questions not entirely answered about the years of the Spanish civil war. But in the context of this study these questions, while not insignificant details, are only of relative importance. In connection with the hypothesis presented at the end of the previous chapter, the basic question we must formulate is whether we can know if any new fact is produced in this period that is susceptible of decisively affecting the history of Opus Dei. Because if there were reasons that led us to question the actual existence of Opus Dei as such before the war, there is no doubt that it did indeed exist immediately after the war. And beginning in February 1941 its name starts to appear in official documents. Unlike the situation in the previous chapter, on this occasion our hypothesis will point to the "clue paradigm" as the most appropriate model: that is to say, in this case we lean toward the model of Sherlock Holmes rather than that of Father Brown. It is true that with respect to the secondary questions mentioned above, it might be that the "official" literature provides too many pieces for the puzzle on occasion; but what prevails is a scarcity of clues relative to events of the war years that might have decisively conditioned the future history of Opus Dei.

In its boldest form, we might state the hypothesis by saying that Opus Dei was born, not in Madrid in 1928, but in Burgos in 1938, in the middle of the war. But we cannot supply sufficient proof to support this. More modestly, and more reasonably, we will limit ourselves to stressing the growing importance of the figure of José María Albareda throughout this period. Albareda, who had no relationship with Escrivá until 1937, became a key person during this period. Above all, and unlike most of the boys who gathered around the Padre, he was not a student, but "Professor Albareda" (Sastre, 211). He was exactly the same age as Escrivá; furthermore, he had been abroad, was a doctor, and was recognized as a distinguished scientist. Albareda played a decisive role in the organization of the flight from Madrid; through his family (a brother had already crossed the border, and his mother was in Barcelona), contacts were established that facilitated the secret journey across the Pyrenees (Vázquez, 177ff.). Once in Burgos, he not only stayed with Escrivá in the hotel, but he was the only one of them who was earning money (Gondrand, 134).

Albareda came from the Asociación de Propagandistas (Sáez Alba, Prologue, 74). Before the war was over he performed "consulting services at the Dirección General de Enseñanza Media" (Gondrand, 134). He was a good friend of José Ibáñez Martín, also a member of the Asociación Católica Nacional de Propagandistas. In July 1939 Ibáñez Martín was named minister of education. In November of the same year, when the Consejo Superior de Investigaciones Científicas was created, Albareda was made its secretary general.

Pursuing certain objectives explicitly inspired by the idea of the "re-Christianization of Spain," the Consejo enjoyed a considerable amount of autonomy in research projects and the granting of fellowships for research. It was, consequently, the ideal place for the training of competent professionals who planned to go on to occupy university chairs. Finally, the early selection of these professionals could be accomplished through the creation of a network of university residences, a task to which Escrivá devoted his efforts as soon as the war was over (on July 14, 1939, the contract was signed establishing the first residence in Madrid; Sastre, 241).

We are not in a position to say whether this set of initiatives was part of an overall plan. Nor would we dare to assert that these are the only "authentic pieces" of the puzzle of Opus in that period. But it seems undeniable that these pieces do belong to the puzzle. Whether it started in 1938 as a project drawn up in Burgos, or in 1939 as the more or less deliberate result of the conjunction of all these factors, this was the incipient Opus Dei that began to function in the Spain of the "Year of Victory." For the moment it was no more than that: a very small group of persons close to the Padre, but this time they really existed and had a concrete project, although this project was limited for the time being to the university community. The enthusiasm and energy of this early little group, blessed by the availability of resources as well as by the dominant ideology, gave rise to one of those situations where the degree of success comes as a surprise to the enterprise itself. But it also led to the first serious confrontations with the Society of Jesus. Unexpected success and open conflict with the Jesuits would be the two major characteristic features of Opus Dei's period of consolidation and expansion in Spain, which we will analyze shortly.

But first we will interrupt the chronological sequence in order to devote a few pages to what might almost be considered a symbol of the "boom" of Opus Dei: Escrivá's book *The Way*, published in 1939.

6

1939: *The Way*

General Observations

(*a*) One of the things that most profoundly displeases the authors belonging to Opus Dei is, precisely, to have anyone juxtapose the appearance of the most famous of Padre Escrivá's writings with the end of the Spanish civil war and the installation of Franco's political regime. This is probably the reason that they insist so strongly on emphasizing the direct connection between *The Way* and the *Consideraciones espirituales* of 1934, generally without even mentioning the fact that the 999 points in *The Way* are, in the earlier book, reduced to less than half. *The Way,* it says in the editorial preface to the volume, "appeared for the first time in 1934, under the title *Consideraciones espirituales;* in 1939, the date of the second, enlarged edition, it was given its present and definitive title." Notwithstanding this, the press itself counts as the first Spanish edition the one published in Valencia in 1939, with a printing of two thousand copies, while the official second edition is the one published in 1944.

Rafael Gómez writes, "They have tried a few times to relate *The Way* to the Spanish political atmosphere after 1939. In reality, if a book that deals with the interior life had any kind of link with a political situation, it would have to be the republican atmosphere of the thirties, the epoch of the gestation and publication of *Consideraciones espirituales,* of which *The Way* is the second edition" (Gómez Pérez, 1976, 253). But it does not seem easy to maintain that a language dealing with caudillos, not to mention the direct allusions to the

Spanish civil war, could have existed prior to 1936. It is even a little ridiculous to argue that "*The Way* was conceived in an epoch in which no one even mentioned General Franco as a possible head of government," as well as to "remind us" that after 1946 Msgr. Escrivá lived outside Spain, in order to conclude that the supposed political inspiration of the book is "a lie that must be denounced," and that "it contradicts the history" (Coverdale, 1964, 492).

The goal of *The Way,* writes Msgr. del Portillo, "was simply to place in the hands of the people who surrounded him, to whom he was a spiritual leader—mostly young university students, workers, and patients—some points for meditation that would help them to improve their Christian lives" (in Morales, 45f.). A little further on we find, in this same volume, a statement similar to that of Jesús Urteaga: "*The Way* was written for everyone. Professors, journalists, politicians, diplomats, students, workers, men, and women, appear in its pages as the intended recipients of its penetrating lines" (in Morales, 88). We have already had occasion to comment on this preoccupation with putting "workers" and "laborers" on the list, and we know that in 1939 Opus Dei had "as many members as the number of years that had passed since 1928" (Helming, 44), and that they were all academics, as Escrivá himself, furthermore, declares in the preliminary foreword to *Consideraciones:* "notes written in response to the needs of young secular university students who are under the author's direction."

The facts are intractable, and the desire to deny the evidence often leads to little contradictions. Thus on the one hand we are told that it makes no sense at all to relate *The Way* to the political atmosphere of the Spanish civil war; but on the other hand Salvador Bernal specifies that in Burgos in 1938, the Padre was in a rush to get the pages of *The Way* "printed as soon as possible, in order to facilitate the meditation of those who were at the front or in the navy," and that if the book "was not published until after the war," it was "for lack of economic means" (Bernal, 133). In fact, the book was printed on September 29, 1939, "The year of Victory." It contained a preface by Msgr. Xavier Lauzurica, a bishop known for his Francoist sympathies, who was at that time "apostolic administrator" of Vitoria, where he had replaced Msgr. Mateo Múgica, one of the few who had refused to bless Franco's "Crusade." Monsignor Lauzurica concluded his introduction to *The Way* with these words: "Reader, do not rest; always stay vigilant and alert, because the enemy does not sleep. If you make these maxims your life, you will be a perfect imitator of Jesus Christ and a gentleman without blemish. And with Christs like you Spain will return to the old grandeur of its saints, wise men, and heroes."

(*b*) The success of *The Way* is indisputable, and quite extraordinary, and Opus Dei keeps it constantly in view. In spite of Escrivá's admonition against "spectacles," including "pictures, charts and statistics" (*The Way,* no. 649), many publications contain statistics about the editions of *The Way*. When the forty-ninth Spanish edition appeared (1990), the press listed 235 editions in 38 languages, with a total of more than 3.5 million copies.

This success did not happen overnight, however. The 1939 edition had a printing of 2,000 copies, and the next one, five years later, 5,000. The size of the printings did not begin to increase significantly until 1950 (1950: 15,000;

1951: 15,000; 1952: 21,000; 1953: 20,000; 1954: 20,000). It was first translated into Portuguese (1946), then Italian (1949), English (Dublin, 1953; Chicago, 1954), Catalan (1955), and German (1956). By that time, then, there were editions in five languages total, compared to thirty-eight today, and about 150,000 copies copies published during the first sixteen years, compared to 3.5 million by 1990. Furthermore, of those first 150,000 copies, only 20,000 were published outside the Iberian Peninsula.

Thus while the book's success grew gradually and, at first, almost exclusively within Spain (as the growth of Opus Dei was gradual, first in Spain, then in other countries), it is essential to add that the controversy surrounding *The Way* was also fairly slow to develop. In spite of the fact that at present everything written about Opus Dei tends to quote the maxims of *The Way,* and in spite of the fact that it tends to produce a notable polarization which does not accept the slightest critical remark in the case of the "official" literature, and which tends to adopt a systematically adverse position in the case of the "unofficial" literature, the truth is that during the first years the reception of *The Way* was much less critical and generally positive. As is to be expected, the historical context of Francoist Spain weighed heavily not only on the author of *The Way,* but also on its readers.

If, as we have just seen, during the 1950s there were annual printings of between fifteen and twenty thousand copies, this means that the book had many more readers than there were current or potential members of Opus Dei. In the Catholic world of the period, *The Way* was in fact widely read and appreciated. In movements such as the Cursillos de Cristiandad, for example, it was required reading. Even many of the representatives of what a few years later would become Catholic "progressivism" thought well of it. Josep Dalmau declares that in the early postwar years *The Way* represented "the first breath of fresh air. It was a banner which, both in the university and outside it, sought in its way the routes of sincerity within the framework of that euphoric and victorious climate for the Church. As for myself, that book spoke profoundly to me" (quoted in Casañas, 146).

But the historical context was not the only thing that weighed upon *The Way*'s author and its readers. The phenomenon of *alternation* (understood as the reinterpretation and modification of the past in light of the circumstances of the present) was in no way the exclusive monopoly of Opus Dei ideologues. A person like Alfonso Carlos Comín, "a communist in the Church, and a Christian in the Party," as he defined himself, one of the principal forces behind the Christians for Socialism movement of the seventies, wrote in the prologue to Carandell's book about Msgr. Escrivá: "For me reading *The Way* was an unforgettable experience. I went from amazement to holy indignation" (Comín, in Carandell, 7). What he forgets is to specify how many years it took him to get from the first stage to the second: while in 1975 he found *The Way* to be "a vulgar, egotistical, and mediocre vision," twenty years earlier he was recommending it and making his friends read it (Casañas, 147).

However difficult it may be to admit it today, the truth is that during the forties and fifties *The Way* and Opus Dei seemed less triumphalist than the official Church, and less Francoist than the Jesuits (Hermet, 1980, 1:266). And here lies precisely—and paradoxically—one of the keys to its success and its rapid implantation after the Spanish civil war, as we will see in the next chapter.

(*c*) The controversy around *The Way,* while timidly initiated somewhat earlier, acquired a substantially different tone after the publication of the article "Integralismus" by the theologian Hans Urs von Balthasar in 1963. The international renown and prestige of the great Swiss theologian made his diagnosis especially serious for Opus Dei—a diagnosis that was prejudicial to the group and did them harm, and that they interpreted as a frontal attack. Von Balthasar put Opus Dei on the same level as other manifestations of Catholic integrism, and declared that "Opus Dei is today, without doubt, the most powerful of them all" (von Balthasar, 1963, 742).

In the German-speaking world, the line of thought thus opened by von Balthasar was pursued, and it would in fact be German-speaking authors who would be most emphatic about the parallels between Opus Dei and the whole set of trends and movements comprised by Catholic integrism; one of the most recent examples is the volume edited by the theologian Wolfgang Beinert, published in 1991, on Catholic fundamentalism (see Neuner, 1991, 426).

Speaking specifically of *The Way,* von Balthasar, after admitting that he found it "alarming" reading, quotes a group of fragments from it and categorizes it as a "little boy scout manual" (von Balthasar, 1963, 743).

Two years later, another distinguished writer, the Spanish thinker José Luis Aranguren, took up the debate unleashed by von Balthasar (which provoked angry replies from several Opus Dei members), and in an article published in the French review *Esprit* he stated that there are original things and good things in *The Way,* but the problem is that those that are original are no good, while those that are good are not original (Aranguren, 1965, 763). He concluded that *The Way* is ultimately a book of "Ignatian asceticism," but "unbelievably trivialized" (ibid., 765).

After this, Opus Dei seems to have reacted in two different directions. First—and on this point we can surmise a certain strategic retreat—they began to assert that Opus Dei could not be reduced to the contents of *The Way,* since Escrivá's little book "did not cover *the whole* spirituality of Opus Dei" (Coverdale, 490, in an article that was, furthermore, a direct reply to—and an attack on—von Balthasar). After twenty years, they continue to return to exactly the same argument: *The Way* is not "a systematic or exhaustive exposition of the spirit of Opus Dei" (Mateo Seco, in Rodríguez et al., 475).

At the same time, on the other hand, a wave of exaggerated eulogies of Escrivá's book began to appear, in an apparent contest to see who could win the prize for the highest flights of imagination. Thus those "notes written without literary pretensions or need for publicity," of which Escrivá spoke in 1934 (*Consideraciones espirituales,* foreword), were transformed into "a classic of spiritual literature" and a "Kempis of modern times" (Editor's note in recent editions of *The Way*). While some speak of the "manual of the clerical-authoritarian" (Ynfante, 137), and someone once in a while makes an effort at balance and thoughtfulness in his opinion (Artigues, 88f.), the "official" literature compares *The Way* to works like *Romeo and Juliet, La Vida es Sueño,* and *The Divine Comedy* (Cejas, 105), and even ventures to assert that "it is, without doubt, one of the most transcendental books of spiritual literature of all time" (Cardona, 1988, 174). If the "unofficial" literature finds it superficial, primitive, and often in bad taste (see, for example, Walsh, 116ff.), or even elitist, macho, and egotistical (Lannon, 226), the biographers of Msgr. Escrivá declare that "since Pius XII, all the Popes have read *The Way*" (Sastre, 233) and that "for many years Paul VI used *The Way* for his own personal meditation" (ibid., 333).

(*d*) The trouble is that meanwhile we still do not have a serious analysis, lucid and critical at the same time, of the contents of Padre Escrivá's book and its 999 points of meditation. We know that the number (999 = 3 x 333) is an homage to the number 3, an expression of "the author's firm devotion to the Holy Trinity" (Saranyana, 63; Le Tourneau, 98). And we also know that Msgr. Escrivá wanted his next book, *Surco,* published posthumously in 1986, to have 1,000 maxims instead of 999, "so that you and I could finish the book smiling, and the blessed readers not have to worry about whether, through simplicity or malice, they were seeing cabalistic meaning in the 999 points of *The Way*" (*Surco,* no. 1000).

But we don't know much more than this. The volume *Estudios sobre Camino* (1988), edited by José Morales, contributes some items of information and some interesting commentary, but it should be complemented by a study of a set of other aspects that are not even mentioned there. Dalmau's book *Contrapunts al Camí de l'Opus Dei* (1969) is an interesting and well-crafted exercise illustrating the bias of Escrivá's text and emphasizing the extent to which every coin has two faces; but although it offers several points of analysis, it is not really, nor does it pretend to be, an analysis properly speaking. Ynfante alludes to several studies done in Madrid in 1967 on the language of *The Way* (Ynfante, 383f.), but he provides no specific bibliographic references.

It would be interesting to see a comparative study of *The Way* and the two posthumous volumes, *Surco* and *Forge*. The latter are presented as the result of notes made by Escrivá during the thirties, published after his death just as he had left them (*Forge,* introduction, 17), since only lack of time had prevented him from giving them a final editing (*Surco,* introduction, 16). But sometimes the reader has the impression that he is confronted with a language and a set of concerns that are somewhat different from those of *The Way*. For example, when Msgr. Escrivá writes that "there are two capital points in the lives of peoples: the laws on marriage and the laws on education" (*Forge,* no. 104), or that "freely, and in accordance with your likes or talents, take an active and effective part in the honorable official or private associations of your country" (*Forge,* no. 717), he seems to be reflecting some uneasiness and to be using a language characteristic of certain sectors of a Catholicism from a time somewhat later than the Spain of the thirties.

Likewise, *Surco* contains a series of points that seem to reflect problems of Catholicism during and after the Second Vatican Council, for example, the distinction between "freedom of conscience" and "freedom of the consciences" (no. 389); or the reference to the false understanding that some Catholics have of notions like "ecumenism, pluralism, democracy" (no. 359). It is difficult to place statements like the following in a context prior to the council: "It is a sad ecumenism that we hear from the mouths of Catholics who abuse other Catholics!" (no. 643). And it is hard to see how, from the Spain of the thirties, a priest could write that "perhaps in other times the persecutions were done openly, and now they are often organized in an underhanded way; but, today as yesterday, they keep on attacking the Church" (*Forge,* no. 852).

It would also be enormously useful—we have said this before—if we had a comparative study of *The Way* and the text that preceded it in 1934, *Consideraciones espirituales.*

From a theological point of view, it would be very interesting to find out what kind of relations are established between man and God or between man and the Virgin in the pages of *The Way.* What is his conception of the Church, what is the role of the "guardian angels," and so on? What connections exist between the spirituality promulgated by Msgr. Escrivá and that of Saint Ignatius? To what extent is his asceticism an "Ignatian asceticism," as Aranguren maintains (1965), or a "trivialized" Ignatian asceticism? How does this presumed "triviality" manifest itself?

From a linguistic point of view, it would definitely be interesting to analyze Escrivá's vocabulary and style, such as his use of the familiar second-person singular, his question marks and exclamation points, his adjectives, the symbolism of fire that we mentioned earlier, the language of war and military usage, the frequent use of strongly negative expressions—"sticky, disgusting, obscene, filthy, offal, pus, putrefaction, swine, fetid carrion, piles of excrement," and so on.

Finally, although Opus Dei is never mentioned in the text of *The Way,* it would be interesting to analyze when and under what conditions a "we" is mentioned that could be interpreted as a reference to the incipient organization, and how this "we" relates to an external world of "them" and "others."

(*e*) Since this last aspect, from a sociological point of view, is the most relevant of all the points raised in the foregoing section, let us pause on it briefly, not to lead into an exhaustive analysis, but to provide at least a few examples that illustrate both the content of the "we" and its contrast to "them," and to observe in passing several of the characteristic features of the language employed by Escrivá.

The first time the pronoun "we" or "us" appears in *The Way,* it is to assert that "among us there is no place for the lukewarm": this is the reason why "you" (Escrivá uses the familiar "tu") may not "be just commonplace" or "a sheeplike follower" because "you were born to be a *audillo*"(*The Way,* no. 16). Here we have colloquial language, military references, and connotations of what has been called Opus's "elitism," or, at the very least, a marked contrast between "us = caudillos" and "the others = commonplace."

"Children, many children, and a lasting trail of light shall we leave behind if we sacrifice the selfishness of the flesh," he writes on the second occasion where he uses the first-person plural (no. 28). Here indirect and metaphorical language referring to sexuality and the commitment to celibacy initially required of every member of Opus Dei figure in a maxim which begins with the famous "marriage is for the rank and file and not for the officers of Christ's army" (no. 28), and which again used terminology belonging to the military world, as does number 311, "war has been for us," on which we commented earlier.

"And how shall I acquire 'our formation,' and how shall I keep 'our spirit'? By fulfilling for me the specific norms that your Director gave you and explained to you and made you love: fulfill them and you will be an apostle" (no. 377). In this case the quotation marks, put there by Escrivá himself to emphasize "our formation" and "our spirit," seem to constitute an unequivocal allusion to Opus Dei, whether real, incipient, or in the planning stages, in the years 1938–1939. No less manifest is the hierarchical model of the organization, real or projected, with the accent placed on obedience to the director ("laymen can only be disciples," he had written in no. 61). And, from the point of view of language, another distinctive feature appears here: it is not "fulfilling the norms," but "fulfilling the norms *for me*." A form of salutation that Escrivá used habitually in his letters and writings used the reflexive pronoun "me," as in "May Jesus watch over you *for me*" (Que Jesús te *me* guarde) (see for example *The Way*, no. 312).

We have another unmistakable reference to Opus Dei, according to the biographers, in the references to "madness" and the "madhouse." "Lord, make us mad, with that contagious madness that attracts many to your apostolate" (no. 916). "Good news: a new madman for the madhouse" (no. 808) is, according to Gondrand, the comment Escrivá made in the early days when someone asked to join the Work (Gondrand, 90). "Answer, decisively, that you thank God for the honor of belonging to the 'madhouse'" (no. 910).

One final example, perhaps the most illustrative of all, both for the type of language used and the application to which that language is put, in the service of the contrast between "us" and "the others," is point number 914 of *The Way*. The Padre writes:

> How pitiful are those crowds—high and low and middle-class—without an ideal! They give the impression that they don't know they have a soul: they are . . . a drove, a flock . . . , a herd.
>
> Jesus: we, with the help of your Merciful Love, will convert the drove into a company of soldiers, the flock into an army . . . , and from the herd we will extract, purified, those who no longer want to be filthy.

Another Small Study of *The Way*

As you can see, there are many studies of *The Way* to be done, that could be done, and that are worth the trouble to do. Because what interests us most in this whole part of the work is to understand the historical evolution of Opus Dei, here we are going to focus exclusively on only one of these possible studies: the presumed manifestations of this evolution in the text of *The Way* itself.

In other words, if in spite of the brusque affirmations of the "official" literature that Msgr. Escrivá "engendered Opus Dei as we know it today" (Helming, 4), and that from the first moment he "saw" it as it would be down through the centuries (Vázquez, 113), with a profile that already contained all the fundamental features of present-day Opus Dei (Fuenmayor et al., 39ff.), so that it would never be necessary to bring it up to date because "God Our Lord brought the Work up to date once and for all" (Escrivá, *Conversations,* no. 72); if in spite of all this our thesis, basic and simply based on common sense, affirms that Opus Dei is the fruit of a specific epoch and specific circumstances, and that over the years it has continued to evolve—in its objectives and in its style—adapting itself to different epochs and contexts, the question we are going to pose here is whether we can detect in the text itself of *The Way* certain indications of such an evolution and such adaptations.

(*a*) We must consider as secondary, although sometimes significant, those modifications that affect only the presentation of the book. For example, in later editions of *The Way* the reference to the year of its original publication as "The year of Victory" disappears. Monsignor Escrivá's name keeps changing, on the title page, in accordance with the successive mutations we discussed in Chapter 2: it goes from "José María Escrivá" in the first Spanish edition to "Escrivá" or "Monsignor José María Escrivá" in the first English editions; and from "Josep Mª. Escrivà de Balaguer" in the first Catalan edition to "Josemaría Escrivá" or "Josemaría Escrivá de Balaguer" in the most recent ones.

As the years passed introductory "notes" were added to the new editions signed by the author (note to the Spanish third edition, 1945; note to the seventh edition, 1950, etc.). The most recent editions also include a biographical profile of the author, which also went through modifications, to reflect, for example, the vicissitudes of his cause for beatification. This same kind of updating is found in a "publisher's note," which gives the number of editions that have appeared and the number of copies circulated, and contains a list of all the languages into which *Camino* has been translated, as well as those in which a translation is being prepared.

Some translations include an introductory note expressly drafted for that specific edition; not all the editions reproduce the original introduction written by Msgr. Xavier Lauzurica in 1939. Finally, the most recent editions usually include an unsigned preface, or editorial note, which gives a synopsis of contents of *The Way* and an assessment of its significance ("a classic of spiritual literature," a "Kempis of modern times," and so on).

(*b*) Dispensing, then, with these aspects which are somewhat external to the book itself, our first line of inquiry to detect possible alterations in the text would be a comparative analysis of the original to the various translations.

It is clear that the language of Msgr. Escrivá, which is colloquial, full of expressions that are unique to the Spanish language, and which frequently resorts to frankly "typical" expressions, is not easy to translate. For example, that reflexive "me" in "May Jesus watch over you [for me]" or "fulfilling the norms [for me]" disappears in English;* in the "official" translation (*The Way,* Scepter, New York and London, 1987) these phrases appear as "being faithful to the norms" and "may Jesus watch over you."

But on certain occasions the translation appears to deliberately avoid the use of the equivalent word in the original, even when that word exists in fact in

* Unless reinserted, as here, in a rather awkward manner—Trans.

the other language. Perhaps this is not true in the case of the celebrated "caudillo," weakly translated as "leader" in the English version, surely because the English language in this case has nothing better to offer, and has never enjoyed the privilege of knowing exactly what a "caudillo" is; in the German edition, by contrast, the word is translated as *Führer,* which is in fact a word capable of awakening resonances in the reader quite similar to those of the original term.

Camino has a whole chapter dedicated to "proselitismo," but this word, which has an exact cognate in English, proselytization, is converted nevertheless into a gentle "winning new apostles," so as to win new apostles. "This zeal to proselytize that eats at your entrails is a sure sign of your surrender" (no. 810) becomes, in English, "That burning desire to win fellow apostles is a sure sign that you have really given yourself to God." "Proselytization. It is the sure sign of true zeal" (no. 703) becomes "The search for fellow apostles. . . ." In German as well "proselytization" is converted to *Apostolsuche,* that is, "a search for apostles."

Another chapter of *Camino* is devoted to the virtue of "discretion." In the German translation this is rendered as "authentic discretion" or "true discretion" (*echte Diskretion*). And the chapter entitled "tactics" in Spanish (and also in English) is rendered "apostolic tactics" in German.

"The plane of holiness that our Lord asks of us is determined by three points: holy intransigence, holy coercion and holy shamelessness" (no. 387). "Shamelessness" could have been given a fairly literal translation in German (*Schamlosigkeit*); but they preferred to use a much weaker word, *Unbefangenheit,* which the dictionary translates as "freedom from or lack of prejudice." Immediately after this, Escrivá writes, "holy shamelessness is one thing, and lay impertinence another" (no. 388); lay impertinence, or "lay unscrupulousness," in the Catalan version, becomes "mundane impertinence" in German and "plain cheekiness" in English.

These examples should be enough—we could find others, and we could consult editions in other languages—to indicate that the modifications are not always due to the inevitable problems of translation, but that in some cases they result from a deliberate intent to avoid the strong connotations of some of the original expressions, in an exercise that we don't know whether to call "holy shamelessness" or "lay unscrupulousness" or just "plain cheekiness."

(*c*) As for the original text, properly speaking, of the Spanish edition of *Camino,* have any changes been made over the years?

Given that this analysis has already been done, and furthermore in a text belonging to the "official" literature, it will be sufficient to reproduce here the observations of Josep Ignasi Saranyana (in the volume *Estudios sobre Camino,* Morales, ed., 59–65).

According to Saranyana, "the text of *Camino* has remained substantially unaltered, except for four variations. Very few, if one keeps in mind that the book has 999 points" (Saranyana, 61f.). The first of these four changes belongs to 1950. Until then, point 381 and point 940 were repetitious: "Don't forget [no. 940; "let us not forget," no. 381] that unity is a symptom of life: to disunite is putrefaction, the sure sign of being a corpse." This text is retained in the sixth Spanish edition in the chapter "The Apostle" (no. 940), while in the chapter "Formation" (no. 381), the new text says: "Don't let it matter if they tell you that you have esprit de corps. What do they want? A fragile instrument that falls to pieces when you grasp it?"

The third change pointed out by Saranyana dates from 1965 and is the consequence of a change in the liturgical norms of the Church. In point number 750 of *Camino* Escrivá said that "my Mother, the Holy Church, makes Priests invoke Saint Michael at the foot of the altar every day against the evils and malice of the enemy." The present text says "my Mother, the Holy Church—for many years: and it is also a laudable private devotion—used to make priests. . . ."

The most recent alteration is from the year 1974, and it affects point 145, which alludes to a group of officers "in noble and joyful camaraderie" on the front in Madrid during the Spanish civil war. Saranyana writes: "In 1974 (28th edition) the first and second paragraph of point 145 were slightly modified, perhaps to avoid references to the Spanish civil war of 1936, which at the time might have caused puzzlement in certain circles" (Saranyana, 62). It is a little hard to imagine specifically what "circles" would have found this puzzling in 1974 and not before. But in any case, the author goes on to say that "in 1983 (37th edition), since the circumstances that made the modification prudent had changed, they were able to restore the original text, much more expressive and poetic," so this alteration of the text has now disappeared from the most recent editions of *Camino*.

The last of these changes, and undoubtedly the most significant, is, however, the one which Saranyana mentions as the second, since chronologically it was made in 1957.

"Point 115, taken from page 16 of *Consideraciones,* was revised first in 1957 (14th edition) and later in 1958 (15th edition). Since then it reads as follows: '"Minutes of silence."—leave them for those who have a dry heart. We Catholics, children of God, talk with Our Father who is in heaven'" (Saranyana, 62). The author adds no comment, and says nothing about the possible circumstances that recommended this modification. And, above all, he does not explain what point 115 of *Camino* said before 1957. Let us look at it, since as in the preceding case the original text might equally be "much more expressive and poetic."

In the first edition of *Camino* we find this text: "'Minutes of silence.'—*leave this for atheists, Masons, and Protestants, who have a dry heart.* We Catholics, children of God, talk with Our Father who is in heaven" (emphasis mine). The line "leave this for atheists, Masons, and Protestants, who have a dry heart" is changed, in the first German edition (1956—a little before the change in the Spanish version, in fact), to "leave it for the atheists and those who have a dry heart," and later to the present "for those who have a dry heart," without specifying whether it is Protestants or atheists who find themselves in this lamentable condition.

The original mixture of Protestants, Masons, and atheists is quite characteristic of Franco's Spain in the forties and fifties. In the headquarters of the Barcelona superintendent of police, for example, at the beginning of the sixties there was still a so-called Bureau of Masonry and Communism, which was precisely the place where the files on the Protestants in the city were kept. It is not too surprising, then, that initially Msgr. Escrivá would be likewise partial to this kind of mixture: it matches perfectly with the image we evoked above of the Escrivá of the years of the republic and the war; if it is not a perfect match with the image presented by the "official" biographies, the problem is obviously a problem with the biographers.

A little more surprising, however, is that the modification of the text of *The Way* was not made until the second half of the 1950s, after Msgr. Escrivá had already spent ten years away from the Spanish atmosphere of the era, living in Rome. This is even more surprising if we recall that Opus Dei boasts about having admitted Protestant co-workers since 1950. In that year, in fact, the "Founder—after a 'filial struggle'—got the Holy See to admit as 'cooperators' all those persons (Catholics, non-Catholics, and even non-Christians) who wanted to collaborate, materially or spiritually, in the apostolate of the Work" (Vázquez, 258, who comments that the event "was quite unheard of in the pastoral history of the Church"). The "filial struggle" (an expression also used by Bernal, 295) is explained by Escrivá himself in these words: "When we officially asked the Holy See for authorization to accept non-Catholics and even non-Christians as 'cooperators' of our Work, the first answer was that it was impossible. I asked again and the answer was a *delay,* which was to recognize the legitimacy of our petition, although advising us to wait. Finally, in 1950, the affirmative response: the Work was thus the first association of the Catholic Church to fraternally open its arms to all men, without distinction of creed or confession" (quoted in Sastre, 456f.).

This does not explain, however, how the position of these non-Catholics is reflected in the Constitutions of the Opus Dei of 1950. They are those "who are alien from the paternal house or who do not profess the Catholic truth," but who help Opus Dei with their work or their donations. All members should work with these "cooperators," so they might attain "the light of the faith" and be attracted "gently and effectively to Christian ways" (*Constitutions,* 1950, art. 29.2). But only Catholics may become full members, to the extent that if a member later joins—and the language is significant!—"a non-Catholic sect," his admission is declared invalid (*Constitutions,* 1950, art. 36.1). After the transformation of Opus Dei into a personal prelature, Article 29 of the old Constitutions was incorporated intact into the new Statutes, and is therefore still current today (*Estatutos,* 1982, art. 16.2; Fuenmayor et al., 630f.).

In sum, then, from 1950 on Opus Dei admitted some co-workers who were not Catholics, at the same time trying to convert them to Catholicism; but until 1956–1957 one could still read in *The Way* that "Protestants have a dry heart." In 1958 the Padre, walking in the rain through the old university town of Oxford, "kept on saying: God must be brought into these places" (Sastre, 373), as if students had been meeting in the chapels of all the Oxford colleges for centuries only to keep "minutes of silence." Finally, in 1960, at the first audience granted to him by John XXIII (Gondrand, 214), Msgr. Escrivá had the audacity to tell him that in the Work everyone, Catholic or not, had always been suitable, and that "I did not learn ecumenism from Your Holiness" (*Conversations,* nos. 22, 46). "The Holy Father laughed, moved" (*Conversa-*

tions, no. 46), "he laughed out loud" (Gondrand, 215), "he smiled, pleased" (Vázquez, 258).

We might ask whether we are looking here at a practical demonstration of "holy shamelessness" by the Padre, which provoked "the hilarity of John XXIII" (Sastre, 457). Or perhaps it is rather an original contribution of the literature of Opus Dei to the widely diffused legend of John XXIII as good-natured and amiable, but a little foolish and irresponsible, in light of the consequences of the council he decided to convoke. The biographers talk of the "affable and paternal charm of his manner" (Bernal, 296), but at the same time the pope's audience with Escrivá becomes, in the text of Vázquez de Prada, an informal session: "Chatting one day with John XXIII, the president general commented . . ." (Vázquez, 258). While this is not the time to divert our attention to the figure of Pope Roncalli, I cannot resist the temptation to reproduce the wonderful anecdote related by López Rodó in the first volume of his memoirs, immediately before turning to an interview he had with Msgr. Escrivá. It refers to an audience given by John XXIII to an "official Spanish mission," led by the minister of industry, in 1959. The wife of the minister "took advantage of the first pause to ask him if he would bless a stack of medals and rosaries she had in her purse and John XXIII blessed them immediately without giving her time to take them out of her purse, and she was not satisified and asked the Pope again for the blessing, which he repeated immediately. But the medals and rosaries were wrapped in plastic; she finally took them out of their wrapping and asked the Pope to bless them a third time. John XXIII said to her, smiling, "Signora, do you believe my blessing cannot pass through plastic?" López Rodó comments, "He looked very amused and his legs were swinging like a pendulum because he was short and, seated on the chair, his feet did not reach the floor" (López Rodó, 1990, 158f.; see also Helming, 61, and Gondrand, 216).

(*d*) There is another approach to detecting possible alterations in *The Way,* not in the text properly speaking, but in *the guide that suggests a specific reading* of that text. We refer to the long subject index at the end of the volume, which takes up thirty pages in the Spanish edition of 1988 (pp. 301–30; it is only fifteen pages in the 1987 English edition, but this is because the type is much smaller; the content is identical). In fact, although we have just said the text of *The Way* has seen only four changes, and only one of them really significant, the subject index of the modern editions is, by contrast, very different from the one in the original 1939 edition. (*Consideraciones espirituales* did not contain a subject index.)

The difference is not only that it has been made much larger overall. New concepts have appeared that were not included in the earlier index, and new subsections have been created that did not exist before; at the same time certain sections grew larger, while others shrank; and some of the concepts that were in the 1939 index have simply disappeared. To the extent that the subject index of a book like *The Way* constitutes a kind of reading guide, its evolution might be considered as symptomatic of changes not in the text itself but in the way in which we are meant to read and understand that text.

(1) In the original edition, the titles of the forty-six chapters all appeared in

the subject index, except for "supernatural life" and "more on interior life," which, being consecutive chapters, were grouped together in the index under the heading "supernatural life," just as "spiritual childhood" and "life of childhood," also consecutive, were grouped under "spiritual childhood" in the index. Except for these two entries, the correspondence between chapter headings and index headings is exact, keeping in mind that the chapter "Our Lady" appears in the index under "Blessed Virgin Mary" and the chapter "Holy Mass" as "Mass," while "other virtues" is entered as "virtues" and the chapter "afterlife" as "death." In the modern editions we find some changes:

- Three partial additions: where before it said "direction" (both in the title of the chapter and the index heading), the index now says "spiritual direction"; "examination" is retained as the chapter title, but the index entry is "examination of conscience"; and "formation" is changed in the index to "doctrinal formation."
- Two substitutions: the chapter "internal struggle," appearing in the 1939 index in identical form, now appears as "struggle, ascetical" while "calling" has been changed, in the index, to "vocation."
- Two surprising disappearances: "discretion" and "tactics" are concepts that have been *removed* from the index. In the Spanish editions of the 1950s they were still there; now, however, the points from the chapter on "discretion" have been incorporated under the index entry "naturalness" (which originally had only nine references and now has many more), and the points in the chapter on "tactics" are included (in English as well as in Spanish) in the index with those referring to "apostolate."

In a way we are given some insight here into a process of evolution that parallels the one we observed earlier in reference to the translations: while the German translator softened the terminology in the chapter titles ("authentic discretion" and "apostolic tactics," respectively), in the preparation of the indexes the editors have gone further and produced a more radical alteration.

(2) In the original edition of 1939, the subject index had a total of 136 concepts. The 1988 edition has 168. Originally there were no subheadings under each entry, but now these are abundant; overall, under most of the index headings, more points in the book are cited. Even setting aside these data, which would involve us in interminable calculations, and concentrating only on the set of main headings in the index, we see that of the 168 present headings, more than half are identical to those of the original edition.

To say this in reverse: although we are concentrating on the concepts, and not on the specific number of references in each instance, nearly half the present subject index of *The Way* has changed with respect to the original edition. Between the addition of new headings, partial modifications, and the elimination of concepts which have now disappeared, the changes affect about one hundred headings in all.

Let us look sequentially at these modifications, suppressions, and additions, pausing only on those that hold some special interest for us.

(3) The changes are small, and some of them have to do with an effort to

make the titles of the chapters more precise: "spiritual direction" where before it was simply "direction"; "interior struggle" converted into "struggle, ascetical"; and so on. Further than that, perhaps the only element to be underlined is that "ambitions" is today presented as "noble ambitions" [in the English edition, the entry reads, "AMBITION (noble ambition)"—Trans.], and what appears in the original edition as "intransigence" has been changed to "Intransigence, holy." (We should add, in passing, that neither then nor now do the other two points of the "plane of holiness" of the members of Opus appear in the index; they are, as you will recall, "holy coercion and holy shamelessness.")

(4) Of more interest, of course, are the concepts that have disappeared from current editions of *The Way* (twenty-eight in all). In some cases the references have been incorporated under other headings that already existed ("communion" and "eucharist," for example), or under new headings (the twelve references formerly under the heading "children of God" are now part of the twenty-eight references under the previously nonexistent "divine filiation").

A whole group of headings pertaining to topics that refer to defects has been omitted from the modern editions; for example, "confusion," "contradictions," "disputes," "indifference," "rumors," "idleness," "pedantry," and "ridicule."

Undoubtedly even more important is the disappearance of the heading "caudillos," which was clearly important in the first edition, with a total of nineteen references. Another one was "zeal" (twenty-six points in the book, according to the 1939 edition), as well as the two that have already been mentioned as chapter titles: "discretion" and "tactics."

Finally, another group of concepts whose disappearance seems to me extremely important is the trio of headings "evangelical counsels," "imitation of Jesus Christ," and "perfection." Together they provided a total of nearly forty references, present in the original edition and retained in the subject index for many years. They were obviously suddenly eliminated at the moment when Opus Dei, seeking new juridical status in the Church, wanted to exhibit more distance from certain virtues considered appropriate to monastic orders. If in the previous cases one might argue that they were eliminating unnecessary headings, or else ones that were not flattering to Opus Dei's image ("caudillos," "tactics," etc.), here we have an alteration that reflects a modification of the conception that Opus Dei had of itself, its objectives, and its spirituality. The "evangelical counsels" were equivalent to the three classical "vows" of poverty, chastity, and obedience of the religious orders: "You know that there are 'evangelical counsels.' To follow them is a refinement of Love. They say it is the path of the few. Sometimes, I think it could be the path of the many" (*The Way,* no. 323). Furthermore, it is said that the religious lead a "life of perfection." The *Constitutions* of 1950 define Opus Dei as "a secular institute dedicated to the attainment of Christian perfection" (art. 1), which has as its most generic goal the "sanctification of its members through the practice of the evangelical counsels" (art. 3.1) (in Latin, in Fuenmayor et al., 553). The *Estatutos* of 1982 define Opus Dei as "a Prelature which proposes the sanctification of its faithful through the exercise of the Christian virtues . . . according to its own specifically secular spirituality" (art. 2.1) (in Latin, in Fuenmayor et al., 628).

We will address this topic in more detail at a later time. For now we will only confirm that, in fact, some of the changes in the index of *The Way* are indicative of the historical evolution of Opus Dei.

(5) Can we find this same kind of confirmation by examining not only the deletions but the new headings introduced into the subject index? Compared to the original 1939 edition, the new concepts, sixty in all, outnumber those that disappeared.

Just as some headings were eliminated because they were judged relatively unnecessary, others were added that are elementary and basic, but were not included originally, such as "Gospel" and "Jesus Christ." Further, while earlier we observed a tendency to eliminate concepts that connoted defects, the modern editions clearly include a whole set of good qualities and virtues: "comprehension," "fidelity," "generosity," "justice," "freedom," "maturity," "magnanimity," "meekness," "patience," "forgiveness," "service," "sincerity," "truthfulness," and so on. Retrospectively, in every case what is significant is not so much the fact of its inclusion here, but its absence from the early editions.

Another noticeable feature is the incorporation of several curious headings that seem to reflect certain expressions very typical of the Padre. We will discuss only two of these: "good pride" and "mística ojalatera," the first of which has only two references, and the second four. "Mística ojalatera" (an expression I would not like to have to translate into English, but that I would certainly never have translated as the comparatively weak "wishful thinking" that appears in the index of the English edition) consists, according to the Padre, "of vain dreams and false idealism" (*Conversations,* no. 88), exemplified by people who go around saying "if only this," "if only that." The strangest thing is that Escrivá does not actually use the phrase at all in any of the four passages referenced in the index. It is one of the Padre's typical expressions, but he probably began to use it after he wrote *The Way*; in any case it does not appear in *The Way* even once. In other words, by means of the subject index elements are introduced into *The Way* that do not appear in the text.

What happens in a case like that of "mística ojalatera" remains purely anecdotal. But on other occasions it is somewhat more than an anecdote. For example, in accordance with the "official" literature, "divine filiation is the basis of the spirit of Opus Dei," to the extent that according to Escrivá himself, "For the first time in the history of the Church, God wanted Opus Dei to be the group that lived this filiation corporatively" (quoted in Vázquez, 516). Anyone who thinks the phrase "for the first time in the history of the Church" excessive can try to become convinced by the forty pages in Fernando Ocáriz's study "La filiación divina, realidad central en la vida y en la enseñanza de Mons. Escrivá de Balaguer" (in Rodríguez et al., 173–214), where the author offers "a first draft of analysis and systematization, which will help in the understanding of the theological richness—truly impressive—contained in the teachings of Opus Dei on divine filiation" (Ocáriz, 176). Let it be clear that we are not attempting to deny the importance, or even the "centrality" of the topic in Escrivá's teachings. All we are emphasizing is that in the subject index to *The Way* the expression "divine filiation" did not occur before, and now it does.

The same can be said of several other concepts eventually incorporated into the subject index, for example, "supernatural vision," "universality," and "vocation." It is true that in reference to the last, the heading "calling" appeared in the first edition, and it has now been deleted; but while there were 49 points referenced under the heading "calling," there are 194 under "vocation," and 20 subheadings.

(6) If one had to sum up in a single sentence the "official" view of the originality and specificity of Opus Dei in the life the Church, there is not the slightest doubt that this sentence would revolve around holiness and sanctification, sanctification of work, in work, and through work, and around work, and sanctification in worldly life. Unanimity on this point is so complete that one can cite almost anything: even in the pontifical document making Opus Dei into a Personal Prelature in 1982, John Paul II writes that Opus Dei has endeavored to "carry into practice the doctrine of the universal vocation for holiness, and to promote among all social classes sanctification of and through professional work" (Fuenmayor et al., 622).

Well then: in the subject index of *The Way* the heading "holiness" does not appear in the 1939 edition or in the editions of the 1950s (the heading "plane of holiness" does appear, corresponding to the chapter with that title, and contains references only to points in that chapter).

The heading "world" is also new to the index; it was not present in the 1939 edition. As for "work," it had nineteen references in the original edition, thirty-five in a 1955 edition, and sixty-three at present. One of its subheadings, "sanctification of work," is new; and none of the points in *The Way* classified under this rubric are cited under the heading "work" in the original edition. The "sanctification of work" is today perhaps the very essence of Opus's message. But when Father Escrivá wrote *The Way,* it definitely was not.

(*e*) In view of all this, I believe the conclusion that Opus Dei has evolved over time is justified. And it has evolved not just a little. We may conclude that Opus Dei, as Escrivá planned it and "saw" it when he wrote the maxims of *The Way,* was not proposed *primarily* "to carry into practice the doctrine of the universal vocation for holiness, and to promote among all social classes sanctification of and through professional work," to repeat the words of John Paul II's document.

Of course, this does not mean we are denying that this is what Opus is today. We are denying that it was so in 1928, 1939, or even several years later than that. Some of these elements might have been more or less present, in more explicit or more implicit form, but they were not the primary elements. If they had been fundamental, they could not have been forgotten in the subject index of *The Way*.

Consequently, and contrary to what the "official" literature asserts, the point is not that *The Way* is not "a systematic and exhaustive exposition of the spirit of Opus Dei" (Mateo Seco, 475), or that "it does not cover *the whole* spirituality of Opus Dei" (Coverdale, 490); the point is that it does not cover what the "official" view today wants to present as the specific components of the spirituality of Opus Dei. In other words, Opus Dei is not what it was, and it was not what it is today. Monsignor Escrivá did not "engender Opus Dei just as we know it today" (Helming, 4), nor did "Our Lord God bring the Work up to date once and for all" (Escrivá, *Conversations,* no. 72).

Once again the "scheming" method of Father Brown, in suggesting that we proceed to eliminate the ill-fitting pieces in the subject index of *The Way,* has allowed us to prove that the "scheming" disciples of Msgr. Escrivá sometimes help us to track down lost pathways. Thus if Opus Dei was not what it is today, let us return to where we left off, at the end of the Spanish civil war, and ask what it really was during the forties.

7

The Establishment of Opus Dei in Spain (1939–1946)

Synopsis of the Official Version: Establishment of Opus Dei and First Juridical Recognition

We left Padre Escrivá as he was entering Madrid with the first of Franco's troops in the last days of March 1939 (Vázquez, 197), with a "work I don't know if born yet or planned" (Laín, 241) that had approximately as many members as the number of years passed since 1928 (Helming, 44). If there are more than enough reasons to doubt the real and concrete existence of Opus Dei up to this point, from 1939 on things began to happen: its existence as a minimally organized group soon became unquestionable, the name Opus Dei appeared officially, and the process of its establishment and diffusion throughout Spain was under way. All this took place while Europe suffered the tragedy of the second world war and Spain survived under an autocratic regime—in an isolation more miserable than splendid. Franco's government was smart enough to avoid getting directly involved in the war but fascist enough that it did not take much imagination to guess where its sympathies lay.

This period took Opus Dei from a mere project without a clear official designation to formal recognition by the Church as an institution belonging to it. While in its initial moments Opus Dei was, in Escrivá's words, comparable to a "madhouse" with a dozen "madmen" ("That—your ideal, your vocation—is a madness. And the others—your friends, your brothers—madmen . . ."; *The Way,* no. 910), the following point of *The Way* might be considered programmatic for the whole period we are about to analyze:

> The great longing we all have for "this thing" to move forward and grow larger seems to be turning into impatience. . . . When will it leap ahead, when will it break out, when will we see the world ours? . . . The longing will not be useless if we use it in pestering and "coercing" the Lord: then we will have made excellent use of our time. [*The Way,* no. 911]

The years between 1939 and 1946 were, in fact, years of little concern for style ("coerce," "pester"), while Opus Dei passed from a condition of anonymity ("this thing") to one in which the final objective ("to see the world ours") began to be perceived as relatively attainable as the process of international expansion began.

(*a*) The progressive establishment of the first nuclei of Opus Dei in the large university cities of Spain is fairly well documented in the "official" literature. The main center was Madrid, where they opened their first university residence. In addition, there was an apartment for those who had already graduated, a large house for the "intensive formation" of the members (Gondrand, 148), which was also the residence of Escrivá, his mother, and his brothers (a house which today continues to be the headquarters of Opus Dei in Spain), and, a few years later, a new university residence much larger than the first one.

In the first year alone, the Padre's initial dozen followers swelled to about forty members (Gondrand, 146). At the same time the women's section began to function. This section began by renting an apartment, but soon ("it did not seem prudent for a young priest to go regularly to an apartment where no one lived in order to educate a group of equally young girls"; Bernal, 149) they moved to the central building ("with full separation from the men"; ibid., 150), until in 1942 a center exclusively for women was opened. The first women in Opus Dei were for the most part sisters of the first members.

The expansion from Madrid to other cities was accomplished in a variety of ways. In Valencia, which during this period became a notable breeding ground for "vocations" (future priests, future college professors, future Franco politicians, and even one who would later switch from Francoism to anti-Francoism, future tutors to the future king of Spain, etc.), Opus established itself basically through a college preparatory school for selected students. (On the Colegio de Burjasot, see Calvo Serer, in Martí and Ramoneda, 39; Laín, 41ff.) In Barcelona, on the other hand, the first members of Opus emerged from among the select students of the Jesuits, who were organized into the so-called Marian Congregations." The future minister and commissioner of the Plan for Economic Development, López Rodó, tells it this way in his memoirs: thanks to a fellow student, he had read *The Way;* in January 1941 he went to Madrid expressly to meet Escrivá, and he was impressed by the "brief and forceful homily" at the mass that Escrivá celebrated in the oratory of the residence; in a conversation later, Escrivá introduced him to Opus Dei—apparently without any reference to the university "apostolate of penetration" as a means of attaining the goal of the re-Christianization of Spain—and two days later, while he was attending a Jesuit mass in Barcelona, "I saw clearly the call of God and decided to ask for admission into Opus Dei" (López Rodó, 1990, 22f.).

The biographies of the padre even provide the names of many of the persons who joined Opus Dei during this period (except for those who later left it: these people, even though they may have held responsible positions in the organization while they were there, are never mentioned). By the end of the period Opus Dei was established not only in Madrid, Valencia, and Barcelona, but also in Valladolid, Zaragoza, Bilbao, Seville, and Santiago (Berglar, 389).

Opus's sphere of action was in every case strictly academic. In some cases there were small student centers which sooner or later would become residences. In others the initiative came from professors, members of Opus Dei, who joined the faculties of various universities (this was the case in Santiago, for example). This, then, was the period whose main thrust has come to be called "the clear will to take over the university" (Aranguren, 1962, 4), "the conquest of the intellectuals" (Angel Sagarmínaga, 1945, quoted in Ortuño, 46f.), or "the conquest of the university chairs" (Artigues, 52f.).

The French historian Max Gallo sums up the situation by saying that its control over the Consejo Superior de Investigaciones Científicas allowed Opus to get control over the university, where it trained the country's elite members and leaders, and also intervened in the appointment of the committees that chose among competing candidates for university chairs (Gallo, 105, 155). For Laín, "the newly born Opus Dei thus began its penetration of the fortress—so uniformly falangist, viewed from the outside—of the Spanish State of the Victory" (Laín, 280; he makes a specific accusation about the "evil arts" of Opus members in the competitions for university appointments; ibid., 332f.). On the same subject Artigues asserts that thanks to its control over the Consejo Superior, "Opus Dei was able to select grant recipients, send them overseas, thereby facilitate their success in the competition for university chairs, and finally transform them into collaborating members of the Consejo" (Artigues, 50).

Of course, the "official" literature denies all this. Its authors allude to the rumor that "Opus Dei wants to take over the university" in order to add that this rumor "makes the children of Don Josemaría laugh—and suffer," and also that some of them "have the well-founded impression that they failed to win a university appointment precisely because they belonged to Opus Dei" (Gondrand, 161f.). Nothing is said about this in a book called *La crisis de la Universidad en España,* by José Orlandis, an Opus Dei member, made professor at the age of twenty-four and ordained into the priesthood a little later (Orlandis, 1967). And the "rumor" is refuted by Antonio Fontán, Opus Dei member and professor at age twenty-six, in his book *Los católicos en la Universidad Española actual* (Fontán, 1961, 60–66).

The only thing the "official" literature seems willing to admit is that Escrivá "had come little by little to the conclusion that, so as not to lose sight of the universal horizon that he saw on October 2, 1928, and to reach it as soon as possible, it was a good idea to pay more attention to the apostolic labor with academics, promoting among them the call to holiness in the midst of the world, making a commitment to celibacy: then they would have the basis of a nucleus of persons who would extend the labor later to many more—celibate and married—of all social conditions and professions" (Fuenmayor et al., 85).

Pending this subsequent diffusion and extension, Opus Dei's sphere of action during this entire period was strictly confined to the university. This is confirmed by the official documents, reproduced (yes—in Latin!) in the documentary appendix of the volume by Fuenmayor and his colleagues: the decree of official recognition (*decretum laudis*) of the Sacred Congregation of Religions in February 1947, for example, states that "Dr. Escrivá de Balaguer felt powerfully called and gently led to an apostolate among the students of the Civil University and the Secondary Schools of Madrid," and over the years Opus Dei has seen the number of its members increase "not only among doctors, lawyers, architects, career military officers, cultivators of science and the arts, writers, university and high school teachers, etc., but also among the students of the faculties, who perform a fruitful apostolate among their classmates" (Fuenmayor et al., 533).

(*b*) It was precisely the success of this apostolate of professionals and students that provoked an immediate and angry reaction from the Jesuits.

This is without doubt a chapter of transcendent importance for the entire subsequent history of Opus Dei. But the "official" literature is not very explicit on the topic, and it confines itself to providing a few small clues—in this case, the corresponding part of the puzzle has been deliberately left undone.

The persecution to which Opus Dei was subjected after 1941 was "a mixture of misunderstandings, calumny and jealousy"; but "Don Josemaría prohibited his children from talking about these painful events" (Helming, 46). From Gondrand we know that in 1940 Escrivá stopped confessing to the Jesuit priest Sánchez Ruiz, who he felt was under pressure (Gondrand, 155) in an atmosphere that had broken out into "an authentic vendetta of persecution fed by perfidious calumnies" (ibid., 154). Berglar sums up the situation by saying that "during this period immediately following the war the misunderstandings grew worse: some people did not understand the universal call to holiness taught by the Founder of Opus Dei. In spite of the adverse climate he continued his labor serenely, praying for those who attacked him and suffering for the harm these calumnies were causing to souls and to the Church" (Berglar, 389).

No one denies that the campaign existed. (This was especially true in Madrid and Barcelona; one of my interviewees, a Jesuit priest from Madrid, admitted that they had been on the point of expelling a student from school because he had attended Opus Dei meetings, and a former member of the Marian Congregations in Barcelona admitted having taken turns watching the door of the apartment where the Opus boys lived, checking who went in and out.) The disagreement in the literature begins over the motives for the persecution, and the "official" literature probably takes us down some dubious paths in saying that Escrivá was accused of preaching the "sanctification of married persons in the midst of the world" (Vázquez, 224), or that "the fundamental cause of the scandal was the anticipation of the doctrine which would be embraced by the Second Vatican Council in 1965" (Sastre, 263). Escrivá himself, on the rare occasions when he agreed to mention the topic, says it was another kind of accusation: *the accusation of secrecy*.

In fact, to a journalist who asked him about the charges of secrecy that have often surrounded Opus Dei, the Padre answered, "I could tell you, point for point, the historical origin of this calumnious accusation"; then he added that, in any case, he preferred not to talk about the "powerful organization" which had devoted itself for many years to "falsifying what they did not know" and accusing Opus of being "a secret society" (*Conversations,* no. 30). On another occasion, after reiterating that "I don't like to talk about these things," he says more specifically, "There are people who said we were working secretly—perhaps that is what they were doing—that we wanted to occupy high positions, etc. I can tell you, specifically, that this campaign was started by a Spanish religious who later left his order and the Church, contracted a civil marriage, and is now a Protestant pastor" (*Conversations,* no. 64; the Jesuit he alludes to is Ángel Carrillo de Albornoz, in those years head of the Marian Congregations in Madrid, and later in Rome, who at the time of Escrivá's interview [1968] had spent many years as the head of the Secretariat for Religious Freedom of the World Council of Churches whose headquarters is in Geneva).

As for working "secretly" with a wish to occupy "high positions," Escrivá says that "perhaps that is what they were doing." I am inclined to think that the origin of the conflict between the Jesuits and Opus Dei lay precisely in the fact that this is what *both* of them wanted. But we will leave that question for later, because the "official" version does not tell us much more. Note, however, that at least on one occasion an Opus Dei author leads us in this same direction, when he states that "those who did not understand the new thing that had just appeared" attacked Opus Dei because they saw it "as dangerous competition in the field of apostolic work with young people" (Gómez Pérez, 1976, 255).

(*c*) The conjunction of the two factors considered here—the progressive establishment of Opus Dei, which for the first time began to make its existence known in a real and concrete manner, and the hostile reactions, mainly coming from the Jesuit world—yielded as their result the first juridical recognition of the Work of Padre Escrivá (by now Escrivá de Balaguer).

There is no unanimity in the "official" literature on the order of events in this matter. Vázquez de Prada maintained in 1983 that it was the bishop of Madrid, Leopoldo Eijo Garay, who took the initiative in giving written approval to Opus Dei, against the wishes of Escrivá, who did not want any approval at all: "it was a friendly initiative by the prelate, without any petition by the Founder having anything to do with it" (Vázquez, 221). This hypothesis must be definitively discarded, however, once we look at the published official request sent by Escrivá to the bishop. It said that "he is privately directing an apostolic labor, called Opus Dei," and that "it is appropriate for the glory of God and the service of the Holy Church to furnish this Work with the characteristics of stability and official canonical status that would assure the permanence of its fruits." For these reasons he begs that the bishop "will deign to give canonical approval to 'Opus Dei' as a Pious Union," that "he will deign to kindly approve the attached Bylaws, Regime, Order, Customs, Spirit and Ceremonial which inform and govern 'Opus Dei,'" and, finally, that he find it worthy "to designate the persons of this Curia to know the Bylaws of 'Opus Dei,' given the character of the Work" (Fuenmayor et al., 511; the same document was published several years earlier by Rocca, 1985, 130, and it is not impossible that this is why the "official" literature has changed its attitude and admitted its existence).

A month later—on March 19, 1941—the bishop of Madrid signed the approval. Thus we have, twelve and a half years after the "official" date of the founding, the first document in which the name of Opus Dei appears (or "*la* Opus Dei," to follow the form used at the time by Escrivá), a document which also contains the date 1928 as the year of Opus's founding, thus gaining official recognition by the diocesan institution of the bishopric of Madrid. In response to Escrivá's request, the bishop decided that "keeping in mind the discreet reserve that must be kept for the greater glory of God and the effectiveness of the Work," the documents of the "Bylaws, Regime, Order, Customs, Spirit and Cermonial will be kept in Our Secret Archive" (Fuenmayor et al., 514). More than half a century has passed since then, but this set of documents seems never to have left the episcopal secret archive, and it has not been published. The authors of *El itinerario jurídico* thought it appropriate to reproduce only the first one (Reglamento; Fuenmayor et al., 511–13), which is very generic in character and drafted in a language very much like that used today in the "official" literature to present Opus Dei: it speaks of Opus as "a Catholic association of men and women, who live in the midst of the world and who, through the sanctification of ordinary work, seek Christian perfection" (ibid., 511f.). There is no reference, however, to "the university apostolate." Thus if it really is a document from the year 1941, we must wonder whether the rest of the texts that have neven been published have a radically different tone; otherwise it is difficult to understand the reasons for keeping them secret. Their publication, in fact, would help to eliminate many misunderstandings, and to definitively clarify long-standing controversies.

Therefore, with regard to this initial ecclesiastical approval of 1941, the information supplied by the "official" literature is insufficient. While it insists on trying to convince us that Opus Dei "has no secrets," it continues to apply the criterion of "discreet reserve" to documents written more than fifty years ago and kept in a "secret archive." We know, in a word, that in this archive are kept some of the missing pieces of the puzzle.

(*d*) What is no secret at all, on the other hand, is that from this moment on the bishop of Madrid was ready to take up the defense of the "Pious Union," just approved, against all its detractors. Cardinal Pedro Segura himself, archbishop of Seville—who certainly was not accusing Opus Dei of anything more than acting secretly—wrote to the nuncio Gaetano Cicognani that if he wanted information he could ask the bishop of Madrid (letter of 1941; Rocca, 1985, 135). And Federico Sopeña, who in addition draws a very interesting portrait of the bishop, Msgr. Eijo Garay, states that he knew for certain that Eijo Garay "was definitely protecting an experiment like that of Opus Dei" (Sopeña, 35f.).

The "official" literature corroborates this defense of Opus Dei on the part of Eijo Garay (whose only error may have been—according to the nonofficial literature—wanting to enroll some of Escrivá's first followers in the "Blue Division" to fight on the Russian front beside the German troops, an idea that Escrivá opposed vigorously). As proof of this defense all the biographers except Berglar point to a letter from the bishop to the assistant abbot of Montserrat, Aureli Escarré, dated May 1941. This letter brings us back to the topic of the conflict between Opus Dei and the Jesuit Father Vergés and his Marian Congregations in Barcelona.

The letter, written in response to a request for information that Escarré had sent to Eijo, begins in a very peculiar style: "I already know about this commotion that has been stirred up against Opus Dei in Barcelona. It is plain to see the pustule it raises on the evil enemy. The sad thing is when people highly committed to God become instruments for evil; it is clear that they are doing it with the intent to serve God" (Bernal, 279, and Vázquez, 228; Gondrand, 154, and Fuenmayor et al., 92, choose to eliminate the words "pustule" and "evil enemy"). The bishop continues that he knows all about it, because from the beginning he has kept abreast of it, so "believe me, Father abbot, Opus is truly of God, from its first idea and in all its steps and works" (this sentence appears in all the texts cited above, and also in Sastre, 267).

Eijo Garay then breaks out into a grand eulogy on the qualities and virtues of the "model priest" that Escrivá is (in the original letter, which is preserved in the archive of the abbey of Montserrat, the bishop writes—three times—"el Dr. Escri*bá*"; Rocca, 1985, 132), adding that "the secret association, as its denigrators call it," has his express authorization, and "the discreet reserve (no social secret) that Dr. Escribá inculcates in the members" is a sign of collective humility: "and nevertheless, today it is the good who attack him."

"Opus Dei deserves only praise," he says in the next paragraph; Bernal (p. 280), Vázquez de Prada (p. 228), and Sastre (p. 267) reproduce it in part, but only Rocca (p. 132f.) transcribes the whole thing. The bishop makes two statements that *none* of the texts of the "official" literature reproduces: "It is not a work of the multitudes but of *selection*"; and, "to equip all the *intellectual* professions from *chosen* groups, without flags flying in the wind or attention-getting labels, to live a holy life and exert influence on the welfare of the rest; that is the mission of Opus Dei" (emphasis mine).

The bishop closes his letter by saying that "the discredit of good people who persecute good is hurtful" (Gondrand, 154), and casting doubt on the claim by some Jesuits that "it defames the Society of Jesus to assert that it is persecuting Opus Dei and seeks its destruction" (Rocca, 1985, 133).

(*e*) In 1943 an extremely important event took place in the historical evolution of Opus Dei and led to the first modification in the juridical statute it had obtained two years earlier. It was even more important than this very approval, because it constituted the culmination of José María Escrivá's work as a founder.

If on February 14, 1930, the Lord had made him see the women's section of Opus Dei, on exactly the same date in 1943, while Escrivá was celebrating mass in the women's section center, the Lord "enlightened him" again and gave him, "loud and clear, the juridical solution he had been looking for but had not found" (Vázquez, 233): this was the creation of what might loosely be called the priestly section of Opus Dei, which translated into the official language as "The Priestly society of the Holy Cross." Thus the "supernatural vision" of 1928 was made complete, a vision that had already been complete ("I received the enlightenment about *the whole Work*"; in Fuenmayor et al., 26) without being complete ("women, not even as a joke"), and which after fifteen years was made even "more complete" (ibid., 118), in spite of the fact that it was already complete from the beginning. Perhaps all this is a little complicated to understand, but by this time we are familiar enough with the apparent mystery inherent in Msgr. Escrivá and his Work not to be too surprised.

We are also now quite familiar with the flowery style of the biographies of Escrivá so we do not have to entertain ourselves with all the anecdotes adorning the description and explanation of the birth of the "Priestly Society." We will confine ourselves to presenting the chronological sequence of events, based on the more restrained text of *El itinerario jurídico del Opus Dei,* complementing it when necessary with material from other sources.

In spite of its "thoroughly lay mentality" (Vázquez, 235), Escrivá had been very clear from the first that "in Opus Dei there must be not only lay persons, but also priests" (Fuenmayor et al., 29). But although prior to 1943 some priests had helped him, "the performance of these first collaborators was not a positive experience" (ibid., 116; they had even been his "crown of thorns," according to Vázquez, 232).

Escrivá thus reached the conclusion that the priests of Opus Dei ought perforce to come "from the ranks of the lay members of the Work" (Fuenmayor et al., 116). The problem was that, once ordained, they came automatically under the control of the diocesan bishop and were obliged "to accept whatever ministry he assigned them to" (ibid., 117). But Escrivá wanted them to be priests *of* Opus Dei and *for* Opus Dei, and instead of depending on the bishop he wanted them "to depend on the president general [that is, Escrivá himself] for the exercise of their ministry" (ibid., 119f.). In principle, this was feasible only in the case of the religious orders, or in associations provided for in the Code of Canon Law called "societies of common life without vows," more or less comparable to the religious orders but not exactly like them; in some cases, these societies "could enjoy the privilege of accepting priests as members, since the Code gave them a limited equivalence in their juridical regime to that appropriate to the religious orders" (ibid., 120).

Through this narrow chink filtered the light with which God, "interfering once more in his life," illuminated Escrivá on February 14, 1943. The illumination was "loud and clear" (Vazquez, 233), but the problem "was not easy to solve" (Fuenmayor et al., 119). So "Don Josemaría Escrivá de Balaguer began immediately to think" (ibid.). *The next day,* he suggested to Álvaro del Portillo, José María Hernández de Garnica, and José Luis Múzquiz that they become priests. And on June 13 of the same year he asked the bishop of Madrid for the establishment of the Priestly Society of the Holy Cross as a society of common living without vows (after having sent Álvaro del Portillo to Rome to explore—and probably prepare—the territory).

Nine days later Msgr. Eijo Garay asked permission from Rome, specifically from the Sacred Congregation of Religious, whose authorization was necessary for this kind of association. In October Rome granted the authorization, and before the year was over, on December 8, 1943, the bishop of Madrid signed the decree establishing the Priestly Society. This whole set of documents (reproduced in Fuenmayor et al., 515–29) will help us to clarify a series of questions, while at the same time, of course, raising new ones in their place.

Finally, in June 1994 the first three Opus Dei priests were ordained. If anyone asks why Father Escrivá declined to attend the ceremony, I am at a loss for the answer.

A Few Questions

Throughout these pages we have seen that there are many questions in the "official" literature that remain unanswered or unsatisfactorily answered. We have pointed out a few of these, and we will not repeat here the questions that have already been mentioned. We will pause only on two problems: first we will try to explain the issue that seems to us most important, specifically in relation to the approval of the Priestly Society of the Holy Cross; and second we will highlight the curious paucity of references to be found in the "official" literature to the events of the second world war, which coincided precisely with the period under consideration here. Finally, we will conclude the chapter with some complementary information about the conflict between Opus Dei and the Society of Jesus.

The Priestly Society of the Holy Cross

The approval of the Priestly Society of the Holy Cross presumes—for a movement wanting to present itself as eminently laical—an important change in Opus Dei's orientation as well as that of its organization. In fact, for many years Opus Dei did not find it easy to justify this change, and all the conflicts that followed from its desire to differentiate itself from the religious orders, which were not resolved until 1982, stem from this change.

Likewise, for a while there was considerable ambiguity about the relation between the Priestly Society and Opus Dei. According to the official version, it was not that "Opus Dei, as such and in its entirety, be transformed into a society of common living, but that a small part of the Work, made up of priests and a few lay persons, be established in a society of communal living, to which would remain closely linked, under the name Opus Dei, the association of the faithful approved with that name in 1941, of which the members of the Priestly Society of the Holy Cross would continue to be members" (Fuenmayor et al., 131). Notwithstanding this, in the request sent to the bishop of Madrid, the Padre wrote (in Latin) as "President of the Pious Union called Opus Dei" that "all and every one of the members" of this Pious Union request that "an association of the faithful of communal living without vows be established, by diocesan law, with the name Priestly Society of the Holy Cross" (ibid., 516). In the decree of establishment Bishop Eijo Garay wrote (also in Latin) that the increase in the number of members of the institution (Opus Dei) and the expansion of its field of action showed that it could not continue to be a simple association, but that it was appropriate for it to be transformed into a real ecclesiastical society, legitimately established and constituted, and that for this reason he established and constituted the pious association previously approved into a society of common living without vows, with the name the Priestly Society of the Holy Cross (ibid., 526).

This obvious contradiction gives rise to a new round of questions. If the former Opus Dei was now transformed into a Priestly Society, does this mean that Opus Dei disappeared? The "official" literature firmly denies it. Do the society of common living and the Pious Union coexist simultaneously? In approving the Constitutions of 1944, the bishop expressly asserted that they were the constitutions of the Priestly Society, "and of its specific apostolic instrument, called Opus Dei" (Fuenmayor et al., 529). We may assume that the Constitutions specify the nature of the specific connection between the society and "its instrument." But the Constitutions of 1944 have never been published. Were they already in the same form as the 1950 Constitutions? It is not certain, since in 1950 they always refer to the "Priestly Society of the Holy Cross and Opus Dei," while in 1943 and 1944 they refer exclusively to the former. During this period, what was the status of the women's section?

We have proof that noticeable ambiguities exist in this whole matter such that even the members of Opus Dei frequently fall into error and confusion. For example, Dominique Le Tourneau gives 1943 as the year of "the establishment of Opus Dei" instead of the year of the establishment of the Priestly Society (Le Tourneau, 58). Another author states that "in 1943 the Holy See gave first approval to Opus Dei as a reality for the whole Church" (Gómez Pérez, 1976, 254), which is clearly an error, since the institution still remained under diocesan jurisdiction. Berglar also seems to be in error when he writes the Msgr. Escrivá founded the Priestly Society "as an inseparable association, belonging to Opus Dei" (Berglar, 75), when in 1943 the situation was, if anything, the reverse: Opus Dei was the "instrument" belonging to the society. Probably the best approach, however, is to leave the disentangling of this imbroglio to the experts in canon law. Giancarlo Rocca, on one side, and Amadeo de Fuenmayor and his colleagues, on the other, have supplied enough information for the debate, and some degree of clarification is possible. But it is unfortunate that the Opus Dei authors have not wished to enter into a dialogue with Rocca, which creates an obstacle to clarification. Rocca complains about this, and on this point there is no doubt that he has indeed every right to complain (Rocca, 1989, 384).

Some other points, however, are fairly clear, thanks to the documentation reproduced and published by these authors. The request presented by Escrivá on June 13, 1943, to the bishop of Madrid clearly shows that since 1928 he had conceived his task as "an apostolic work among the young people who attend the civil universities" (Fuenmayor et al., 515). In his letter to Rome requesting the *nihil obstat* for the Priestly Society, the bishop reiterates this same argument and presents Escrivá's work as destined to cultivate Christian piety among university youth and to inculcate in them the highest standards of perfection in their professional studies, so that "well-educated and imbued with sincere religiosity, they would exercise effective influence on the community from various levels in civil administration" (ibid., 520).

In the curriculum vitae that Escrivá sent to the assistant bishop, Casimiro Morcillo, which was forwarded to the Holy See as supplementary information, he lists as qualifications his lectures at the summer session at the University of Jaca, reminding the Holy See that on that occasion the Holy See itself "conceded the privilege of a portable altar" (1940); it also points out that Escrivá was the only priest of the secular clergy in the Consejo Nacional de Educación, serving alongside three bishops and a few members of the religious orders (the shadow of Albareda and his Consejo Superior de Investigaciones Científicas was long), and that he was professor of ethics and morals at the Journalism School (the shadow of Herrera Oria and the Asociación Católica Nacional de Propagandistas was also an appreciable factor). The curriculum vitae ends by quoting an excerpt from Eijo Garay's letter to Abbot Escarré mentioned above: "exemplary priest, chosen by God for the sanctification of many souls . . . , apostle of the Christian education of university youth" (Fuenmayor et al., 522f.).

In the rough draft of the future Constitutions of the Priestly Society (why do Fuenmayor et al. publish the rough draft in their documentary index, pp. 516–20, instead of the final text?), the Society of the Holy Cross is presented as a society that is basically, or preferentially, clerical (*praeferenter clericalis*). How does this assertion connect with the fact that at the moment the text was written, there was only one cleric in the society, which, furthermore, was proposing "an eminently lay spirituality"?

The second article of the draft specifies that the general end of the society was "the sanctification of its members through the practice of the evangelical counsels," while its specific goal was "to work with a view toward having the intellectuals and managers in civil society practice the precepts, including the counsels, of Jesus Christ" (Fuenmayor et al., 516). On this point it could not be more explicit, and we are thankful that the "official" literature, after going around in so many circles, offers us such a clear text, even if it is in Latin. Other aspects can also be discerned from this draft of the Constitutions: for example, those baptized as adults and those who lack a minimum of three generations of Catholicism in their family history cannot be members (art. 14); and members must have a university degree (art. 15). But these are secondary in comparison to the fundamental assertion that the goal of Escrivá's Work is to work specifically among "the intellectuals and the managers of civil society."

Thus the official texts themselves corroborate the thesis that the "re-Christianization of Spain" was Escrivá's primary objective during the forties. At the very least, the text of the decree establishing the society confirms this (Fuenmayor et al., 526f.). Finally, in a document from the diocese of Madrid, reproduced in Rocca's book, it is explained that the creation of the Priestly Society of the Holy Cross was motivated by the veering off of the intellectual world toward doctrines and ways of life that were contrary to the Catholic Church, as well as by the desire to bring everyone, but principally intellectuals, to the Christian and religious life. As for the first goal of the society, its members were to attain "Christian perfection through the sanctification of professional work, whether in the exercise of their professions or in the performance, with apostolic spirit, of the public duties entrusted to them by the state." The text also specifies that "the priest members will occupy the principal posts of the Work," and that the Work will obtain its resources from "the incomes of the secular members in their professions as professors, architects, lawyers, etc." (Rocca, 1985, 147).

In the mid-1940s, in spite of all the ambiguities surrounding the existence of Opus Dei and the Priestly Society, about the official status of the former, and about their relation to each other as well as that between both of them and the women's section, one essential piece of information is clear: the objectives of the association as approved in the diocese of Madrid with the authorization of the Holy See are the ones specified in the "official" documents we have just reviewed.

World War II

In comparison to the question of the apearance of the Priestly Society of the Holy Cross, a topic of fundamental importance to our comprehension of the history of Opus Dei, the second theme to which we now turn might seem to be of purely anecdotal interest. But we must keep in mind that if it seems like an anecdote, it is simply because the facts and events of the second world war are never mentioned in the "official" literature except as the background to several anecdotes about Father Escrivá and his "children" during this period.

The war is presented, basically, as an *obstacle* that hindered the process of Opus Dei's expansion. If in 1936, according to Escrivá's biographers, the outbreak of the Spanish civil war had initially prevented the realization of his international projects (totally nonexistent projects, according to our hypothesis), the world war was now another setback. Vázquez de Prada, the only one who devotes nearly *two* entire pages to the war, poses the question in the following terms: "Spain had just emerged from the heavy punishment of a civil war when it found itself on the point of sliding into an international maelstrom. Suddenly there was the brutal invasion of Poland and the Machiavellian division of its territory between Russia and Germany. His hands free, Hitler launched the western campaign" (Vázquez, 218). He concludes a little further on that, for Escrivá, "his dedication was to consecrate himself to the total fulfillment of the Divine Will. Nothing else interested him. . . . There was much to do. The apostolic campaign presented various simultaneous fronts: the training of new members, the expansion in Spain, and the problematical foray into foreign lands" (ibid., 200).

Up to a point we might say that it seems logical for the official literature not to say anything about his attitudes toward the war. It takes someone from the outside, like Vladimir Felzmann (who was an Opus Dei priest), to explain that Escrivá saw the war as a crusade against Marxism, and that "the members of Opus Dei offered to volunteer for the so-called *Blue Division*," eager to fight on the side of the German army against the Soviets, although they did not actually join it. Once the war was over Escrivá denied that Hilter, who had been of so much help to Franco, could have been responsible for the death of six million Jews. "Certainly there were no more than three or four million," Felzmann reports that Escrivá said to him (Hertel, 205f.). Hertel's statements may seem very strong. (Later, the publication of these same statements by Felzmann in *Newsweek* at the beginning of 1992 aroused a considerable stir on an international scale.) It is true that Felzmann did not meet Msgr. Escrivá until quite some time after the end of the war; but it is also true that he maintains that Escrivá himself spoke these words to him. Finally, it is no less true that "philonazism" was widespread in Spain during those years, as an Opus Dei member, Rafael Gómez Pérez, demonstrates with quotes and concrete examples; none of these, of course, originated within Opus Dei (Gómez Pérez, 1986, 16ff.).

The "official" response to this kind of question is that Escrivá systematically refused to take sides in political affairs. The Opus authors thereby run the risk of confusing the principle of freedom of action for Opus members with Escrivá's personal convictions, to which he obviously had a perfect right, even if he had no right to impose them on others. For example, they justify Escrivá's "stubborn refusal to place himself at the service of social regimes or ideologies" as a product of his "deep respect for freedom of conscience" (Vázquez, 290).

It is interesting to note, furthermore, that this author, Vázquez de Prada, is the only one of the Spanish biographers who includes Franco's name in the subject index to his volume (neither Bernal nor Sastre does); one reference is to "Francoist Spain" and the second is to "the Francoist dictatorship" (Vázquez, 240, 290). Neither he nor the others, however, allude to the fact that in 1944 Escrivá led spiritual exercises for Franco (Calvo Serer, in Martí and Ramoneda, 113). Only in the non-Spanish biographies do we find this mentioned, in a very indirect manner in Gondrand (p. 173), so indirect that in fact it would be insufficient if we did not see it explicitly confirmed in Berglar: "When Msgr. Escrivá de Balaguer, in the forties, led several days of spiritual retreat for the head of state and his family . . ." (Berglar, 173).

In the end, then, all the "official" literature offers us about the second world war is pure anecdote. Let us consider a couple of examples at least, which will lead us toward the venue of our next chapter, Rome.

The first members of Opus Dei who went to Rome were José Orlandis, of Mallorca, and Salvador Canals, a Valencian, toward the end of 1942. The former was twenty-four years old and had just been made professor of law; the second, who was not yet twenty-two, had already earned a doctorate in law. Both had obtained scholarships to further their studies in Rome (no one specifies whether they got the scholarships through the Consejo Superior de Investigaciones Científicas, although it might well be). "When José and Salvador arrived, Rome was teeming with German troops. The wartime navy occupied the coastal cities and the Allied attacks were not far off. The climate of the city exuded tension. But nothing could stop them; the pursuit of their studies was very important for their future professional work" (Sastre, 320; let us specify that the two men would be ordained priests; one would become a professor at Zaragoza, and the other would take a job in the Vatican Curia). Furthermore, this "furthering of studies" would also serve to let "the Work begin to be known in Roman circles." Sastre gives a whole list of bishops, abbots, cardinals, and so on, who "began to know Opus Dei," and in January 1943 Orlandis and Canals were granted an audience with Pius XII. The master of ceremonies "was astonished at the youth of the two Spaniards: so young and already professors!" (Sastre, 320).

In May 1943 Álvaro del Portillo made a trip to Rome, prior to the permission for the establishment of the Priestly Society. During the flight, "he became aware that they were flying over a squadron of airplanes. . . . Suddenly the bombing and antiaircraft fire broke out. . . . In the midst of the general turmoil, Don Álvaro remained calm. The papers he had in his briefcase guaranteed the safety of the airplane and also that the frightened passengers would not end up as fish food" (Vázquez, 234; this is another example of the idée fixe held by Escrivá and his followers that "it was Heaven that was endeavoring" to bring about the realization of "the Work"). Portillo was also granted an audience with Pius XII, and he too aroused admiration: "So young and already an admiral! one gentleman said to another on the trolley" (ibid.). In truth Portillo had dressed for this important occasion in "a navy blue uniform with braid, gold buttons, and two-cornered hat," which is the "dress uniform of a highway engineer" (Sastre, 322). The reference to the war here is: "the Romans of those days had already seen a long parade of pomp and colorful plumage" (Vázquez, 234). But the sentinel of the Swiss Guard, seeing him arrive in this outfit, gave orders for the Guard to come to attention, "and Don Álvaro, letting things take their course, passed the troops in review" (ibid., 235). In this manner, Sastre announces, Álvaro del Portillo made "his first entrance into the Vatican."

A month and a half later Rome was hit by "a shattering bombardment"; "whole neighborhoods of Rome were covered with rubble," and the airport "burned from one end to the other"; but "in the midst of this chaotic situation, the process for conceding the *Nihil Obstat* followed its course" (Sastre, 323). As we said earlier, the Work was God's and it was God who "was endeavoring for it to be realized."

When the war was over, the Founder himself traveled to Rome, *Videre Petrum,* to see the Holy Father. And beginning in this year, 1946, Rome became Escrivá's permanent residence.

Opus Dei and the Society of Jesus

We would not like to conclude our analysis of this period (1939–1946) without adding something more about the conflict which, starting during this period and lasting for many years, pitted the Jesuits and Escrivá's followers against one another. We are not attempting to settle the question—quite the contrary. This is a question that will unavoidably remain open as long as so many pieces of the puzzle are still missing. Furthermore, it is obvious that in this case the problem is lack of information; the Padre himself is very explicit in this respect: "I do not wish to speak of these things" (*Conversations,* no. 64); and in the forties, Opus Dei members had specific instructions not to talk about the matter (López Rodó, 1990, 24).

In fact, not only do we know that there are many pieces missing from this part of the puzzle, but we even know where they are hidden, at least a good number of them. In addition to the above-mentioned letter from the bishop of

Madrid to the assistant abbot of Montserrat, which is the only document of its kind that the Opus Dei literature decided to publish (not, however, in its entirety), even before Rocca published it, we are certain that the archives of many Spanish bishops contain more documentation. The same can be said of the nunciature's archives, since the nuncio received denunciations and since, as Rocca has shown, the nuncio himself wrote letters asking for information about Escrivá's little group of men, which was practically unknown at the time and was, according to some, guilty of "secrecy" (Cardinal Segura, in Rocca, 1985, 134) and of being a "heretical sect" according to others (López Rodó, 1990, 24). Finally, of course, many of the puzzle pieces must be located in the archives of the Marian Congregations of the Jesuits, as well as in Opus Dei's own archives. Leaving aside the degree of difficulty one might encounter in trying to get access to all this material, there is no doubt that it would be useful to accomplish this research, which is not a simple matter, but holds undeniable interest.

To understand the history, largely still kept secret, of the tempestuous relations between the Society of Jesus and Opus Dei, it seems essential to take as our point of departure the parallels and the competition between the two institutions. *Parallels and competition*—an apparently dual hypothesis that is really only one, since if there is competition between them, it is in fact because they are so similar; and their extreme similarity could only lead to radical confrontation in an atmosphere like the Spanish Catholicism of the era, which was hardly a situation of religious pluralism.

This is a theme which can never be stressed too strongly: the influence of Anglo-Saxon sociological literature has led us to assume that "pluralism" implies peaceful coexistence, and even fruitful cooperation, between different religions. But the truth is that in the case of the countries of southern Europe the eruption of a situation of religious pluralism has meant something entirely different: the progressive "demonopolization" of a specific religious tradition. In the case of Spanish Catholicism in the early Franco years, the only possible way for an emerging religious group to coexist with already existing groups was to follow the path of ghettoization, of isolation. But Opus Dei was from the first too ambitious and too belligerent to resign itself to occupying an isolated, marginal position in the Catholic picture. In the specific terrain in which it proposed to act ("the university apostolate," "the leading sectors of society"), it inexorably had to face a head-on confrontation with the group that aspired to preserve its privileged monopoly over this terrain, and that group was the Jesuits.

The basic thesis of parallelism and competition between the Jesuits and Opus Dei must be qualified by certain shades of meaning. First, in affirming the existence of this fundamental parallelism we do not deny that there are differences between the two institutions. The differences exist, and the members of Opus Dei have clearly pointed them out; but these are differences of nuance and not of basic nature. Second, both the Society of Jesus and Opus Dei have evolved with the passage of time; however, they have not done so at the same pace or in the same direction. This means that at present the competition between them frequently (though not always) has less to do with their similarities than with the fact that they currently occupy opposing positions in the bosom of the Catholic Church. But when the conflict began to unfold in Spain in the 1940s, there was no such polarization, and if one of the two adopted newer or more daring postures under those circumstances, it was certainly Opus Dei, not the Society of Jesus. Consequently, everything is reduced to something as basic—but as fundamental—as this: any analysis of their similarities and their confrontations must be situated in its specific historical context.

(*a*) Before initiating the process of Opus Dei's establishment Escrivá often referred to Saint Ignatius Loyola, always in eulogistic terms.

> Who are you . . . nobody. Others have done wonders, are doing wonders, in organization, in publishing, in promotion. What means do they have, while you have none? Well: remember Ignatius. Ignorant, among the doctors of Alcalá. Poor, very poor, among the students of Paris. Persecuted, defamed. . . . This is the way: love, and have faith, and . . . suffer!: your Love and your Faith, your Cross, are the unfailing means to make effective and to perpetuate the zeal for apostolate that you carry in your heart. [*The Way,* no. 474]

Saint Ignatius and his first companions appear often as a model or a precedent for Escrivá's small group of early followers. Even in the manner of presentation of the history and vicissitudes of the early years in the biographies, we frequently find surprising parallels (in spite of the fact that the name Ignatius does not appear in the name index of Bernal, Vázquez, or Sastre) with the founding period of the Society of Jesus (see, for example, among many possible sources, the fine summary by Woodrow, 1984, 23–65, or the great work of Lacouture, 1991). Some authors have found similarities between *The Way* and Saint Ignatius's *Spiritual Exercises* (Artigues, 127). Others—we saw this earlier—speak of "trivialized Jesuit asceticism" in the case of Opus Dei. We can certainly say that the more or less legendary histories of the beginnings of all religious foundations have common features (especially when they are told by faithful, fervid disciples), and furthermore that the spirituality of the Jesuits has been so dominant in the modern Catholic Church that even the Code of Canon Law of 1917 expressly states that "when doing spiritual exercises preference shall be given to the Ignatian method" (Moll, 1991, 2). The assertion that without the Jesuits Opus Dei would never have come into existence is unpopular on Opus Dei's side because they don't want to be accused of having such irresponsible 'paternity," and on the Jesuit side because they reject any "filiation" that is so un-divine; but the assertion is obviously true.

Escrivá does, however, borrow from Ignatian spirituality a specific and very characteristic feature: that of *selection;* the idea of the "education of select persons," in accordance with the title of the work of the Jesuit Ángel Ayala, the founder of the National Catholic Association of Propagandists, which has so many parallels to Escrivá's Opus Dei. In his *Spiritual Exercises,* Saint Ignatius expressly prescribes the "selection of those whom you want to practice," considering that "not everyone is fit to follow [our discipline], and good will is not enough" (Moll, 12). If it were even minimally legitimate to effect the transposition, setting aside the logical differences in language, we would say that for

Ignatius Loyola, as for Escrivá, his followers could not be "of the crowd," because "they were born to be caudillos" (*The Way,* no. 16).

In the last chapters we will pose the question of a possible relation between Opus Dei's economic ethic and the Puritan ethic analyzed by Max Weber; this will be an important element to keep in mind. If in fact there are common features between the conception of work as a vocation in both cases (the notion of *Beruf;* Weber, 1904, Chap. 3; *ut operaretur,* a message that Escrivá is said to have rediscovered in Genesis); if this leads in both cases to a perception of success as a guarantee of election; and if this leads to an interpretation of "Christian morality as a strategy for improving the world" (Moll, 21) as a result of which the problem of means and the problem of tactics become questions pertinent in the sphere of spirituality and ethics (and *The Way* dedicates a chapter to each), we will have to ask to what extent is there a direct connection between Puritanism and Opus Dei, or whether this connection is not better established between Puritans and Jesuits, on the one hand, and Jesuits and Opus Dei, on the other. (The debate between Puritan economic ethics and Jesuit ethics is old, and it is well illustrated in Robertson, 1933, criticizing the Weberian thesis, and in the reply by the Jesuit priest Brodrick, 1934.)

It would be convenient if we could confirm this first fundamental similarity between the Society of Jesus and Opus Dei as groups of "select" persons, of those "called to be perfect." We will try to do this from the sociological perspective, although clearly sociology alone is not sufficient, and a complementary theological treatment is necessary, even indispensable, but this is not possible in these pages. Even though Escrivá reminds us the "the whole world is called to be perfect," this does not rule out the distinction made between "the crowd" and "the caudillos," nor can we forget that, in speaking of the "evangelical counsels," he says that "the Lord has asked more of me." Does this have anything to do with the accusations of gnosticism or Pelagianism to which Opus Dei has been subjected at times? From a theological perspective, is it possible to speak of a "life of perfection" without automatically implying that one is professing a life "more perfect than that of the rest"? Do we find ourselves once again—with predestination or without—facing the classic distinction between "the holy ones chosen by God" and "the masses"? In the case of Protestant Puritanism, the call to perfection and its logical consequence, the emergence of groups of "select" persons, withstood the disappearance of the religious orders. In the case of the Society of Jesus, Xavier Moll quotes Kierkegaard, who asserts that in the bosom of Catholicism "the religious orders came to an end with the Jesuits: with them they finished converting themselves into a completely worldly business" (Moll, 12). Is this one of the reasons why Opus Dei, the continuation of the Catholic tradition begun by the Jesuits, systematically rejects any comparison to the religious orders? Would it be theologically correct to say that Opus Dei is, in the words of Kierkegaard (a Protestant thinker whom several Opus Dei authors nevertheless quote frequently) "a completely worldly business"?

(*b*) But while Escrivá spoke reverently of Saint Ignatius before the process of Opus Dei's establishment in Spain began, after the civil war his relations with the Jesuits deteriorated rapidly. It was not long before he broke with the priest who had been his confessor for many years, the Jesuit Sánchez Ruiz (Gondrand, 155), during the unleashing of the persecution and the campaign (of "calumnies" according to the "official" literature) against himself and the nascent Opus Dei. It is not impossible that these events had a decisive influence on Opus Dei's adoption of "secrecy" (according to its detractors) or "discreet reserve" (according to its defenders).

Today one may read declarations like those of the Dutch priest Kolvenbach, present general of the Jesuits. When asked about the strained relations between the Society of Jesus and Opus, he answered that

> it's only a rumor, constantly fed by the press. Probably it is due to the fact that when Opus Dei began there was some trouble with the Jesuits, because, among other things, of the politically very tense times. At present, I can say that my relations with the prelate of Opus Dei are very cordial, when there are problems we deal with them frankly. Furthermore, relations between the Society and Opus Dei cannot be frequent or constant, because our respective missions in the Church are different. [Kolvenbach, 1990, 153]

The prelate of Opus Dei, Msgr. Álvaro del Portillo, has made similar declarations (see also Escrivá, *Conversations,* no. 54).

Given all this courtesy and diplomacy, it is hard to imagine that Msgr. Escrivá could have said, even in a moment of pique, "that it would be better to die without the last sacraments than to receive them from a Jesuit" (Walsh, 125). But no less difficult to imagine is the fierceness of the attack on Escrivá in the forties: the Jesuits persecuted him, abused him, denounced him. The families of the boys who joined Opus, or who sympathized with it, were subjected to pressure and threats. And Escrivá was obliged to write to those families, frequently on the official stationery of the bishop of Pamplona, to try to demonstrate that it had official protection and blessing. Much has been said about the role played by the bishop of Madrid as defender of Opus Dei, while the "official" literature speaks relatively little of the bishop of Pamplona, Marcelino Olaechea, in whose house Escrivá found refuge. (If Álvaro del Portillo wanted to, he could explain it, because he was there too.) Olaechea is the one who went to see Father Vergés of the Marian Congregations in Barcelona; and it was Olaechea who spoke with the bishop of Barcelona, Gregorio Modrego, about the conflict and the denunciations. The formal argument given by the Jesuit Vergés was of the legal sort: according to the statutes of the Marian Congregations, no one can belong to another religious organization without telling his superiors. And the boys Escrivá was "fishing" (the Jesuits say "stealing") from the Marian Congregations did not want to say anything (or they were not authorized to talk about it).

All this happened at a time when the Jesuits—their schools had been closed during the republic and they had been officially expelled and gone into exile, mostly to Belgium—had just returned to Franco's Spain, triumphant, without having suffered the casualties that many other religious had endured, ready to exercise a kind of monopoly over the best young minds.

In a doctoral thesis on "the socialization of Catholic leaders" in Barcelona during the forties, Francisco J. Carmona analyzes the functioning of the College of San Ignacio and the Marian Congregations, basing his research mainly on documents of the period. Although he never mentions Opus Dei, his work is very useful to us in providing a foundation for the hypothesis concerning the similarities then existing between the objectives proposed by the powerful Society of Jesus and those of the incipient Work of Padre Escrivá (Carmona, 1991).

The author defends the thesis that the Society proposed, through its preparatory schools, "to educate a small number of Catholic leaders who, inserted into the elites of Barcelona, will make it possible for the Catholic Organization to increase its influence and control over Barcelona society" (Carmona, 31f.). Opus Dei, similarly, states in its Constitutions that "one of its apostolic means is public positions, particularly those requiring management functions; and in order to get them, it is important for its members to distinguish themselves professionally and concern themselves with the acquisition of scientific education and training" (*Constitutions,* 1950, arts. 202, 203).

The section Carmona dedicates to the theme of the "Kingdom of Jesus Christ" and the "Message of the Sacred Heart of Jesus" can equally be read as an explication of the sources of inspiration for Padre Escrivá's thinking. He refers to "a project of control and influence on all social institutions." Attacked for its role as legitimizer of the ancien régime, the Catholic Church reorganized its plan of action "under the banner of the social kingdom of the Sacred Heart." To confront all the errors of modernism, it proposed "a program of struggle to impose the Catholic social program on society" (Carmona, 207). By coincidence, one of the works repeatedly cited by Carmona in this section is the *Social Catechism* of the Jesuit priest Valentín Sánchez Ruiz, the spiritual director and confessor of José María Escrivá from the official date of the founding of Opus Dei until 1940. Another coincidence is that the final approval of Opus Dei as a secular institute of the Catholic Church bears the date of the "festival of the Sacred Heart of Jesus" in the year 1950 (Berglar, 390). The whole book *The Way* is based on this concrete vision of the function of the Church in the world.

The Marian Congregations were not open to just any student of the Jesuits but only to the best, the "choicest," who aspired to a "higher perfection." (The bishop of Madrid likewise defined Opus Dei as a work "not of the multitudes but of the choice"; Rocca, 1985, 132.) The first of their rules defines them as associations destined to make of their members "true Christians, who sincerely pursue their own sanctification in their own lives, and work very hard, as their social condition permits, to save and sanctify others" (Carmona, 288f.). The text of the document for the cause of Msgr. Escrivá's beatification similarly describes the members of Opus Dei as Christians who "endeavor to find holiness and practice the apostolate among their associates and friends, each in his own sphere, profession and work in the world, without changing his state" (Berglar, 76). This means sanctification of oneself and of others, work, and preservation of one's state (meaning not to become a religious, but to remain secular). If this is also the principle of every Marian "congregant," then where is the radical innovation and originality of Escrivá de Balaguer's message?

(*c*) Some of the most interesting pages in Carmona's study are those he devotes to relations between the Society of Jesus and Francoism, General Franco, and the army during the forties. "Within the framework of the proliferation of spiritual exercises of the era, it became the norm for high-ranking civilian and military authorities, including General Franco, to participate in annual spiritual exercises" (Carmona, 84). We know that in 1944 Escrivá directed the Caudillo's spiritual exercises, but usually these were conducted by the Jesuit fathers, "the experts in these methods of Ignatian spirituality" (ibid., 85).

> The collaboration and sympathy between the high authorities of the Franco regime (especially the military) and the Society of Jesus were great. In fact, these figures and their families were looked upon by the Jesuit fathers as human material that needed to be molded to the good of the Church and the Fatherland; in turn, for many of these individuals their communication and relationships with the Jesuit fathers were either a product of family history or a social practice attached to their new status. [Carmona, 85]

A reading of the *Revista* of Saint Ignatius College in Barcelona reveals veneration and cordial loyalty to Franco on the part of the Jesuits. Carmona gives a few concrete examples, and concludes by saying that "for the Jesuits, as for the Catholic hierarchy and the great majority of the Catalan clergy, Francoism appeared as the savior of the Church and the destroyer of the laicism and anticlericalism of the republican era" (Carmona, 88). In Opus Dei, the adhesion to and veneration for the figure of Franco were just as great. But here as in other matters, it was less ostensible, less noisy, and more "discreet." Here, paradoxically, lies one of the keys to Opus Dei's success in the forties (especially in Catalonia). Nearly all the non-Opus authors who have written about the Opus Dei of those years agree on this point: Escrivá's men seemed less openly committed to the regime, and the tone they adopted—presenting themselves as an emerging movement, and in opposition to the Jesuits—made them seem less triumphant, less openly totalitarian, and thus apparently more liberal.

In contrast to the Falange with its totalitarian zeal or to Acción Católica, mortgaged to its earlier political commitments, "Opus Dei looked like something new, different from both Acción Católica and the Falange" (Hermet, 1981, 2:182). In Catalonia especially, Opus looked like an advanced movement compared to the traditional Catholicism of the Jesuits. Its establishment in Barcelona was aided by a posture that was apparently not closely committed to Francoism and not necessarily anti-Catalan (Hermet, 1980, 1:266). Opus Dei offered to the flower of "the bourgeois intellectual youth a framework and a mystique that were more attractive than the Jesuits or the Falange" (Gallo, 105). In Catalonia, "the first group to form that was eager to purify that hodgepodge of Acción Católica's slogans, mottos, impositions, and challenges of every sort, was Opus Dei" (J. Dalmau, 1979, quoted in Casañas, 146).

Finally, if during the first years of Franco's regime Opus Dei seemed like an innovative movement, it was really because—as Carmona says—even though "in those postwar years the national Catholic pastoral tendency was the majority tendency in the Spanish church, . . . the members of the Society of Jesus were the ones who most effectively supported it" (Carmona, 102). It is particularly important to keep this in mind, since the Jesuits' subsequent evolution (especially in the sixties) tends to make us see, from our present perspective, only what sets Opus Dei and the Jesuits apart from or even against each other, rather than what made them so similar for many years. When one speaks today of competition or conflict between these two institutions, most often it is a consequence of the fact that today they are defending opposing ideologies within the Catholic Church. In the epoch we are analyzing, on the contrary, the competition was due to the fact that they were maintaining positions that were fundamentally identical. Today there are many authors who present Opus Dei as the continuation of the historical role played for so long by the Society of Jesus: an organization stressing educational institutions, the "training of the select few," social works, influence in the world of politics and business, a presence in the world of the communications media, and, on a more strictly ecclesiastical level, positions of power in the Vatican bureaucracy. Assuming that we take all this as true (the "official" literature of Opus Dei tends rather to deny it, although the facts, which are very stubborn, point exactly in this direction), we would doubtless have to attribute it to Opus Dei's successful and deft strategy. ("What beautiful houses are being prepared for us!" exclaimed Escrivá when contemplating how the others "were erecting magnificent buildings and building sumptuous palaces"; *The Way,* no. 844.) But we would also have to see it partly as a consequence of the recent evolution of the Society of Jesus, which has ceded several buildings and palaces to anyone who wants to occupy them.

In other words: there have been those from the Society who have written a "Requiem for Jesuiticalism" (Coy, 1974). Opus Dei has evolved too, and it is not today what it was in the forties; but it has not evolved in the same way, nor has it admitted it out loud. Up until now, no one from Opus Dei has chanted a requiem for the spirit of Opus Dei of those years.

(*d*) I basically agree with Artigues's thesis—at least as it relates to this initial period—that "the Institución Libre de Enseñanza, on the one hand, and the Asociación Católica Nacional de Propagandistas [ACNP], on the other, represent in a way two prefigurations of Opus Dei or, if you prefer, two archetypes, two models on the basis of which Father Escrivá and his disciples conceived their own enterprise, giving it, of course, special features" (Artigues, 23).

Defined by Hermet as "a prototype of the lay associations conceived for the Catholic elite" (Hermet, 1980, 1:245), the ACNP was founded in 1908 by the Jesuit Ángel Ayala as an extension of the model of the Marian Congregations (the same author calls it "the lay filial of the Society of Jesus"; ibid., 262). It identifies as its primary objective "to prepare a few select Catholic university students capable of playing a role in public affairs" (ibid., 246). Up to this point the similarity to Opus Dei is clear and undeniable (*Constitutions,* 1950, art. 202: the performance of "public jobs as a means of apostolate"). Where then are the differences, and what are the "special features"?

According to Msgr. Escrivá,

> I have never stopped worrying about that attitude which makes a profession of calling oneself a Catholic, as well as the attitude of people who wish to deny the principle of personal responsibility on which all of Christian morality is based. The spirit of the Work and of its members is to serve the Church and all creatures, not to use the Church to serve themselves. I want the Catholics to carry Christ, not in name, but in their conduct, bearing real witness to Christian life. Clericalism is repugnant to me, and I understand that—besides an evil kind of anticlericalism—there is also a good kind of anticlericalism which comes from love of the priesthood, which opposes the use of a sacred mission for earthly ends by either the priest or the simple believer. [*Conversations,* no. 47]

Escrivá's language is not as cryptic as it seems at first sight. As "lay filial of the Society of Jesus," the Asociación Católica Nacional de Propagandistas "carries Christ in name," and it could be considered a camouflaged form of "clericalism." According to its founder, Ayala, the ACNP wants Catholics to exert influence in public life, and it wants to supply men for public positions. It says—just like Escrivá—that the ACNP itself takes no political stand, but that its members as individuals may and ought to do so. The difference rests in the fact that these men, who are "needed to act in leading circles of power and of society" (Hermet, 1980, 1:247), are provided by the ACNP "to the Church" and more specifically to the ecclesiastical hierarchy, something which does not, by contrast, happen in the case of Opus Dei.

Opus Dei would tend to argue that it is precisely the "good kind of anticlericalism" which informs the principle of personal freedom and responsibility. ("Be free, my children," Msgr. Escrivá liked to say.) The Jesuits would defend their position by asserting that "the ANCP never tried to act outside the Spanish hierarchy, as Opus Dei would do later" (Sáez Alba, 29). "At no time did it occur to Msgr. Escrivá de Balaguer to separate the members of Opus Dei from the jurisdiction of the diocesan bishops. Anyone who states the contrary would be spreading lies" (Berglar, 17). But in spite of Berglar's vehement protests, the truth is that Opus Dei was accused of trying to bypass the diocesan hierarchy. The same could be said of the discretion recommended in the Constitutions of 1950 (especially arts. 189–91), which stipulate that the members "shall never reveal to anyone that they belong to Opus Dei without express permission of the local director" (art. 191). Official documents have been published, signed by Álvaro del Portillo in 1949, in which he consults the Vatican to find out if it is necessary to obtain the authorization of the bishop of the diocese in order to undertake various initiatives (Rocca, 1985, 170f.). As late as 1979 Cardinal Baggio, whose job it was to coordinate the studies preparatory to the transformation of Opus Dei into a prelature, asked Msgr. del Portillo for an "information supplement" and explicitly asked him to state "what specific criteria" they intended to adopt in order to prevent Opus Dei from being able to convert itself into "a parallel Church, in the bosom of the territorial jurisdictions" (Fuenmayor et al., 613).

Still less cryptic than Escrivá's language is that of another Opus Dei priest, Juan B. Torelló, when he states that "Jesuit spirituality," with its principle of absolute and total obedience, could never lead to "an authentically lay spirituality." By not being based on personal freedom and responsibility, in fact, it easily leads to "esprit de corps," and to an "instrumentalization of temporal values" whereby the laymen end up being only the "worldly *longa manus*" of the Society of Jesus (Torelló, 1965, 23).

Even the Jesuits cannot deny that the Society of Jesus has been characterized from the beginning by its esprit de corps. (As for Escrivá, this was something that did not bother him at all: "don't worry if people say you have too much 'esprit de corps'"; *The Way,* no. 381.) But while there may be real differences between "Jesuit spirituality" and "the spirituality of Opus Dei," we must not forget that in 1950 Opus Dei defined itself as "a family, but also a militia" (*Constitutions,* art. 197). According to Otto Roegele, an author cited in the bibliographies of the "official" literature (see, for example, Le Tourneau, 123; Mateo Seco, 529), Opus Dei, which "tries to influence the masses by occupying key positions in society," is characterized basically by its "aristocratic and gentlemanly spirit" (Roegele, 398, 392).

On the topic of esprit de corps, however, we will leave the last word to Giampaolo Bonani, who in an article published in 1971 in *Nuestro Tiempo,* a magazine put out by the University of Navarre, describes in detail all the accomplishments of Opus Dei in Italy, and states: "Using another graphic comparison, one might say that the overwhelming majority of the members of the Work, while they are '*soldiers of Christ*'—as is anybody who is baptized and confirmed—have no vocation at all to act as an *army corps;* on the other hand, they are well versed in the '*guerrilla*' *tactics* by means of which in daily life, *with supernatural violence,* we gain the road to heaven" (Bonani, 63; emphasis mine).

(*e*) In spite of all the possible differences we still insist that the history of the rivalry between the Society of Jesus and Opus Dei arose from a situation of competition provoked by their similarities, not their differences.

In his study on the Asociación de Propagandistas, Sáez Alba states that "relations between ACNP and Opus Dei are not so bad as people think," and that in both organizations "one finds the same elitism concerning a managerial vocation and the same objectives of taking over positions from which they can exert influence." The ideological foundations are identical, and both institutions turn to the same arguments to support the idea that it is the individual members who intervene in politics, and not the institutions as such (Sáez Alba, 97f.; for a recent formulation of this thesis by a distinguished politician of Francoism and member of Opus Dei, see the repeated assertions of López Rodó, 1990, 1991).

The very Consejo Superior de Investigaciones Científicas would be, according to the author of the long prologue to this study of ACNP, a joint project of the "propagandists" and the members of Opus; during the forties there would have been many instances of dual militancy, that is to say of individuals belonging simultaneously to both institutions (Sáez Alba, 44–46), a thesis that would logically weaken the clear differentiation between them established by Msgr. Escrivá (*Conversations,* no. 47).

However, in spite of the indubitable interest of Sáez Alba's study, in this case we also prefer to remain faithful to our own method of using, insofar as possible, sources closer to Opus Dei itself. An analysis of the ACNP by Antonio Fontán, a member of Opus, in his 1961 book *Los católicos en la Universidad española actual,* provides more than enough support for our hypothesis.

Fontán bases his work principally on statements by three key figures of ACNP: its founder, the Jesuit Ángel Ayala; Ángel Herrera, who was its president from 1909 to 1935; and his successor, Fernando Martín-Sánchez Juliá, who headed the association from 1935 to 1953. In 1925, three years before the official date of the foundation of Opus Dei, Herrera stated that education must be a priority concern, and that "education and the press are the two great enemy fortresses" (Fontán, 1961, 43f.). The propagandists therefore proposed a strategy of "penetration in university departments," at the same time proclaiming the need for a "University of the Church."

Education and the press, then, were the top priorities. Significantly, in the report that Msgr. del Portillo wrote in 1979 to request the transformation of Opus into a prelature, the professional activities of the members of Opus that he mentions first are the work "in 479 universities and secondary schools on five continents; in 604 newspapers, magazines and scientific publications; in 52 radio and television stations, 38 news and publicity agencies, and 12 movie producers and distributors" (in Fuenmayor et al., 609). Education and the press. . . .

Herrera Oria himself asserted in 1935 that "whoever in a society becomes master of the university is in the end master of the whole society"; "the day we have accomplished the apostolate of the university, we have accomplished all the rest of the apostolates"; "give us the university and all the rest will be given to us besides" (Fontán, 1961, 45f.). When the Spanish civil war was over, Martín-Sánchez said that "the Catholics have a right to be present everywhere and the propagandists can and should, therefore, go to the university chairs" (ibid., 47).

"The conquest of the university" and "the preparation of university chairs" were thus the watchwords of the ACNP. Fontán concludes that the propagandists, in fact, "carried out an active, conscious, and organized penetration of the university faculties" (Fontán, 1961, 52). We hardly need remind ourselves again that after the war Escrivá also devoted himself fundamentally to "the university apostolate," or to what extent the watchwords of the ACNP applied equally to the activities of the members of Opus Dei during the forties. Paradoxically, however, when Fontán begins to speak about Opus Dei, his tone suddenly changes. He becomes defensive; criticizes the legend that accuses Opus of doing the same thing as ACNP (Fontán was appointed to a university chair at the age of twenty-six); and confines himself to saying that Escrivá's "program" was "simple, supernatural, and clear: the spiritual life of the children of God and work" and that in the case of Opus Dei there was no "tactic of apostolate," but "a fundamental principle of asceticism, child of a profound theological inspiration" (ibid., 56, 57).

Since this author resists recognizing the parallels between the organization founded by the Jesuit Ayala and the one founded by Escrivá, let us turn to other texts, also written by members of Opus Dei. Jesús Urteaga observed in 1948 that "many Christians have not understood what Christianity is." They only understand passive Christianity, the kind that means going to mass, not "a Christianity that is influential in the public and private lives of men. They understand the more zealous Christians who daily frequent the temple of God, but they hardly imagine these same men participating as Catholics in the country's politics, education, and business" (Urteaga, 1948, 19f.).

Writing much later, Rafael Gómez Pérez states that "although it seems strange, I seek converts. And not only among 'the masses' . . . but also in that embroidered border formed by the intellectuals in every society" (Gómez Pérez, 1976a, 94f.). "To seek converts for the supernatural among the intellectuals" (ibid., 95)—is this not the same thing as Herrera's search for the "heights"? And if Martín-Sánchez claimed the right for Catholics to be present "everywhere," Florentino Pérez Embid, a member of Opus Dei, is even more explicit. Speaking in the name of "Catholics who want to leave a mark on collective life and contribute to the configuration of the national future," he writes that it is a matter of "simply occupying, by right, the leadership position that belongs to the Christian in collective life. In Spain, specifically, other causes will join this one," and he lists the first two, which are "victory in a just war" and "the thrust of the Catholic intellectual minority" (Pérez Embid, 1955, 153f.).

Some researcher should do a comparative study of Ayala's *Formación de selectos* and Escrivá's *The Way*. The two books were written at practically the same time. Although Ayala's volume is much longer, some of its chapters, for example, Chapter 6, "Qualities of the Select: Tenacity and Character," clearly allow us to see what "variety" of Jesuit spirituality Escrivá shared. Near the end of his book Ayala considers that "one cannot separate religion from politics. If politics is the art of good government, how can religion take no part in that government on which material and religious prosperity, the temporal and eternal good of men, depend? In this sense, religion is essentially political; it cannot do without it; just as citizens cannot do without it, not only lay persons, but also priests and religious" (Ayala, 1940, 407f.).

In spite of his wish to separate himself from Ayala's philosophy, Fontán states in the concluding chapter of his book that "Catholic Spain of today is not, as has incorrectly been said, the result of 'political coercion' or of 'a group of Catholics who have ascended to power.' On the contrary, the official Catholicism of the state is a consequence of the Catholic reality of Spain and a condition *sine qua non* for the maintenance of that same state and of the public peace" (Fontán, 1961, 156; unlike Msgr. Escrivá, Fontán could hardly be considered "a precursor of the Second Vatican Council"). He continues: "it is a reality, demonstrated by experience, that in Spain one cannot govern *against* the Catholics, and what is more, one cannot govern *without* the Catholics" (ibid., 156).

Earlier we referred to articles 202 and 203 of the Constitutions of Opus Dei (1950), on the necessity of professional distinction and acquiring a solid scientific background in order to acquire the means of Opus Dei's apostolate, namely, public positions, and especially those involving the exercise of leadership. In a speech given before the ACNP general assembly in 1944, its president asked, "Don't we say and think that the Association wants to be a select minority of apostolic men with the ability to manage? Well then, what more managerial positions can there be than those with the august function of governing others?" (Martín-Sánchez, 392).

(*f*) Independently of any differences that may have separated them, in the Spain of the forties Opus Dei and the Society of Jesus followed strictly parallel paths. The Jesuits still had more means, and much more power, than the nascent Opus Dei. But even so they were aware that Escrivá's men were competitors to be feared ("dangerous competition in the field of the apostolate among the young," according to Gómez Pérez, 1976, 255). The Jesuits came out fighting and the members of Opus Dei responded in kind, because neither of them had the "vocation of martyrs." As Jesús Urteaga wrote at the time,

> We Christians of today do not have the vocation of martyrs, but of warriors. It is not because of the mere fact of standing in the truth that we have to bite those who do not know it; but it is because of the mere fact of having been graced with it that we have to defend it to the teeth against anyone who offends it. If the declared enemies of God believe they are going to find in Christianity cowardly and slippery spirits who would raise their hands so the enemy can aim better at their hearts, they are mistaken. If the men of this bruised sea, reddened by envy and rancor, seek blood to quench their lies and their hate, let them know that the Christians of this generation are ready to die killing. [Urteaga, 1948, 98f.]

A Spanish Jesuit of the forties could hardly have expressed it better. . . .

We began this chapter about the establishment of Opus Dei in Spain (1939–1946) by quoting a passage from *The Way* which we considered programmatic for this period: "The great longing we all have for 'this thing' to move forward and grow larger seems to be turning into impatience. . . . When will it leap ahead, when will it break out, when will we see the world ours?" (no. 911). For the Opus Dei of the forties, as for the Jesuit ideologues of the "formation of select persons," the first conquest had to be the university, the "heights." But this was only the first stage, the stage that gave access to "key positions in society" (Roegele, 398). Those who want to "see the world ours" must go further: "You can—having Christ with you—feel the pride of a conqueror of a hundred worlds. The world is ours because it is Christ's" (Urteaga, 1948, 113).

In 1946, Escrivá de Balaguer and his most "select" followers, as they prepared for the start of Opus Dei's international expansion, set out to conquer a new world. It was a difficult world, and a world in which the Society of Jesus had preceded them and had a big advantage over them: Rome. In 1946, the city of Rome would become the place of Msgr. Escrivá's residence, Opus Dei's center of operations, and the arena of new battles in the conflict that would permanently set it against the Society of Jesus.

8

1946: Rome

General Observations

The year 1946 was absolutely decisive in the history of Opus Dei. It was the year of the *grand tournant,* by the end of which nothing would be the same as before, the year after which Opus Dei would begin to resemble, in many respects, what it is today.

At the start of 1946, Opus Dei had "centers" (student houses, or residences) in eight or nine Spanish cities and had about two hundred and fifty members and four priests. By the end of the year, there were already ten ordained priests, and Opus had begun—or was about to begin—its penetration into Portugal, Italy, England, Ireland, and France. At the beginning of 1946, Father Escrivá had never left the Iberian Peninsula (except for his brief crossing of the northern border in the Pyrenees during the war, in order to pass from the republican to the Francoist zone); by the time the year was over, he had made two trips to Italy, had been received twice by the Holy Father, had established his residence in Rome, and was about to be named "domestic prelate," which would give him the right to be addressed as Monsignor. Álvaro del Portillo was also living in Rome, as were Salvador Canals and several others (Sastre, 329). During the final days of 1946 five women from the women's section arrived, "to take care of the apartment," carrying a great deal of luggage and "a potato *tortilla,* to grace the supper table" (Vázquez, 245). In 1947, the magazine *Arbor* reported that "they are planning to install a delegation of the Consejo Superior de Investigaciones Científicas in Rome, so that the Eternal City may be intimately linked with the cultural zeal of our men of science" (*Arbor,* no. 21, 1947, 569; the notice is amplified in no. 23, 1947, 324f.).

The internationalization of Opus Dei was, then, a process that had its beginnings in 1946. If until then the official documents specified that *"Opus Dei unicum habet domicilium nationale"* (Fuenmayor et al., 513), from now on the "official" literature could begin to proclaim that "Opus Dei was born with a universal Catholic nature." The boldest writers have even asserted that Escrivá was abandoning "a Spain in which the necessary freedom did not exist, which for him was a natural right and an indispensable condition to enable him to lead Opus Dei into the whole world" (Berglar, 235). Rather than an example of *alternation,* this sentence is a sheer distortion of Escrivá's point of view; but it allows Berglar to characterize the Padre as a "universal apostle," comparing him to Saint Paul, Saint Ignatius Loyola, Saint Francis Xavier, Saint Augustine, Saint Thomas Aquinas, and Saint Francis de Sales (ibid.).

On another level, at the beginning of 1946 Father Escrivá was "a sincere, honest, simple, ingenuous priest," who in Rome was going to discover "intrigue" and a style of action and governing, in the bosom of the Church, that would persuade him that even he could leap over or sidestep certain norms of behavior when it was convenient to do so (Walsh, 58). I hasten to stipulate that in my judgment Walsh's adjectives for Escrivá—more or less reflecting the statements of Raimon Panikkar—are hardly appropriate and extremely ambiguous. If one interprets the words ironically, they could imply that with all the accumulated experience of a man who had been a priest for twenty years, had gone through the war, had founded Opus, had withstood persecution, and had conducted spiritual exercises for Franco, the Padre was not in the least "ingenuous." Nor is "simple" the most suitable epithet to describe a man who had ennobled his name and become Padre Escrivá de Balaguer. On the other hand, to deny him "honesty" and "sincerity" in 1946 seems excessive and probably erroneous: in earlier chapters we said that Escrivá did indeed give the impression of being perfectly "sincere" in his passions and his ambitions.

Given these ambiguities, let us forget about all those epithets and keep the only textual affirmation of Panikkar, also reproduced in Walsh's book. Panikkar commented on several occasions that the Padre, after his first visit to Rome, told a small group of his followers: *"My children, in Rome I lost my innocence."* Raimon Panikkar was one of the six Opus Dei priests ordained in 1946; it is curious to note that while *all* the biographies of Escrivá mention the names of the first three priests ordained in 1944, *none* of them mentions the names of the ones ordained in 1946, in spite of the fact that three of them were disciples "from the very first" (Francisco Botella, Pedro Casciaro, and Ricardo Fernández Vallespín); it would not be at all improbable if this silence were motivated by the fact that not all of them remained in Opus. (On Panikkar's present view of Opus Dei—a topic of which he is so reluctant to speak that, respecting his feelings, I will refer to him much less than I would have liked—see what are perhaps the best pages in the whole book *Historia oral del Opus Dei;* Moncada, 1987, 129–38.)

"My children, in Rome I lost my innocence"; in spite of his accumulated experience, Escrivá still had a particular, local, and even parochial view of the

Church and society. All he knew was Spanish society and the Spanish church of the era—and what a society, what a church!

In Rome, things were quite different. If Escrivá came there more or less convinced of the relative originality of Opus Dei, in Rome he quickly discovered that there was a proliferation in the Catholic Church of movements and initiatives that were to some extent just like Opus (Father Gemelli, the missionaries of la Regalità, Milites Christi, groups of Notre-Dame-de-Vie, the Compagnia di San Paolo, etc.; see Rocca, 1985, 34–47). He also discovered what Walsh calls "intrigue": the scheming that went on in the Vatican Curia, not because it was the Vatican but because it was the Curia, a bureaucracy which functioned sociologically like all bureaucracies. He discovered, in other words, what Max Weber called "the trade secrets" of any bureaucracy.

This is the sense in which we must understand Escrivá's statement "I lost my innocence": the experience of Rome was a hard experience, but at the same time it was profoundly liberating. Beginning in 1946 he no longer had to repress his ambitions as before; the means employed by the Vatican could also be used by him. "When it comes to God Our Lord it is my duty to use every supernatural and human means in order to fulfill the Holy Will of God, in matters concerning the establishment of his Work, as He has given me to understand" (Fuenmayor et al., 345). If the world belonged to the *schemers,* and even in the Church people acted in a scheming fashion, that meant that God was asking us to be schemers. The "loss of innocence" for Escrivá meant not the acquisition but the *sanctification* of *scheming:* in Rome, from 1946 on, the "scheming" of the Padre became a "holy scheming," since God himself "wrote law with crooked lines" (ibid., 295).

Finally, 1946 was decisive in the history of Opus Dei because while at the beginning of that year it was just an institution under diocesan jurisdiction, established under the bishopric of Madrid, by February 1947 it had become an institution of pontifical law, that is, of the whole church. Furthermore, as we saw in the previous chapter, at the beginning of 1946 the juridical status of Opus Dei was not even clearly established, nor was its precise relationship to the Priestly Society of the Holy Cross, which was the entity actually approved in 1943 as a "society of common life without vows" under diocesan authority. And as we shall see below, this difficulty was an important issue throughout Portillo's and Escrivá's negotiations in Rome during 1946. When the year was half over, however, the official documents began finally to refer to the "Priestly Society of the Holy Cross and Opus Dei" as a single institution, officially recognized as such.

Chronology of a Decisive Year

Very schematically, the chronology of the "Roman" events of the year 1946 is as follows.

January. Salvador Canals returned to Rome, having spent Christmas in Spain.

February. Álvaro del Portillo and José Orlandis came back to Rome. Portillo brought the documents for the petition for a status of pontifical, rather than diocesan, law, as well as about sixty letters of recommendation. "The reception was cordial," but "the initial balance of the negotiations was not positive" (Fuenmayor et al., 148).

February to June. They lived in a furnished apartment in the center of Rome ("all the balconies looked out over the beautiful Piazza Navona"; Sastre, 325), obtained through the Spanish consul. In May Orlandis returned to Spain.

June. On June 10 Portillo wrote to Escrivá asking him to come to Rome; two days later he again urged, in a second letter, "the need for his presence in the Eternal City" (Fuenmayor et al., 156f.). On June 20 and 21 Escrivá was in Barcelona, where he embarked with Orlandis for Genoa. He reached Rome on June 23 and was received at the new apartment that had been rented near the Vatican ("the dining room looked out over Saint Peter's square . . . and, nearby, the lighted window of the Pope's private library could be seen"; Sastre, 329).

July. On July 16 Padre Escrivá was received at an audience with Pius XII.

August to September. On August 31, Escrivá went to Madrid with a document "in which the goals of the Work were approved," and with "the relics of Saint Mercuriana and Saint Sínfero, two child martyrs of the second century" (Vázquez, 244). At the end of September the six new priests of Opus Dei (or of the Society of the Holy Cross) were ordained.

November. On November 8, after passing through Barcelona again, Escrivá returned definitively to Rome.

December. Escrivá was received again in an audience with Pius XII. Before the end of the year, "the [five women] of the Work who would be taking care of the apartment" arrived, in spite of the fact that it had "very little space" (Vázquez, 245); arrangements were made for the women "to sleep at a nearby residence" (Sastre, 335).

While there are pages and pages of anecdotes in the various biographies of Msgr. Escrivá, here we will limit ourselves exclusively to trying to find some kind of answers to these two questions: what kind of activities did Álvaro del Portillo engage in from March to June 1946, and why did he demand the Padre's presence in Rome?

Álvaro del Portillo's Activities in Rome

On the topic of Álvaro del Portillo's journey to and stay in Rome, our usual sources of reference are not only brief; they are also confusing.

According to Vázquez de Prada, the "secretary general of Opus Dei" arrived in Rome "well provided with letters of recommendation from nearly every Spanish prelate," he "boldly and persistently" approached bishops and cardinals, he presented himself at the Vatican every day, he was very insistent, but "negotiations stalled," and that was when he wrote to the Padre asking him to come (Vázquez, 240). He does not specify what exactly "the final approval" was that the Founder was seeking, and above all he does not seem to realize that officially Portillo's job was "secretary general and attorney of the Priestly Society of the Holy Cross"—not "of Opus Dei."

Ana Sastre adds some supplementary detail: the letters of recommendation came from sixty Spanish bishops, and they accompanied "the request for the *decretum laudis* of the Holy See for Opus Dei." She adds that Portillo, on disembarking in Genoa, was in a hurry to get to Rome before the departure of several bishops who had just been named cardinal, so as to get letters of recommendation from them as well. Even so, "the negotiations were not going to be easy," "it was going to be a difficult business," and in the end it looked as if "it was leading into a dead end" (Sastre, 325). Leaving aside the minor question of whether the sixty letters left Madrid with Portillo or whether "they had been arriving at the Holy See over a period of several months" (Fuenmayor et al., 154), there still remains considerable ambiguity of a more essential nature: the request for the *decretum laudis*—a requisite for conversion into an institution of pontifical law. Was this for Opus Dei, or for the Priestly Society? According to a document reproduced in part by Rocca, dated January 23, 1946, Escrivá, acting in his role as "president general of the Priestly Society of the Holy Cross," asked the Holy Father to "deign to concede the decree, as well as the approval of the constitutions of this society, which was founded on October 2, 1928, and canonically approved as a Pious Union on March 19, 1941" (Rocca, 1989, 390; 1985, 10f.).

Here, according to all indications, is the crux of the problem. It is not a matter of the legal impossibility of approving an institution made up of both laymen and priests (according to the thesis of Gondrand, 177; this is denied by Rocca, 1985, 33, who asserts that there were precedents in Rome). It is, rather, a matter of the relationship between the Priestly Society and Opus Dei. Escrivá's request asks expressly for approval of the *society;* but there is something else: either he made an error when he said the Priestly Society was founded in 1928 and approved as a Pious Union in 1941—when all the "official" literature dates its foundation, "by divine illumination," to 1943—or else he presumed that the juridical transformation of 1943 establishing the "society of common life" affected the entire earlier foundation, which would mean that by 1946 Opus Dei simply would have had no legal existence at all.

Here in fact we seem to be looking at the "dead end" that caused the negotiations to "stall," that is, that the Holy See cannot grant the requested approval, because the confusion of names and dates makes it unclear exactly which is the organization for which such approval is requested. The documents presented in *El itinerario jurídico del Opus Dei* seem to confirm this, although in a very fragmentary and indirect way. The authors of this study explain the unfavorable "initial balance" of Álvaro del Portillo's negotiations (Fuenmayor et al., 148) as based on the conclusions reached, after studying the case, by two of the jurists who were helping Escrivá's men at the time (both Spaniards, and both Claretians: Father Larraona and Father Goyeneche). In a report which seems fundamental for a clear understanding of the problem—but which, inexplicably, Fuenmayor, Gómez-Iglesias, and Illanes fail to reproduce in its entirety in their documentary appendix, limiting themselves to quoting a few fragments in the text of the book—Father Goyeneche begins by saying that

"Opus Dei was the origin of the present society" and "is inseparable from it" (ibid., 152). It certainly seems legitimate to wonder whether it was this duplication of institutions that provoked the lack of agreement among the members of the Sacred Congregation of Religious with respect to the approval requested by Escrivá. All three Opus authors conclude that Father Larraona referred the question of Opus Dei to a new regulation that was still being studied and that would pave the way for the creation of the category of the secular institutes (ibid., 155), and that this also led Álvaro del Portillo to demand the Founder's presence in Rome (ibid., 156).

On the other hand, these same authors curiously say nothing about a series of other activities simultaneously undertaken by Portillo when he reached Rome, which we know about thanks to a set of documents reproduced in Giancarlo Rocca's study.

Between March 19 and June 10, 1946, Álvaro del Portillo brought a total of nine petitions to the Sacra Penitenzieria Apostolica (Rocca, 1985, 148–56, documents 15–23), requesting in sequence:

1. that the priests of the Society of the Holy Cross be able to bless, with the sign of the cross, rosaries and crucifixes, with the usual indulgences in such cases;
2. that they be able to erect the Via Crucis in all the oratories of the society;
3. that they be able to impose the scapular of the Virgen del Carmen on all the members;
4. that, like the religious in their perpetual profession, both in Opus Dei and in the Society of the Holy Cross, plenary indulgence be granted for admission, "oblation," and "fidelity" (note the mention here of Opus Dei!);
5. that the priests of the society be able to impart the apostolic benediction, with plenary indulgence, to those who perform spiritual exercises under their direction;
6. that indulgence of five hundred days be granted each time they kiss or venerate with prayer the cross erected in the oratories of the society, with plenary indulgence for those who visit the oratory on the days of the "Invention" and "Exaltation" of the Holy Cross;
7. that various indulgences be granted for the hours devoted to study;
8. that plenary indulgence be granted on certain holidays of the year, on the day of the emission or renewal of vows ("Opus Dei is not interested in vows, nor promises, nor any form of consecration," Msgr. Escrivá would say in 1967; *Conversations,* no. 20), and on the holidays of the Patrons of the Work; and
9. that the members of the two branches of the society be able to receive general absolution on specific holidays.

Focusing strictly on the content of the requests, leaving other considerations aside for the time being, I believe that these documents produced by Rocca are very important—even more important than the author himself concedes—for a variety of reasons:

1. They seem to radically discredit any assertions to the effect that Escrivá's foundation was, in 1946, an institution of "common Christians," imbued with secular spirit, laical mentality, and so on. The model of "sanctification of ordinary life" proposed here is perfectly legitimate and respectable; but to try to present it as a model comparable only to that of the first Christians, and at the same time as an anticipation of the Second Vatican Council, is simply—let us not mince words—a fraud.
2. Some of these documents again illustrate something that, even though it is obvious, the "official literature" goes to lengths to deny: in 1946 there is no talk of "men and women of every profession and social class," but only of "intellectuals and university people" (document 19), "of intellectuals and educated men who by nature take on the leading roles in civil society" (document 21).
3. Contrary to the protestations of Opus Dei, which would later redouble every time someone likened or compared it to a religious order, in 1946 Álvaro del Portillo established the parallel with the religious orders (document 18) and spoke explicitly of "the emission and renewal of vows" (document 22).
4. With respect to "secrecy," which according to Msgr. Escrivá and the "official" literature never existed, one of these documents (22) offers an interesting illustration. Álvaro del Portillo wrote—for reasons that we shall later attempt to discover—that "the Sacred Congregation of Religious approved an outline of the constitutions of the Society (of the Holy Cross), of which I attach a copy." The Latin of the text leaves no room for confusion: the copy he appends is that of the outline, not of the Constitutions. In December 1943 Bishop Eijo Garay had written to Escrivá asking him "to write and present Us with the Constitutions of the Priestly Society of the Holy Cross canonically established by Us today, developing as best suits you the outline of constitutions we sent to the Holy See" (Fuenmayor et al., 527). After a month and a half (January 25, 1944), the bishop of Madrid signed the decree of approval of these Constitutions (ibid., 529). If we were surprised earlier that *El itinerario jurídico* published the outline and not the definitive text in its documentary appendix (ibid., 516–20), we now see that in June 1946 Portillo sent to the Sacra Penitenzieria Apostolica a copy of the outline, not the Constitutions.

In spite of all this, there remains the possibility of forming an alternative hypothesis: what if these more or less interesting data are not the only really interesting data? What if they are, in fact, red herrings? Could it be that Álvaro del Portillo, in asking for all these "indulgences," was really seeking something other than indulgences? Could he have been seeking, for example, an official document from the Vatican that could be presented as evidence of benevolence, of sympathy, even—for an uninformed public—of approval? There are at least two clues in favor of this hypothesis: first, the response to all of Portillo's petitions could be made, as Rocca explains, either through the "normal channels" of the Sacra Penitenzieria itself, or else "in a more solemn manner," in the form of a document issued by the Vatican secretary of state, and the latter was the form chosen (Rocca, 1985, 37). Therefore, we might say that the indulgences were of interest not in and of themselves, but rather for the formal appearances with which they were conceded. The second clue is that for many years the "official" literature mentioned this document (referred to as *Cum Societatis,* June 28, 1946) as if it had far-reaching consequences for the juridical recognition of Opus Dei, rather than being just a simple concession of indulgences. Thus Le Tourneau asserts that with the concession of the indulgences the Holy Father "implicitly approved the spirit and the ends of Opus Dei" (Le Tourneau, 58).

It should be apparent that we are returning to Father Brown's epistemological model. Letting ourselves be guided by the *scheming* figure of Chesterton, we can take one more step in the formulation of the alternative hypothesis. We could ask, in fact—without exceeding the bounds of pure inquiry—whether the whole string of indulgences and scapulars that seemed to be Portillo's main preoccupation was merely a screen for what we see as *the* problem of Opus Dei at the time: that is, the problem of its status and its relationship to the Priestly Society.

Read from this new vantage point, the documents reproduced by Rocca become interesting for reasons different from those mentioned before, and render our questions plausible. The first four documents delivered to the Sacra Penitenzieria all bear the same date (March 19). The first three are written on the official stationery of the Sociedad Sacerdotal de la Santa Cruz, and in them Portillo presents himself as "a priest, secretary general and attorney of the Priestly Society of the Holy Cross." Typewritten below the letterhead of the fourth, however (document 18), are the words "and Opus Dei": the secretary general and attorney acts for "the partnership called the Priestly Society of the Holy Cross and the Work of God," and he speaks of "our Society . . . and its own work, called Opus Dei."

In all the requests that followed he once again presented himself exclusively as the general counsel of the Priestly Society of the Holy Cross. Two new documents are dated May 3: in one he speaks only of the society, and in the other of "our Institution" and "our Partnership" (documents 19 and 20). In the next, dated May 27, he refers to "our Institute." He prepared another document on June 5, in which he mentions "our Society," "our Institution," and "the Patrons of the Work" (document 22). In the last of the series, drafted on June 19—the same day he wrote to Escrivá and asked him to come to Rome—he refers to the "members of both branches of the Society," and to February 14 as the "anniversary of the foundation."

Our hypothesis is that this whole tangle of names and dates *is not accidental.* And our question is whether it might not be, on the contrary, a carefully calculated *strategy.* The 14th of February is the date of the founding of the Priestly Society. (It is not, therefore, the famous 2nd of October, 1928.) What are the "two branches of the Society"? Which one has "branches": the "Society" or "Opus"? Why is he suddenly calling the "Society" an "Institution," an "Institute," or a "Partnership"? Why does he refer to the "Society" *and* "Opus Dei" on only one occasion, but never again?

We will address these questions in a moment, but first we must bring a new character on stage: the Founder arrives in Rome in person.

Father Escrivá in Rome

On June 10, 1846, Álvaro del Portillo wrote to the Padre. "He could not find his way through this labyrinth, and he feared the matter had reached an impasse" (Vázquez, 240). "His personal presence in Rome was necessary, to try to move forward what seemed humanly impossible" (Bernal, 256). Two days later he wrote again, insisting (Fuenmayor et al., 157). The Padre received the first letter on June 16 (Vázquez, 240).

All the "official" literature inserts an interlude here, sometimes a long one, to explain that the Padre was suffering from diabetes and that his doctor had formally advised against the trip, not wanting to assume "the responsibility of what might happen" (Bernal, 257); "he was not responsible for his life" (Sastre, 326). Nevertheless, the Padre got his things together and obtained a visa. In Madrid he met the members of the Consejo General, "who with him were exercising the leadership of Opus Dei, always collegial" (Gondrand, 178). "He read them the letter from Don Álvaro and asked their opinion" (Sastre, 326). The Consejo gave a "favorable opinion for the trip" (Fuenmayor et al., 157). Escrivá "thanked them, but let them know that he had already decided to go, because that was what God wanted" (Vázquez, 240). "I thank you; but I would have gone anyway: one does what one must do" (Gondrand, 178). The marvels of an "always collegial leadership"!

Only *three days* had passed. On June 19 the Padre left Madrid by car with his "chauffeur" (Sastre, 673). He was carrying a letter from the superior of the Claretians for the Claretian Father Larraona, which said that Don José María Escrivá was "the founder of the Priestly Society, Opus Dei." Unless it was "as a matter of holy obedience," as those in religious orders say, Father Larraona, who knew the puzzle of 1946 better than anyone, could hardly accept that the "Priestly Society" *was* "Opus Dei." "He is a good friend of ours," the letter continued. "His Work is growing fast. It gives great glory to God . . ." (Rocca, 158). The Padre spent the night in Zaragoza. On the 20th he went up to Monserrat, "to say hello to our lady" (Vázquez, 240), and then went on to Barcelona. On the 21st, after celebrating mass at an Opus Dei center, and visiting the basilica of la Mercè, he embarked for Genoa, accompanied by José Orlandis.

Here there is another long interlude in the "official" literature to describe what an authentic "odyssey" the voyage was. He had to go by sea because "air travel between Madrid and Rome had been suspended and the French border closed as a result of diplomatic pressures against Francoist Spain" (Vázquez, 240). The ship was "an old tub" (Berglar, 196), "old enough to be scrapped"

(Vázquez, 241), and "in spite of the good offices of the Transmediterranean Company, they could find nothing but an inside berth for the Founder's voyage to Italy" (Sastre, 328). As if that were not enough, there was a tempest, "a furious tempest, not characteristic of the Mediterranean" (Bernal, 259). Vázquez de Prada's spectacular account outdoes all the others: "the wind and the waves brought uproar and presages of the storm. . . . Children and women were crying amid the clatter of dishes breaking and bundles being hurled from one side to the other. The water flowed over the sides and into the boat, flooding the rooms . . ." (Vázquez, 241). The result was that the Padre passed a very bad night; "he suffered the unspeakable" (Bernal, 259) but "never lost his good humor for a single moment" (Sastre, 329).

Escrivá's comment was: "You should have seen how the devil swished his tail in the Gulf of León! It was plain that he was not at all pleased about our going to Rome!" (Sastre, 329). I don't know why, but this remark reminds me of what Freud said to Jung, also on board a ship, moments before arriving in New York: "These people have no idea that we are going to infect them with the plague!"

On the 22nd, at nightfall, the ship docked in Genoa. (Today, preserved in the central house of Opus Dei in Madrid, one can find "the steering wheel and the compass that guided their route toward Rome; a difficult route that was, however, the path of God"; Sastre, 331). Álvaro del Portillo and Salvador Canals were waiting for them in the port of Genoa. The next day they went by car to Rome, which Padre Escrivá saw for the first time on Sunday, June 23, 1946. In spite of his fatigue after the journey, and in spite of vain attempts to make him rest, "he spent the whole night alternately praying and contemplating the cupola of the basilica of Saint Peter's, beneath which was the tomb of the first Pope, and the windows behind which lived his successor, the vicar of Christ on earth" (Gondrand, 176).

Let us now return to the basic question. Why did he go to Rome? "In 1946, the founder moved to Rome in order—among other things—to direct and drive forward the long process destined to find a legal structure appropriate to Opus Dei within the Church," wrote William West (1989, 52). "Providence carried him to Rome, because Opus Dei was born Roman," concludes Vázquez de Prada (p. 248), quoting an expression from Escrivá himself.

Let us make the question a little more specific, then. What was the hurry? After receiving Portillo's first letter on June 16, in spite of illness and in spite of difficulties, why did he leave Madrid on the 19th, arriving in Rome by the 23rd?

Again we find ourselves amid a sea of questions which have to be interwoven with the questions about the background of the documents sent by Álvaro del Portillo to the Sacra Penitenzieria Apostolica, with the whole tangle of "Society," "Institution," "Institute," and "Partnership," of "branches" and "dates of the anniversary of the founding," of the mixup between the "Priestly Society of the Holy Cross" and "Opus Dei."

It is mainly this last mixup—which in the chronological order occurs first (Rocca, 1985, document 18, March 19, 1946)—that seemed to provoke the most consternation, since the Sacra Penitenzieria requested information from the Congregation of Religious about *the exact name* of the institution that had been approved in 1943 (ibid., 37).

However, before a reply was received from the Congregation of Religious, the Office of Indulgences of the Apostolic Penitentiary sent a resolution (on June 18) to the secretary of state of the Vatican in response to several petitions from Álvaro del Portillo, asking the secretary of state to prepare, officially and solemnly, the document of concession of the indulgences. In this resolution the penitentiary assigns the title "Sanctae Crucis *et* Operis Dei" to the society.

Ten days later, the secretary of state issued the document conceding the indulgences (*Cum Societatis,* June 28, 1946), referring to the "Priestly Society of the Holy Cross *and* Opus Dei," and dating its foundation as October 2, 1928 (Fuenmayor et al., 529f.).

During the following week, the Sacred Congregation of Religious answered the Apostolic Penitentiary, saying that the authorization granted in 1943 was for the establishment "in Institute of diocesan law of the Priestly Society of the Holy Cross, which, according to the outline of the Constitutions, is united with a work called Opus Dei, which is the means used by this Society in its apostolate." The same note adds that the official nomenclature "*cannot be modified* without previous authorization from the Congregation itself" (Rocca, 1985, document 26, July 2, 1946, p. 159; emphasis mine).

Meanwhile Padre Escrivá had arrived in Rome. Just before leaving for Italy, during the mass he had said the same morning in Barcelona, he uttered some words that all the biographies quote:

> Lord, have You been able to allow me in good faith to deceive so many souls? Everything I have done is for your Glory and knowing that it is your Will! Is it possible that the Holy See will say we have come a century too soon? . . . I never had the will to deceive anyone. I have only had the will to serve you. Will I end up a swindler, then? [Fuenmayor et al., 157; Bernal, 258; Gondrand, 178; Vázquez, 241; Sastre, 327]

A "swindler" is, according to the dictionary, a trickster, someone who shrewdly tries to defraud or deceive. In Escrivá's passage, then, the word "deceive" appears three times in a row. What supposed deceit is he talking about? In all the documents we have examined in this chapter, who has used the word deceive? *Who* accused him, *when,* and of deception *in what?* Could there be perhaps some reference to this in Portillo's letter demanding his presence in Rome? Is there any possibility that "the deception" has something to do with the whole matter of the indulgences, interpreted as a pretext to obtain recognition for the existence of Opus Dei together with the Society of the Holy Cross, when only the latter was officially approved? Could this explain the urgency of Escrivá's trip to Rome? Could it be that the accusation of deceit, formulated by some entity in the Holy See, was an expression of disapproval of an exhibition of "holy scheming" on the part of Álvaro del Portillo?

There are many questions, and few verifiable resources for answering them, naturally. What does seem unquestionable, in any case, is that the Congregation of Religious itself, which on July 2, 1946, had declared that "the official nomenclature cannot be modified without prior authorization," a month and a half later drafted a document addressed to the "Most Reverend Father José María Escrivá de Balaguer y Albás, Founder and President General of the Priestly Society of the Holy Cross and of Opus Dei," in which it says that "some time after having conceded the canonical establishment of the Priestly Society of the Holy Cross *and* of Opus Dei there arrived various documents in abundance from different places which are true and authentic and which extol and recommend Your Institute," and it urges him to continue working on "such a noble and holy Work" (Fuenmayor et al., 532).

If there was a scheme, it was a roaring success, judging by all appearances. Where at first there had been an "Opus Dei" (1941), later transformed into a "Priestly Society" (1943), from now on there would be an entity called "Priestly Society *and* Opus Dei."

After a stay of a little more than two months in Rome, the Padre returned to Madrid, on August 31, 1946, with this document of "approval of the goals of the Work" and with the "relics of Saint Mercuriana and Saint Sínfero" (Vázquez, 244). And when he reached Madrid he said, *"My children, in Rome I lost my innocence."*

9

The International Expansion of the Secular Institute of Opus Dei (1947–1958)

Synopsis of the Official Version: Approval of Opus Dei as a Secular Institute and International Expansion

The period we are about to consider is characterized basically by the approval, at first provisional (1947) and soon afterward final (1950), of Opus Dei as a secular institute of the Catholic Church; by its notable international expansion (Europe, North and South America, and, toward the end of the period, Japan and Kenya); and by a definite broadening of its sphere of action, involving the full inclusion within Opus Dei of married people, diocesan priests, and even non-Catholic participants, the creation of its own institutions in the education field (principally the Roman colleges and the University of Navarre), and the appearance of members of the Work in very high places in politics and business, especially in Spain

The Pontifical Approval of Opus Dei as a Secular Institute

For chronological reasons and also for more thematic proximity to the questions analyzed in the previous chapter, let us first consider the most juridical aspect of the official recognition of Opus Dei as an institute of pontifical law.

At the beginning of 1947 the Holy See, through the document called *Provida Mater Ecclesia,* approved the creation of the secular institutes, which are "societies, clerical or laical, whose members, in order to achieve Christian perfection and exercise the apostolate fully, profess the evangelical counsels in secular life" (art. 1). Three weeks later, a decree of the Sacred Congregation of Religious established "the Priestly Society of the Holy Cross and Opus Dei, called for short Opus Dei," as a secular institute (*Primum Institutum Saeculare,* reproduced in Fuenmayor et al., 532–35).

(*a*) Since we have available the text (published in 1949, but not always paid much attention to in the "official" literature) of a lecture delivered in Madrid in 1948 by Msgr. Escrivá de Balaguer on the *Provida Mater Ecclesia* and on "the nature of Opus Dei as it relates to the state of evangelical perfection" (Escrivá, 1949, 5), we will use it here as basic source material. Escrivá delivered the lecture at the headquarters of the Asociación Católica Nacional de Propagandistas and addressed his audience by saying that "it is my great pleasure to spread the knowledge of the Work among the good children of our mother the Church" (ibid., 5). He began by presenting the pontifical document (*Provida Mater Ecclesia*) as a recognition of a new form of the life of perfection, different from that of the religious.

The Church is, according to Msgr. Escrivá, an organism that demonstrates its vitality as a movement that is not a mere "adaptation to the environment: it is an intrusion in it, with positive and lordly spirit" (Escrivá, 1949, 7). This is an old idea of Opus's founder, who in *The Way* had contrasted the idea that "the environment is such a strong influence!" against the assertion that "we are such a strong influence on the environment!" (*The Way,* no. 376). The Church, guided by the Holy Spirit, had to go into the world "with a firm and sure step, opening the road, conscious, furthermore, that she carries in her breast the sign of contradiction for the ruin and the salvation of many" (Escrivá, 1949, 7). This brief introduction is not only "a profession of faith in the presence and action of the Holy Spirit in the Christian community" (Fuenmayor et al., 218), but also a good synthesis of the Founder's view of the Church in the world.

The text continues with a short historical sketch of the successive ways of living "the state of perfection" in the Church (asceticism, monastic life, orders and religious congregations, societies of common life without vows). These last, he says, are clearly distinct from associations of a laical type; on the other hand, they are very similar to the religious congregations (common life, practice of the evangelical counsels of poverty, chastity, and obedience, apostolate form, and centralized internal organization). The only significant difference between these societies and the congregations of religious hangs on the question of vows, whether they exist or not, or are promises rather than vows, or are vows that lack a public character (Escrivá, 1949, 14). All these observations are particularly interesting when we recall that between 1943 and 1947 Father Escrivá's foundation had enjoyed precisely the status of a "society of common life without vows."

The novelty of the secular institutes lies in the fact that, with the *Provida Mater Ecclesia,* "a new state of perfection is recognized, different from those that existed juridically until now" (Escrivá, 1949, 16). For the first time the Church recognizes the possibility of a state of perfection whose members are not members of a religious order. The distinctive feature of the religious is either "the contemplative life dedicated to prayer and sacrifice" or "the active life dedicated to remedying the evils and necessities of the world from outside it" (ibid.). With the secular institutes there appears, on the other hand, "a new form of the life of perfection, whose members are not religious"; on the contrary, "now these apostles who dare to sanctify all the everyday activities of men come from the world itself" (ibid., 16f.).

This became, after 1947, a good definition of what, according to Escrivá, the members of Opus Dei were: *"apostles who dare to sanctify all the everyday activities of men."* In this sense the secular institutes, of which Opus Dei was the first, mark "the end of the evolution of the forms of life of perfection in the Church" (Escrivá, 1949, 17). The secular institutes "constitute in secular life a true state of perfection," undoubtedly distinct from the religious state, but no less distinct "from the mere secular state" (Canals, 1954, 85). When Msgr. Escrivá asserted years later that "Opus Dei is not and cannot be considered a reality linked to the evolutionary process of the state of perfection in the Church," and that "our Association does not pretend at all that its members change their state, that they cease to be the simple faithful just like everyone else, in order to acquire the peculiar state of perfection" (*Conversations,* no. 20), the "official" literature could say what it wanted, but it seems undeniable that Msgr. Escrivá was literally *retracting* what he had said in his 1949 lecture.

In the second part of his lecture, in fact, Msgr. Escrivá presents Opus Dei as the first secular institute approved in accordance with the norms of the *Provida Mater Ecclesia,* "which has been set up as a model for this new type of life of perfection by the Holy Father Pius XII" (Escrivá, 1949, 18). The text is fairly short, and it covers the principal "characteristic notes" of Opus Dei, with respect to which it might well be considered an official authorized version of the period. In the part dedicated to "the goals of the Institute" (a part curiously excluded from Fuenmayor's summary of the lecture; Fuenmayor et al., 219), Msgr. Escrivá stresses two points.

First, "the general goal of the Institute is the sanctification of its members, through the practice of the evangelical counsels and the observance of the Constitutions themselves. The specific goal is for them to work as hard as they can so that intellectuals will adhere to the precepts and even the counsels of Christ Our Lord, and put them into practice, and in this way to foment and diffuse the life of perfection in secular life among the other classes of civil society, and to prepare men and women for the exercise of the apostolate in the world" (Escrivá, 1949, 19). Although by this time references to "the other classes of civil society" begin to appear, it is again repeated that Opus Dei works specifically among the intellectuals, who have to exert influence over (foment and diffuse) "the other classes," in accordance with the old principle—espoused by the Propagandists to whom Escrivá was addressing this lecture—that whoever attained "control of the heights is master of the whole society" (Fontán, 1961, 45).

Second, "the members who consecrate themselves temporarily or in perpetuity swear private vows, as any other faithful person may do . . ." (Escrivá, 1949, 19). Twenty years later, Msgr. Escrivá would say that Opus Dei "was not interested in vows, or promises, or any form of consecration for its members, other than the consecration that all have already received through Baptism" (*Conversations,* no. 20).

(*b*) After this summary, there is not much more to be said. The decree of provisional approval (*Primum Institutum Saeculare,* February 24, 1947) in fact consecrates the unity of the Priestly Society of the Holy Cross and of Opus Dei, and even recognizes that the second name is the short name usually used to designate the Institute, which is furthermore presented as a "model for the secular institutes" (Fuenmayor et al., 534). The document also specifies that although because of the ordinary condition of its members it ought to be a laical institute, because of the Society of the Holy Cross (whose members exercise all the management functions) it is still defined as predominantly clerical, comparable from a juridical point of view to the clerical institutes (ibid.). If at first glance it might seem that Opus Dei would have made every effort to be recognized as laical—given its members' character as "ordinary laymen," and given the existence of a women's branch—in fact its definition as "clerical" confers on it a whole series of advantages from the point of view of its autonomy vis-à-vis the diocesan authorities (Rocca, 1985, 40ff., 52ff.).

Furthermore, the 1947 decree of approval gives a brief historical presentation of Opus Dei which, although it contributes nothing new, is extremely useful for confirming all our hypotheses about the exclusively university-based character of Escrivá's apostolic work. "Dr. Escrivá de Balaguer felt powerfully called and gently nudged to the apostolate among the students of the lay University and the Madrid superior schools," the decree reads. In the "difficult times" before the Spanish civil war, "he worked in the apostolate among university students and educated men, and through them among the whole society of intellectuals and leaders" (Fuenmayor et al., 533; note that there is no reference at all here to "the other classes of civil society"). With its consolidation and the beginning of its international expansion, the number of members multiplied; the decree explicitly mentions "doctors, lawyers, architects, military officers, scientists, artists, writers, professors, and students" (ibid.). The decree goes on to specify the goals of the institute in the words repeated by Escrivá in his 1949 lecture, adding that for this reason it is appropriate for all the members to possess a doctorate in some discipline (ibid., 534). Only after having made this important point does it mention "the sanctification of the ordinary work of the members." This should be kept in mind, since what it really means is that each time the "official" literature mentions the "sanctification of ordinary work" of the members of Opus Dei, we must remember that this refers to the "ordinary" work of individuals who have earned a university doctorate.

(*c*) In 1947 Opus Dei became the first secular institute to be provisionally approved by the Holy See, and exactly three years later (really quite a short time, considering the habits of the Curia, as the "official" sources admit), Msgr. Escrivá requested final approval. Here we have three equally important elements: Msgr. Escrivá's *initiative;* the *hurry* he was in ("hurry to take this new step, so important, in the juridical itinerary of the Work"; Fuenmayor et al., 200); and the *definitive* character of the approval of Opus Dei as a secular institute, which was in fact conceded by the Holy See in June 1950 (decree *Primum inter Institute Saecularia,* reproduced in Fuenmayor et al., 544–53).

We said that the three elements are equally important because they make manifest, in addition to Escrivá's probable desire to again be the first to receive approval, his wish to have Opus continue to appear as a "model." That is, these three factors also make manifest Opus Dei's conformity with the juridical model of the secular institute, which continues to appear to its leaders as appropriate and fully satisfactory. In sum, in 1950 Opus Dei was apparently considered the embodiment of "the perfect type of secular institute" (Canals, 1954, 82). This is why they were trying to obtain "this new and final approval as soon as possible" (Fuenmayor et al., 221), which would be equivalent to "a definitive pontifical sanction." According to Escrivá himself, "the final approval confirmed again the fundamental principles of the Work" (ibid.). In the more recent "official" literature one reads, however, that in 1950 "the Holy See granted Opus Dei the necessary approvals, although the juridical framework attributed to it was not what the Founder was thinking of" (Helming, 52); or that "the juridical status of secular institute was always considered by Opus Dei as a suit that was not made to measure" (West, 187).

In itself, the decree "of final approval of Opus Dei and its constitutions" contributed little that was new (see the long commentary on it by Rocca, 1985, 66ff., and Fuenmayor et al., 237–44, as well as the entire text, reproduced in Latin in both volumes). It says that the request had been filed by the Founder and his general counsel, with the support of 110 letters or recommendation from bishops (among them twelve cardinals and twenty-six archbishops). Again the objectives of Opus Dei are explained. Several categories of membership are distinguished among both men and women (regular members in the strict sense, oblates, and associate members; there is a supplementary distinction in the case of women between regular members and auxiliary regular members, who are those who devote themselves to domestic manual labor). An account is given of the principal works of apostolate, the basic spiritual character of the Institute is stressed, and finally there is mention of the training of members and the regimen of the Institute.

In this last part, the decree specifies that the president general, whose job is vital, must enjoy full authority: "since all the regular members are intellectuals," the authority of the president would be in jeopardy if it could be challenged (Fuenmayor et al., 554). Furthermore, the decree is almost a summary of the Constitutions of 1950, except perhaps as it relates to the "works of apostolate of the members." We will emphasize here in particular those points which are mentioned less often in the "official" literature.

Everyone, especially the men, practices the apostolate:

- Through the sanctification of professional work, procuring the edification and the welfare of the souls of their professional colleagues, and of those who work with them or under their orders. They are to undertake authentic social action with workers, assistants, and collaborators who are their subordinates.

- Through exemplary action in the civil and political jobs they are asked to perform by the public authorities.
- Through the religious, scientific, and professional education of the young, especially university students. In spite of the fact that in certain cases Opus could have its own institutions, it prefers "as far as possible to proffer its anonymous participation in public establishments." This form of apostolate includes teaching in the universities ("where possible, public ones"), moral and religious education in university residences, and social, artistic, physical, and other types of education in all kinds of youth organizations.
- Through the diffusion of Christian culture, resorting to the "most modern means of transmission and reproduction, oral and written, in word and in image."
- Through scientific research, the publication of books and articles, and participation in scientific meetings.
- Through finding a foothold in a special way in the apostolate among those who are in ignorance and in error, who find themselves outside the paternal house (read, Protestants), and who show themselves hostile to the Church. Such are the "associates," who can participate professionally and economically in the works of the Institute, pending their conversion.
- And, finally, through the obedient readiness to work in those regions where the Church suffers persecution (Fuenmayor et al., 548f.).

As for the women, in addition to participating in these same apostolic works, as befits their condition of being female, the decree distinguishes as special the following apostolic works:

- the direction and administration of houses for spiritual exercises;
- work in publishing houses, bookstores, and libraries;
- the apostolic training of other women;
- the promotion and defense of Christian modesty in women;
- the direction of student residences for women;
- the foundation of agrarian schools and centers for the training of household help;
- the administration and financial management of all Opus Dei houses (Fuenmayor et al., 549).

(*d*) As for the Constitutions of 1950, likewise an object of the final approval of the Holy See, which, with a few subsequent modifications, remained in force until 1982, they are a text at once fundamental and peculiar.

The Constitutions are peculiar more than anything because of the "secrecy" in which they have always been shrouded. In spite of the repeated protestations of Opus Dei members that it is not correct to call it secrecy ("any moderately well-informed person knows that there are no secrets"; "informing oneself about Opus Dei is very easy"; obstacles have never been placed in the way of information, "answering all your questions or giving out the appropriate information"; Escrivá, *Conversations,* no. 30), the truth is that "the Constitutions cannot be divulged, nor can they even be translated from the Latin into vulgar tongues without the permission of the Padre" (*Constitutions,* 1950, art. 193). Even some people who have been members of Opus Dei for many years frequently complain that they have never been able to read the Constitutions. Valdimir Felzmann: "I have been in Opus Dei for twenty-two years. I went to Rome, I earned my doctorate, I was ordained a priest. But I have never seen the Constitutions" (in Hertel, 1990, 227; see also Moreno, 1976, 25f.; Steigleder, 1983, 261ff.). The Constitutions were published for the first time by Ynfante in 1970, in the documentary appendix of his book (Ynfante, 397–452), provoking quite an uproar; for a while it was even insinuated that they were not really Opus Dei's Constitutions. In 1986 a bilingual edition was published (Latin-Spanish), translated—history is full of ironies—by the secretary of the Philological Institute of the Consejo Superior de Investigaciones Científicas, the home for so many years of many young Opus Dei university students. Let us point out, finally, that the volme *El itinerario jurídico del Opus Dei* includes an appendix containing its first twelve articles—out of a total of 479—in Latin (Fuenmayor et al., 553–55).

Among the biographies of Msgr. Escrivá, the one by Salvador Bernal does not include "Constitution" or "Secular Institute" in its index, and in its chronology there is no reference whatever to the final approval of 1950 (Bernal, 366). Gondrand, in a section entitled "A Decisive Approval," says that in 1950 "the Pope signed the decree of final and solemn approval of Opus Dei," and adds that "Vatican Radio gave an ample commentary on the decree in all its thirty broadcasts in different languages" (Gondrand, 197), but without mentioning the juridical configuration of the secular institutes, and without a single word about the Constitutions. Vázquez de Prada does the same thing: two brief allusions to "the final approval" (Vázquez, 257, 259), and no mention of the Constitutions or the secular institutes in either the chronology or the subject index. Berglar, for whom "the approval of Opus Dei was based on a new, special juridical formula that had to be created *ex profeso,*" devotes half a page to a discussion of the secular institutes in general (Berglar, 237), although neither they nor the Constitutions appear in the index. In the most recent biography, by Ana Sastre, the pattern remains unchanged.

Thus the only text in the "official" literature that pays any attention at all to the Constitutions of 1950 is the study by Fuenmayor, Gómez-Iglesias, and Illanes. The first element they emphasize as significant is the unity of the Institute: they say that while the Society of the Holy Cross seemed to take precedence before, now the name "Opus Dei" belonged to the whole entity, within which there existed a group formed of priests and those lay members who, according to the padre's criteria, were likely to become priests (*Constitutions,* 1950, art. 1). "We can see that things have literally been turned upside down when it comes to the text of 1943 and that of later years" (Fuenmayor et al., 248), thereby indirectly confirming that this was in fact—as our hypothesis from the previous chapter states—the big problem behind the difficulties of 1946.

In the second place, and in equally euphemistic terms, they assert that with the final approval came the end of the period "during which [Msgr. Escrivá] felt it necessary to give preference to apostolic activity in the university" (Fuenmayor et al., 248). Seldom has the "official" literature offered an interpretation so close to the reality of the Opus Dei of the early years. Even so, the fact remains that the Constitutions (art. 3) continued to speak of work among the intellectuals as the specific goal of Opus Dei, in an article which the authors reproduce in its entirety in Latin in the appendix (p. 553). In the text itself (p. 250), however, an ellipsis replaces the passage in which it is declared that the intellectual class "is the one which directs civil society, because of its greater education, because of the positions it holds, and because of its prestige."

Now, continue the authors of *El itinerario jurídico,* with the authorization conceded by the Holy See to admit men and women "single or married, of any profession, class, or social condition," Opus Dei is described in the Constitutions "as a path of holiness and apostolate, in the midst of the world, without establishing any limitation or specification, and open therefore to men and women of any condition" (Fuenmayor et al., 248). Indirectly, this amounts to an admission that this kind of description of Opus Dei, in spite of all the repeated assertions to the contrary in the "official" literature, becomes valid only *after* this date. Even so, one has to wonder to what extent Opus could have lived up to the claim that after 1950 it was open to persons of any "profession, class, or social condition," without "any limitation," if the regular members "had to have a university degree" (art. 35), "had to pursue ecclesiastical studies at the advanced level" "of Thomist orientation" (art. 135, 136; Fuenmayor et al., 252), and that these were the only members *strictu sensu* of the Institute (art. 16). As the decree of approval to which we referred earlier said, "all the regular members are intellectuals" (Fuenmayor et al., 554); and thus it continued to be after 1950.

It is indeed true that the Constitutions themselves specify that the distinction between various categories of members was made so that each kind of member would know his or her obligations, and so that all the members could be equally committed to the pursuit of perfection (Fuenmayor et al., 255). But it is no less true that this same article of the Constitutions says—and Fuenmayor et al. do not mention it—that the associate members must know that "the vocation of the regular members is higher and more complete" (art. 44.1). Furthermore, nonregulars can be recruited "among persons of any social condition," including those who "are suffering from some chronic illness" (art. 41). By contrast, the regular members of Opus Dei, who have "precedence over the nonregulars" (art. 31), cannot be chronically ill, married, or of "any social condition." Of those regular members who are not priests, the Constitutions expect the following professional activities: "jobs in public administration and university teaching; lawyers, doctors and the like; business and the like; business and economic activities" (art. 15).

Even among the regular members of Opus Dei there are different categories: some, called "registered" members, are destined for "management positions in the Institute" (art. 16.3), by "direct appointment by the Padre" (art. 19). Among these "registered" members, some can vote in the election of the president general: the president general himself is the one who appoints them (art. 22), and they are given the title of "electors" (art. 16.3). Finally, some regular members are at the same time regular members of the Priestly Society of the Holy Cross, either because they are priests or because "the Padre considers them to be candidates for the priesthood" (art. 1); also within this group some constitute a special category (art. 70). It is exclusively the Padre who determines which persons will be ordained; those who "desire it, can express this desire to the Padre, but submitting to his decision" (art. 273).

The Constitutions of 1950, meticulously detailed, raise many other questions on which we shall not linger here. On all matters relating to the obligations of members (obedience, chastity, poverty), to customs and devotions, to the regimen of governing, and to the women's section, we will turn to the text of the Constitutions whenever the specific topic under consideration warrants it. (For a general discussion of the Constitutions, more synthetic but perhaps more precise than Fuenmayor's, see Rocca, 1985, 68–74.)

The International Expansion

From a different perspective, let us consider a second major characteristic feature of this period (1947–1958), which is just as important as the final approval of Opus Dei: its international expansion.

During these years, Opus Dei ceased to be a phenomenon that was exclusively—and typically—Spanish. It is true that most of its members and practically all of its leaders continued to be Spanish. But even though the first persons to introduce Opus in other countries were generally Spanish, the fact is that the Institute became international, both in its sphere of action and in the members who joined it.

On very rare occasions the Opus authors provide us with concrete statistics. In 1950 there were 3,000 members in all, (more than 80 percent men); naturally most of them were regulars, since the admission of associates into the Institute was still a very recent feature. Twenty-three of the total number were priests (Fuenmayor et al., 1950f.). Of the nearly 2,400 men, 1,500 were Spanish (260 Portuguese, about 100 each of Italians and Mexicans, etc.) (Rocca, 1985, 63f.). Eleven years later, the number of members had grown to more than 30,000, with men only slightly in the majority, and with a dramatic reversal of the ratio of regular to nonregular members. In 1961 there were 3,000 women and 3,500 men (300 of them priests) as regular members (Rocca, 1985, 81).

At the beginning of the period (1947) Opus Dei had a few centers in Portugal, Italy, and England. In this same year they began to establish themselves in Ireland and France, and during subsequent years Opus spread to Mexico, the United States, Chile, and Argentina (Le Tourneau, 11). In 1950 Opus Dei had about seventy men's centers, fifty of them in Spain, and about twenty women's centers (Rocca, 1985, 63). At the end of the period (1958), centers had been opened in other European countries (Germany, Switzerland, Austria), in Canada, and especially in Latin America (Venezuela, Colombia, Peru, Guatemala, Ecuador, Uruguay, Brazil, and El Salvador, in addition to the previously mentioned Mexico, Chile, and Argentina), and was beginning to operate in Africa (Kenya) and Japan (Le Tourneau, 11; Fuenmayor et al., 301).

The process, particularly well described in Ana Sastre's book, always follows the same pattern: initially a member of Opus Dei goes to further his studies in another country, or else someone (usually a priest) is sent directly by the Padre to start "The Work." From the first apartment they proceed, in most cases, to the organization of a student residence. Then the moment comes for the incorporation of a small women's group, to take care of the household. Only in a few cases, especially toward the end of the period, did Opus begin to directly create its own institutions: a study center in Kenya, a language institute in Japan. (See Sastre, 363–471, who gives no figures but includes many names.)

In 1948 Msgr. Escrivá established the College of the Holy Cross in Rome, an international center for the formation of members of the men's section. In this manner regular members and those who were preparing for the priesthood could pursue their ecclesiastical studies in common, without ever leaving the atmosphere of the secular institute. This seems to have been an old idea of Escrivá's; he had not even wanted the first three Opus Dei priests (ordained in Madrid in 1944) to mix with other seminarians during their studies.

Another of the Padre's fixed ideas was the radical separation of men and women. Hence five years after the establishment of the Roman College of the Holy Cross (1953), the College of Saint Mary was established, "a center analogous to the previous one, but for the women's section" (Fuenmayor et al., 301f.). On this question of the separation of men and women, Rocca reproduces a curious document, dated 1947 and titled "Internal Regulation of the Administration," which stipulates among other things that "the male members of our Institute shall never go, even to visit, to the houses of the female section"; they shall never see "the servants who form part of the Administration (the 'auxiliary regular members'), or know their names, or ever speak with the help"; and in the case that there was only one oratory "the women shall attend the religious ceremony behind a screen, such as that used by the cloistered nuns when their churches are open to the public"; further, the cleaning of the residence will be done when the men are not there, "and the servants shall always come in a group to do the cleaning, never alone" (Rocca, 1985, 163–65).

The most significant feature of this period was, in any case, that the process of international expansion was accompanied by the possibility for married persons to join Opus Dei (always as nonregular members), as could diocesan priests. The conjunction of these two facts is what provoked, during the fifties, a notable transformation in the physiognomy of the Secular Institute. Although the majority of its members were still Spanish, Opus gradually began to be perceived as no longer exclusively Spanish, but rather universal. Although strictly speaking the members of the Institute were the regulars, who were committed to celibacy and were the only ones who could assume leadership positions, quantitatively they became more and more of a minority ("the officer corps") in relation to the growing majority of associate members ("the troops," to use the terminology of *The Way,* no. 28). In this way Opus's sphere of action gradually widened and, while it had more "limitations" than Fuenmayor, Gómez-Iglesias, and Illanes admit (p. 248), one could at last really begin to speak of persons of "diverse social conditions" and of "ordinary work."

At the same time, however, a new problem began to surface all across

Spain, where Opus Dei was largest and had been in existence the longest. First, whereas in countries where the process of expansion was just beginning the primary apostolate was still in the university community, in Spain many of the first members were already in the full swing of their professional careers. Second, the admission of married persons into Opus Dei rapidly created a problem that had not existed before: the socialization of a second generation, that is, the education of the children of Opus Dei members.

The Broadening of the Sphere of Action

Sociologically, the processes of institutionalization are a fine arena for the verification of the principle of the unforeseen consequences of any action. When we use this formula of the principle of the unforeseen consequences of action, we sociologists generally mean one or the other of two phenomena normally expressed in popular language as follows: on some occasions one experiences "success beyond one's wildest dreams," while on others everything "backfires."

The period from 1939 to 1946 was the beginning of Opus Dei's process of institutionalization in Spain. During the period from 1947 to 1958 this process reached its culmination, and the institutionalization was completed within the framework of a political regime which was likewise in the process of institutionalization, and which was progressing from economic autarchy to a certain degree of development, and from the international isolation of the immediate post–World War II years to gradual international acceptance, the two most spectacular manifestations of which occurred in 1953: the signing of a treaty with the United States, and a concordat with the Holy See.

How did the unforeseen consequences of action manifest themselves (and generally they were "successes beyond one's wildest dreams" rather than "backfires") in the fifties in Opus Dei's case?

(*a*) Once he had been named domestic prelate of the Holy Father in 1947, the Founder became Monsignor Escrivá de Balaguer, and he kept saying—or his biographers make him say, or they say he said—that "my wish is to hide and disappear." Nevertheless, he was named favorite son of Barbastro, his native town (1947), and was decorated with the Grand Cross of Alfonso X the Wise (1951), the Grand Cross of San Raimundo de Peñafort (1954), and the Grand Cross of Isabel the Catholic (1956).

(*b*) As for the achievement of his apostolic objectives, some authors estimate that at the end of this period Opus Dei members occupied as many as 30 percent of the university chairs in Spain (Valverde, 546). Simultaneously, as might be expected, Opus members were slowly penetrating into the organs of government administration, while others went on to occupy distinguished positions from which they influenced the worlds of business and politics.

The official version of this dual phenomenon is frankly ambiguous, and sometimes barely coherent when compared against the text of the Constitutions of 1950. For example, when Msgr. Escrivá defends himself against the accusation that "we want to occupy high positions" (*Conversations,* no. 64), he does not seem to remember the article in the Constitutions that speaks of "public jobs, and especially those that imply the exercise of leadership" (art. 202), or the general content of the pontifical decree of approval of Opus Dei as a secular institute.

A river of ink flowed on this topic in the seventies. But what set it off were the events of 1957, when two Opus Dei members were appointed government ministers (Commerce and Finance), two months after Laureano López Rodó had become technical secretary general of the presidency, at the same time that other Opus members were assuming positions of responsibility in the ministries of Information, Public Works, and Education (Tamames, 1973, 512). Immediately after the constitution of the first of Franco's governments to contain ministers belonging to Opus Dei, an Opus priest, Julián Herranz, published an article titled *El Opus Dei y la política,* which for a long time was the model for the "official version" (Herranz, 1957). This "official" version revolved around the dual argument that the Work did not intervene at all in the political choices of its members, who acted with complete freedom ("in everything temporal the members of the Work are free," declared the Padre; *Conversations,* no. 48) and had no commitment, consequently, to Opus Dei. In the second place, the argument went, members in fact adopted diverse positions in this field, which was perfectly legitimate and Opus of course had nothing to do with the matter.

As for the first point, that Opus Dei never intervened in politics—that only its members did so, under their own personal responsibility and without commitment to the Institution—the most problematical aspect is raised by article 58 of the Constitutions of 1950, which declares that members (regulars and associates) swear that "they will consult the Superiors on every important question of a professional or social nature, even when it is not a matter directly concerning the vow of obedience." The reply—that "on political questions, belonging to Opus Dei is like being a member of a tennis club" (López Rodó, 1991, 18; see also Escrivá, *Conversations,* no. 49)—is not very convincing, since in principle a tennis club does not have as one of its goals the exercising of an apostolate through public service, nor does it oblige its members to consult it on professional matters. Equally unconvincing is the line of reasoning that the members of Opus continue to enjoy total liberty in their decisions after having heard the advice of their superiors on the matter: "in fact, and by right, the commitment obliged them to seek counsel," but not necessarily to follow it (Fuenmayor et al., 244 n. 32). In 1971, Álvaro del Portillo informed the Holy Congregation of Religious that this article had been deleted from the Constitutions, explaining that experience had shown that "these oaths are not necessary to preserve our peculiar foundational charisma" (ibid., 586).

As for the second question—the pluralism of the political choices of the members of the Work—our only caution is that one must not lose sight of the specific historical context of the fifties, so as not to become confused about the dimensions and real scope of this so-called pluralism. Rafael Calvo Serer, one of the first members of Opus who participated directly in the world of politics (Calvo Serer, 1947), said years later that "people begin with the prejudice that Opus Dei is a political organization that acts in a planned way in public life," when "the reality is that, in this matter, one must forget about Opus Dei if one wishes to understand things" (Pániker, 85; see, in the same volume and on the same point, López Rodó, in Pániker, 327f.). "One must forget about Opus Dei": this is what Rafael Gómez did literally in his *El franquismo y la Iglesia*. After his first mention of the ministers who joined the government in 1957, he adds a note in which he says: "Ullastres and Navarro Rubio were members of Opus Dei, the well-known Catholic institution. On earlier pages the names of others who also are members of Opus Dei have appeared: Pérez Embid, Calvo Serer, Rodríguez Casado, López Rodó, and López Bravo. I did not mention this circumstance because it had no specific political relevance" (Gómez Pérez, 1986, 74 n. 5).

Certainly a claim can be made that there were different points of view (and that other names have to be mentioned: Albareda, Fontán, López Amo, Suárez Verdeguer, etc.) and therefore that there was a certain amount of "pluralism." We can also admit the thesis that Opus Dei did not function "as a political organization which acted in a planned way in public life." We can even accept the hypothesis that in Opus's Roman center this highly visible implication of certain distinguished members aroused a bit of anxiety and concern (Artigues, 170). But even so, we must insist again that the situation continued to be a logical and foreseeable consequence of the process of the institutionalization of Opus Dei in Spain, and of its success in achieving the objectives it had set for itself. The paradox is that this very success seems to have caught the leaders of the Work by surprise and obliged them to design a strategy whose goal was not to have to acknowledge as institutional the success of certain objectives which had nevertheless been defined as institutional. We must also keep in mind, furthermore, that the political pluralism of the members of Opus Dei was, during the 1950s, a frankly limited pluralism: for the most part they followed the common platform of the magazine *Arbor*. (On *Arbor,* especially the figures of Calvo Serer and Pérez Embid, see Artigues, 147–77, and Casanova, 1982, 251–79, as well as the monographic issue of 1985 devoted by the magazine to its own history; Pasamar et al., 13–137.) Several of the members played a direct and important role in the education of the then prince Juan Carlos; all of them agreed on the affirmation of Catholicism as an essential element and "backbone" of Spain, as well as of Francoism, which was simply taken for granted.

Within these narrow confines one can—if one wants to—speak of pluralism. But, if one pleases, one could also speak of some very basic similarities and areas of agreement. To give just one example: in 1952 Ángel López Amo, a historian and tutor of the prince and future king, wrote that "freedom of expression, free suffrage, and freedom of religion are the freedoms of destruction and debasement" (López Amo, 313). Five years later Rafael Calvo Serer, by now become a kind of "enfant terrible" of Francoism through his polemical and supposedly nonconformist writings, asserted in a newspaper article that "freedom of conscience leads to loss of faith, freedom of expression to demagoguery, mental confusion and pornography, freedom of association to anarchy and totalitarianism" (quoted by Artigues, 189f.).

(*c*) On the institutional level, the growth of Opus Dei, the success of its establishment in Spain, and the broadening of its base thanks to the acceptance of nonregular members and above all of married persons, had another unforeseen consequence that slowly transformed the entire physiognomy of Msgr. Escrivá's institute: the creation of its own institutions of education.

In the early years, Padre Escrivá had declared himself against the creation of this kind of establishment. His model was to be quite different from that of so many other religious congregations dedicated to teaching, and also different from the Jesuit model, with its schools for the children of rich families. Escrivá did not even share the idea of the "Propagandists," of Herrera Oria and the Jesuit Ayala, on the appropriateness of Catholic universities (see, for example, the chapter "The Teaching Apostolate," in Ángel Ayala, *Formación de selectos,* 260–82). For Ayala the Catholic university was "the educational institution par excellence for the young": "as long as we Catholics in Spain lack our own university, it will be impossible for us to raise a generation of young people fully educated in the doctrines of the Church, and impossible to create a single national body of thought on problems vital to the nation" (ibid., 269). By contrast, in the documents of approval of Opus Dei (1950) it was still being affirmed that Opus Dei preferred "to proffer its anonymous participation in public establishments," and that the members of the Work who were university professors would exercise their apostolate in the universities, "public ones, where possible" (Fuenmayor et al., 548).

Notwithstanding this, in 1951 the first Opus Dei school was opened near Bilbao. In 1952 the activities of the "General Study of Navarre" began, the initial nucleus of what several years later would become the University of Navarre (which does not officially bear the title of a Catholic university but is nevertheless classified as a Catholic university in the corresponding section of the *Anuario Pontificio*).

Aranguren interpreted the creation of the University of Navarre as a consequence of the failure of Opus Dei in "its effort at Spiritual mastery of the university" (Aranguren, 1962, 15). It is true that after 1951, with the naming of Joaquín Ruiz Giménez as Spanish minister of education, the winds ceased to be systematically favorable to Opus; it is equally true that parallel to the first liberalizing efforts of the new minister, the universities became progressively a source of conflict, and for the first time criticism of the professors and students of Opus Dei was expressed. This might support the possible interpretation, as Aranguren suggests, that the creation of the University of Navarre was a maneuver of strategic retreat.

Even granting this to be true, I believe that the explanation must take into account a second and no less important factor. That is, with the broadening of its base of recruitment, Opus Dei began to create for itself a hitherto unknown problem: the socialization of the next generation. During the whole initial period, the small group of Opus's first members had been fully dedicated to what in Catholic language is usually called "the apostolate of penetration." But when the group ceased to be small, and furthermore began to procreate, since it contained married members, it was faced with the need to educate those who were born into the organization, who did not come from outside it. This was not yet, certainly, the phase in which Msgr. Escrivá would radically reject the apostolate of penetration and boldly assert that "I hope that a moment will come when the phrase 'the Catholics penetrate the social sphere' is no longer uttered, and that everyone will realize that it is a clerical expression" (*Conversations*, no. 66). This remark dates from 1968, but for now we are still closer to that other sentiment, according to which it is necessary "to intrude in the environment, with lordly and positive spirit," quoted at the beginning of the chapter (Escrivá, 1949, 7). But at the minimum Opus Dei began to realize the need for a new, complementary style of apostolate. The "biographical memory" of the first members was no longer sufficient; from now on they would have to start "transmitting a tradition" (Berger and Luckmann, 93). In other words, sociologically there is a big difference between the problem of the "apostle who converts the pagans" and the "apostle who attempts to make his own children believe." In my view, the educational institutions of Opus Dei must also be explained in this manner.

In any case, with the creation of these centers at the beginning of the fifties a new activity was begun, which in turn contributed to the broadening of Opus Dei's range of action. The "General Study of Navarre" was the first—and is still the most important today—of a "broad gamut of university institutions all over the world" (*Conversations,* no. 84). Meanwhile the preparatory school in Bilbao, at first an exception, became the precursor of a vast network of schools—belonging to Opus Dei, directed by Opus Dei members, or entrusting to Opus Dei the tasks of spiritual direction—organized into "societies and cooperatives of parents to promote and direct teaching centers. On the death of Msgr. Escrivá de Balaguer, one of these institutions (there are several, in different countries) had more than twenty boys' and girls' schools, through which thousands of students had already passed" (Sastre, 427).

Notice that in all this activity Opus Dei was still competing with the Society of Jesus. In the preceding chapters we saw that during the initial years this competition—although it was real, and perceived as dangerous by the Jesuits—was frankly unequal, given the power and tradition of the Jesuits and the small size and minimal institutionalization of Opus. From now on, however, events confirmed the validity of the Jesuits' apprehensions. The two groups competed in many areas, but now it was on equal terms. In education—more specifically, schools "for the children of rich families"—Opus was not only not behind, but was actually ahead. And in other areas the Jesuits even began to follow in Opus Dei's footsteps.

For example, in the latter part of the forties a group of university students who clustered around the Jesuits Llanos, Díez Alegría, and others, were planning the establishment of a residence "that would bring together a select group of university students to plan the creation of a secular institute" (González Estefani, 59). We have said from the beginning that Escrivá and Opus Dei were clearly inspired by the Jesuit model; now it is not so clear who is copying whom! And both continue to use the same language: "We want to give our activity above all a line of austerity and *profoundly Catholic and Spanish intolerance";* "we want to give our efforts at perfection the note of *a fanaticism* for the most unifying charity toward all classes, entities and individuals of the Church and of Spain"; "we want, as authentic Catholic laymen and Spaniards of our time, to achieve the most perfect harmony between the two services of Church and State" (ibid., 57; emphasis mine). The author of these programmatic lines (written in 1947) was not Padre Escrivá but the Jesuit Padre Llanos. The subsequent evolution of these two men would be radically different, but the "pluralism" of Catholic Spain of that epoch went no further than this!

A second example of this parallelism between the Society of Jesus and Opus Dei appears in relation to the University of Navarre. The Jesuits had opened a center of higher studies in Bilbao (1886), and there were other ecclesiastical study centers in Spain. The Concordat of 1953 between Spain and the Holy See recognized the right of the Church to create centers of higher education, but it was the state which established the criteria for the accreditation of courses and degrees awarded by the centers. After the establishment of the University of Navarre as a University of the Church (1960), it was Opus Dei, and not the Society of Jesus with its University of Deusto, that in 1962 got the state to accredit the curriculum and the degrees offered by the universities of the Church.

Here is one last, and significant, example of this parallelism: at the end of the period we are considering (1958), the Society of Jesus created in Barcelona a Higher School of Administration and Business Management (ESADE); simultaneously Opus Dei created, also in Barcelona, an Institute of Advanced Business Studies (IESE). We will treat these two institutions in more detail in Part II.

Some Unresolved Questions: From Padre Escrivá's Plan to Abandon Opus Dei (1948–1949) to the Attempt to Oust Him from Opus Dei (1951–1952)

The basic profile of these "ten years of success" in the history of Opus Dei—characterized by its final approval as a secular institute, its internationalization, and the notable broadening of its sphere of action—might lead one to think that everything was going smoothly for Padre Escrivá's followers, and that Opus's Secular Institute was a true oasis of peace. A few facts and a few clues, however, lead us to think that the resolution of the grave crisis of 1946, miraculously achieved with the arrival of the Padre in Rome, was in fact not the final overcoming of certain obstacles that stood in the way of Opus's triumphant trajectory.

During the first half of the period 1947–1958 there were at least two episodes that were, for Opus Dei's leaders, a source of displeasure and headaches. Both are perfect illustrations of the kind of situation—which we have already encountered—in which the "official" literature confines itself to telling us, in euphemistic language, that the puzzle is not complete, and also that for some reason or another it is not prepared to place the missing pieces on the table.

(*a*) In 1948 and 1949 Msgr. Escrivá "felt a great preoccupation for the holiness and the sanctification of the priests," so pressing that he came to the conclusion that "it would be necessary to undertake a new foundation with the end of aiding the diocesan priests, even if this would oblige him to leave Opus Dei" (Fuenmayor et al., 229f.). He even went so far as to pose the question—they say—to his family, to the members of the General Council of Opus Dei, and to certain persons in the Curia: "I went to the Holy See and said I was ready to make a foundation for priests. With great surprise, they told me yes" (quoted in Sastre, 298). According to Berglar, Álvaro del Portillo compared this decision of the Padre's with "the sacrifice of Abraham"; and just as in the case of Abraham, "the Lord gave him the solution," in the form of the incorporation of the secular priests into Opus Dei as nonregular members of the Priestly Society of the Holy Cross (Berglar, 410 n. 53). The final approval of 1950, in addressing the possibility of the integration of priests who came from outside the ranks of Opus Dei, withheld "that new foundation, which would have painfully divided Escrivá's heart of father and of mother" (Vázquez, 257; Bernal, 158).

As can be plainly seen, all our authors mention the subject, but none of them in a forthright manner. What is hidden behind Escrivá's decision to undertake a new foundation, even at the cost of abandoning Opus Dei? How and why did he suddenly find an apparently very simple alternative solution, although he had "sought for it for such a long time," and "no one had suggested it to him" (Gondrand, 196)? This solution consisted simply of including the priests who wanted to join in the already existing foundation. Could all this have something to do with the final approval of Opus in 1950, in the sense that someone had proposed, as a condition for the approval of Opus Dei, that the Padre dedicate himself to something else? Is there any possible connection between this project of a new foundation and the granting of the "episcopal miter" to the Founder, a goal which Álvaro del Portillo had pursued stubbornly for a while, and was on the point of achieving (but the "official" literature never mentions this) when it was frustrated at the last moment by the intervention of several members of the Society of Jesus who were very close to Pius XII?

This first question is clearly unresolved, and one moreover that is posed in the "official" literature in terms that denote a definite desire not to resolve it for the time being.

(*b*) The second episode is equally obscure but seems more serious than that of 1949. This one has to do with an outside initiative that took place in 1951 and 1952, whose objective was "to separate Msgr. Escrivá de Balaguer from Opus Dei, and to divide it into two distinct institutes, one for men and the other for women" (Fuenmayor et al., 317). In spite of spending several pages on it, Gondrand is no more explicit: it was "a project to dismantle the Work . . . a truly diabolical plan: it would have split the two sections of Opus—masculine and feminine—and obliged the Founder not only to give up his job as president general, but to leave the Work," which made Escrivá exclaim, "with tears in his eyes: if they oust me, they kill me; if they oust me, they assassinate me" (Gondrand, 206). What is the root of this project that "posed a threat to the existence of the Work and placed the Founder in a nearly desperate situation" (Berglar, 44)? The starting point might be found with the families of some Italian members of Opus Dei, who were alarmed by the denunciations they were hearing about the Work (Vázquez, 259). Behind these families, however, obviously there had to be people who were organizing "the campaign of denunciations and calumnies," who had begun "the secret transmission of false documentation presented to the Holy See," and who according to all appearances were the same ones as always: "Those blessed men of old argued obstinately, and not for exemplary reasons. They scourged him with lies, sticky as flies when a storm is coming" (ibid., 261).

Vázquez's reference to "men of old" allows us to surmise that the "blessed men" in question perhaps belonged to the Society of Jesus. But it is only a guess, not entirely confirmed by the official literature ("things happened that were analogous to what happened earlier in Spain"; Fuenmayor et al., 303). Above all, we are not told a single word about the reason behind the denunciations, or about the content of the "false documentation." We are only told that Msgr. Escrivá decided to "consecrate Opus Dei to the Most Gentle Heart of Mary" (1951), and later "to the Sacred Heart of Jesus" (1952), until, through the direct intervention of the Holy Father, "the affair was stopped" (ibid., 304).

There can be no conclusion other than that of Berglar: "Only in time, when the process of historical sedimentation has advanced sufficiently and the archives are opened, will we know more specifically what the threats were and how they were overcome" (Berglar, 269). This is probably the most explicit statement in the "official" literature whereby someone who has had access to the unpublished documents admits that, in fact, there are areas of the puzzle that cannot be reconstructed until the appearance of a series of pieces that are being concealed at this time.

The Theory of the *Terzo Piano*: A Little Exercise in Vaticanology

"Vaticanology" is a discipline of notable complexity, which has its excellent and very competent experts as well as an abundance of "amateurs éclairés" plus the inevitable horde of the curious who tend to confuse real knowledge with mere desultory gossip. If you do not belong to either of the first two groups, and you do not want to be accused of belonging to the third, it is wise to tread cautiously.

We will begin, therefore, from an elementary and classical outline which, even without covering all the nuances, has the advantage of clarity and, above all—as several experts we have consulted have confirmed—the advantage of being fundamentally correct. In accordance with this plan, in order to minimally understand how the Vatican works it is necessary to distinguish three levels: (1) the masses ("the troops," in Escrivá's language), the mobilizable and identifiable collectivity, with their cheers and banners, standing amid a great crowd in Saint Peter's square; (2) the Curia, that is, the entire bureaucratic complex of the Roman Church, apparently silent and discreet to anyone looking on it from outside, which functions in general fairly efficiently considering that it is a bureaucracy, a very complex bureaucracy, and a bureaucracy that work in conditions of greater than average precariousness; and finally (3) *il terzo piano,* the third story, or the closed world of limited access in which the pope and his closest colleagues move.

At the time when Pius XII was pope (not so much now, of course), a fourth and no less indispensable element had to be added: the "castle," the Borgo di Santo Spirito, headquarters of the Jesuits, where the man who in those times used to be called "the black pope" lived. During that entire period, in fact, the Jesuits, in addition to having their own mass following (laymen who through many associations revolved in the Society's orbit), were present in practically every organism of the Curia, and had access to the *terzo piano*.

If we wanted to write one page imitating the literary style of the Padre's biographers, we would say that during the night of June 23, 1946, his first night in Rome, which he passed "alternately praying and contemplating the cupola of the basilica of Saint Peter's . . . and the windows behind which lived his successor" (Gondrand, 176), Escrivá meditated on all these things, and reflected on the situation of Opus Dei. The Padre dreamed of the day when his people would be able to mobilize masses: "the first time that the Founder was with a real crowd" would not be until 1960, at the ceremony inaugurating the University of Navarre. During the last years of his life, "the supernatural marathon" (Sastre, 528) of journeys through Spain and different countries of Latin America would be "baths of crowds." After his death, on the occasion of John Paul II's trip to Spain (1982), Opus Dei's capacity to mobilize masses reached one of its most spectacular heights: throughout the pope's visit the cities were so filled with placards bearing Opus's greeting to the pope, *totus tuus,* that it gave the impression that the Spanish were *totus Opus*. And every year for quite some time now several thousand students meet at a conference in Rome during Holy Week and hail the pope while he speaks to them "warmly of Opus Dei and its founder" (West, 17). But on that June night in 1946 Escrivá could not yet muster thousands of Opus youths in Saint Peter's Square. Opus Dei was still "not a work of the many, but of the select few," as the bishop of Madrid, Eijo Garay, had written to the assistant abbot of Montserrat (Rocca, 1985, 132).

In the second place, it is obvious that Padre Escrivá did not have direct access to the *terzo piano*. While he was "contemplating its windows," he was thinking about the abysmal difference between being received in a formal audience by the Holy Father and being able to go up to the third floor by the back stairs. Having the Swiss Guard called to attention to usher Álvaro del Portillo into the Vatican "via the great stairway that leads to the Audience Chamber" (Sastre, 332) doubtless had its charms. But a number of Jesuits could go up and down the back stairs with complete freedom. In his meditation the Padre was perfectly aware that for him this access was barred. He would not attain it even with John XXIII, much less during the endless years of the not-so-long reign of Pope Paul VI. The next pope would be the fleeting John Paul I: if he hadn't died after only one month, who knows . . .

What Escrivá dreamed of on that Roman summer night was perhaps well exemplified by Joaquín Navarro Valls, an Opus Dei member who in 1984 was named director of the Press Office of the Holy See, when he was able to observe that "the Holy Father is so informal when he speaks to his closest colleagues, that I have to make an effort to remind myself frequently that I am talking to the Pope"; offering a specific example of this informality, Navarro Valls related that in order to get a better idea of the scope of a news item which had just been released, "I called the apartment and the secretary said to me: come for supper tonight" (statements to *Catalunya Cristiana,* April 25, 1991, p. 16). This man was summoned to go up, by the back stairs, directly to the *terzo piano!* The only problem—concluded Msgr. Escrivá in his imaginary meditation—was that even when some of his "children" finally had access to it, they would cross paths frequently, too frequently, in fact, with the "children" of Msgr. Giussani and his Communione e Liberazione. This would be especially true if one believes the accuracy of the description of John Paul II as a pope who "listens to many people, speaks with very few, and decides by himself" (Grootaers, 211).

But let us abandon our imitation of the literary style—a mixture of reality and fantasy—of the biographers of the Padre. In the Roman panorama of the last twelve years of Pius XII (1947–1958), with the masses excluded as nonexistent and the third floor inaccessible, there was only one possibility left: the Curia. Consistent with his model of the Church, "which exhibits its vitality with a movement that is not a mere adaptation to the environment," Msgr. Escrivá directed the efforts of the members of Opus who lived in Rome toward "an intrusion, with positive and lordly spirit" (Escrivá, 1949, 7), into the organisms of the Vatican Curia.

Weighing the merits of what for Opus Dei represented the pontifical approval of 1950, Fuenmayor, Gómez-Iglesias, and Illanes conclude that "in summary, the drawbacks of the solution of 1950" were twofold: "the dependence on the Congregation of Religious, and the fact that the definition of the secular institute was placed in the sphere of the concept of the state of perfection" (Fuenmayor et al., 295). We have already spoken of this second drawback: specifically we were able to show that at least initially the Opus Dei authors, beginning with Escrivá himself, seemed rather happy about being included in the "sphere of the concept of the state of perfection." In the next chapter, in any case, we will return to this question; for now we will focus on the first "drawback"—"dependence on the Congregation of Religious"—for it is precisely here that all the "intrusion, with positive and lordly spirit" was exercised by Opus Dei in the Vatican Curia.

Using as a basic source the volumes of the *Anuario Pontificio* (the Vatican Yearbook), we can see that in the edition of 1948 some names related to Opus appear for the first time: Giuseppe Escrivá de Balaguer y Albás figures in the list of the "prelati domestici di Sua Santità," and José María Albareda is a member of the Pontifical Academy of Sciences. In the 1949 volume in the pages devoted to the Sacred Congregation of Religious, we find that (1) in the section of ordinary affairs, there is a subsection devoted to the "Societies without Vows and Secular Institutes," comprising a single name, "Don Álvaro del Portillo, of Opus Dei"; (2) in a second section, called special affairs, there is a list of four names, one of which is Álvaro del Portillo; (3) the juridical committee has its own list of thirteen names, the second of which is Portillo; (4) in a final committee, of "government and religious discipline," the second section "for five-year relations" has a list of seven names, the first of which is Álvaro del Portillo.

In 1950, Portillo is again found on the three committees, and replaced as the only member of the "III Ufficio" of the section of ordinary affairs by Salvador Canals, also of Opus Dei. In the 1951 edition the final approval of Opus Dei appears: it says that Opus Dei's goals are "to diffuse among all classes of civil society, and especially among intellectuals, the life of evangelical perfection." Beside Opus Dei appears the Compagnia di San Paolo, whose goals are "the social apostolate and the apostolate of penetration" (*Anuario Pontificio,* 1951, 793f.). In 1952 the order of presentation is reversed, and after the Compagnia di San Paolo and Opus Dei two other approved secular institutes are added: the Ministering Priests of the Sacred Heart of Jesus (founded 1883), with the "training of aspirants to the priesthood" as their goal, and the Society of the Heart of Jesus (founded 1791), whose goal is to "practice evangelical perfection." As for the Sacred Congregation of Religious, Salvador Canals remains in his post, and there is a reorganization of the various committees; in the committee for the Constitutions of the Secular Institutes the name Álvaro del Portillo appears. In another place in the same *Anuario* (1952, p. 1188) the female secular institutes of pontifical law are enumerated: five in all, among them the Teresian Institute founded by Pedro Poveda. Of Opus it simply says: "Opus Dei. Sezzione Femminile. Madrid."

The 1953 edition reports that in the framework of the Congregation of Religious a Committee for the approval of the Institutes and the Constitutions was created. This committee had two sections: one for congregations, and another for societies without vows and secular institutes; Álvaro del Portillo is listed as a member of the second. There are no changes in subsequent years, except that the *Anuario* of 1955 says that the person in charge of "statistics" for the general section of the Sacred Congregation is the "Reverend Alberto Taboada, of Opus Dei"; his name had already figured in the *Anuario* since 1949, but as "Signore" and not "Reverend," and without mention that he belonged to Opus Dei. In 1956 eight persons were promoted to the rank of "consultants" to the Sacred Congregation of Religious; the eighth was Álvaro del Portillo.

In the 1957 edition of the *Anuario Pontificio,* Salvador Canals is still in the Sacred Congregation of Religious, and he is also listed as a consultant of the Pontifical Committee for Cinematography, Radio, and Television. In the 1958 volume, Msgr. Escrivá appears as consultant for the Sacred Congregation of Seminaries and Universities (remember that the decree establishing the University of Navarre was in 1960). And in the 1959 *Anuario*—which shows, as usual, the corresponding figures from the previous year—we find another member of Opus Dei, Javier de Silió, listed as belonging to the Congregation of Religious as "commissioner," while Salvador Canals continues working there, and Álvaro del Portillo is still consultant.

After this date, which coincides with the beginning of John XXIII's reign as pope and the announcement of the Second Vatican Council—but which is outside the period under analysis here—there would be a growing dissemination of Opus Dei members throughout the other curial organisms, now begun in earnest with the appointments of Canals and Msgr. Escrivá. But let us not get ahead of ourselves; for now we only want to register the existence of an apparent contradiction between the claim that "the dependence on the Congregation of Religious was one of the drawbacks" of the juridical solution reached in 1950 (Fuenmayor et al., 295) and the active participation of a number of distinguished members of Opus Dei in the tasks of this very Sacred Congregation. This seeming contradiction could be resolved only if it were possible to clear up the basic underlying question: was the approval of Opus Dei in 1950, officially categorized as "final approval," really considered as final by the directors of Opus Dei during the fifties, or, rather, was it considered, even then, to be just another stage in the juridical itinerary of the Work? This is the question we will consider in the next chapter.

10

1958: *Non Ignoratis,* A Letter from Monsignor Escrivá

General Observations

On October 2, 1958, the day commemorating the thirtieth anniversary of the founding of Opus Dei, Msgr. Escrivá de Balaguer wrote a surprising letter to his "dearest daughters and sons," which is reproduced in the documentary appendix of Fuenmayor et al. (pp. 563–65). The authors describe it as a document that has, "in a way, the flavor of a declaration or exposition of motives and intentions" (Fuenmayor et al., 321). They give a brief summary of it—much briefer, in fact than their summaries of two later letters having similar content, dated 1961 and 1962 (ibid., 327–31, 339–47), perhaps because the latter two are not reproduced in their entirety in the appendix, while the first is.

But there might also be another reason for this brevity, specifically, the fact that the 1958 letter had already been published by Rocca (1985, 184–87), who pointed out that it had not been made public until 1983, together with the Statutes of the new Prelature of Opus Dei. Curiously, while Rocca reproduces the letter in Latin, for once the three Opus authors publish it in Spanish (in contrast to the Statutes, which appear in Latin; Fuenmayor et al., 628–57). We have no way of knowing whether the letter is presented in Spanish because it was originally written in that language, or whether in this case the authors simply wanted everyone to be able to read it.

Whatever the case, the most surprising thing is the actual content of the letter, in which Msgr. Escrivá wrote, among other things, "Our only zeal is to serve the Church, as she wants to be served, in the particular vocation we have received from God. *Because of this, we do not desire for ourselves the state of perfection.* We love it, for the religious and for those who belong to *what are now called secular institutes,* because it is proper to their vocation" (Fuenmayor et al., 564; emphasis mine).

"The peculiar characteristics of the spirit and the apostolic life of the Work of God . . . confer on our Work a personality that is surely very special . . . that clearly differentiates it from the present secular institutes" (Fuenmayor et al., 564). Escrivá mentions in this context the *Decretum Laudis* of February 24, 1947, but without expressly recalling that the document begins: "The first secular institute which has merited the *Decretum Laudis* is the Priestly Society of the Holy Cross and Opus Dei, called, for short, Opus Dei"; Escrivá also cites the Decree of June 16, 1950, which is the decree of final approval of Opus Dei and its Constitutions as a secular institute of pontifical law; and he cites the apostolic briefs *Cum Societatis* (June 28, 1946) and *Mirifice de Ecclesia* (July 20, 1947), which are really nothing more than documents conceding indulgences to the members of the Priestly Society of the Holy Cross and Opus Dei.

The letter goes on to say that *"in fact we are not a secular institute,* nor can that name be applied to us in the future" (Fuenmayor et al., 564; emphasis mine). The *Anuario Pontificio* for 1958, however, and for succeeding years, does list Opus Dei as a secular institute, and its objective is officially defined as "to diffuse among all classes of civil society, and especially among the intellectuals, the life of evangelical perfection." This text continues to be reprinted without modification in the *Anuario Pontificio* until 1965 (p. 683), when it becomes: "to diffuse among all classes of civil society, and especially among the intellectuals, the life of Christian perfection in the midst of the world." In 1967, the objective of Opus Dei, still listed as a secular institute (pp. 905f.), is phrased in these terms: "to promote among all classes of civil society, and especially among the intellectuals, the search for Christian perfection in the midst of the world." This is the last year in which the *Anuario Pontificio* details the objectives of the various secular institutes; starting in 1968 it lists only the title, date of foundation, and address of the central offices. Opus Dei is found among the secular institutes in the pages of the *Anuario Pontificio* until the 1982 edition; in 1983 it is listed as the first, and only, Personal Prelature (p. 1012).

Let us return to the Padre's letter of 1958: "For the same reason and with the same desire, so that no one may originate or diffuse any false opinion about our specific vocation, we never wanted—as the Holy See knows—to form part of the federations of religious, or to attend the congresses or assemblies of those who say they are in a state of perfection" (Fuenmayor et al., 565).

An Important Congress

Exactly two years before this interesting letter was written, from September 23 to October 3, 1956, a National Congress of Perfection and Apostolate was held in Madrid, and its acts were published in several volumes. We shall concern ourselves here only with the first of these (*Introducción histórica y sesiones comunes,* 1957).

In the definitive program of the congress, Salvador Canals is listed as the moderator of the third session, which was dedicated to the secular institutes. Amadeo de Fuenmayor was the speaker in the ninth session, on secular apostolate.

In 1989, Ediciones Rialp published the nineteenth Spanish edition of a book by Salvador Canals, titled *Ascética meditada* (first Spanish edition, 1962). On the dust jacket it says that Salvador Canals, born in Valencia in 1920, joined Opus Dei as a very young man, before 1942, and was ordained into the priesthood in 1948. It also says that in addition to this book he also wrote several works of a juridical nature. However, it does not mention the fact that two of these works were published in the same series by the same publisher: one was *Los institutos seculares,* and the other *Institutos seculares y estado de perfección.*

In this last volume (1954; here we are using the second edition, 1961) Canals asserts, for example, that when a suitable juridical framework was being sought for those associations which after 1947 were to become the secular institutes, "no small contribution was made to the definitive solution through the study of that vigorous and exuberant Institute, the Priestly Society of the Holy Cross and Opus Dei," since the studies done by the Sacred Congregation showed "clearly that Opus Dei embodied the perfect type of secular institute" (Canals, 1954, 81, 82). The final result was that Opus Dei became "the Spanish Institute which enjoyed the glory of being the first secular institute of pontifical law" (ibid., 81).

There is another important detail in Ediciones Rialp's presentation of their author. The dust jacket also says that Salvador Canals "performed important jobs in the Dicasteries of the Holy See: he was Auditor of the Sacred Roman Rota, and consultant to the Congregation of the Clergy, the Congregation of the Discipline of the Sacraments, and the Papal Committee on Social Communications." The *Anuario Pontificio* confirms that until his death (in 1975, a month before the death of Padre Escrivá), he held all these jobs. And he also held another one: he was consultant to the Congregation of Religious, a job he took on in 1968 (or 1967, but it is recorded for the first time in the *Anuario* of 1968), which had until then belonged to Álvaro del Portillo, present prelate of Opus Dei. In fact, until 1960 Canals had worked in the juridical office of this Congregation of Religious, which was in charge of the secular institutes. In Canals's other book, cited above, *Institutos seculares y estado de perfección* (2d ed., 1961, 11), the author is presented as "Rotal lawyer and Officer of the Sacred Congregation of Religious."

These preliminary items seem to denote, then, certain differences in perception concerning the appropriateness of the juridical format of the secular institute for Opus Dei. For this reason we find it particularly interesting to contrast these two texts, written very close in time to one another: the letter by Padre Escrivá written October 2, 1958, and the *Acts of the Congress* which closed exactly two years before on October 3, 1956.

From the first of these documents the important passages to consider are the assertion that "we never wanted . . . to attend the congresses or assemblies of those who say they are in a state of perfection" and the pair of statements, "in fact we are not a secular institute, nor in the future may that name be applied to us" and "nor can Opus Dei be confused with the so-called movements of apostolate" (Fuenmayor et al., 565, 564).

(*a*) In the first volume of the *Acts of the National Congress of Perfection and Apostolate,* there are twelve papers written by members of Opus Dei: three by Excelentísimo Señor Don Amadeo de Fuenmayor (the coauthor of *El itinerario jurídico*), two by Msgr. Álvaro del Portillo, two by Salvador Canals, and one each by the following: Antonio Pérez (then general administrator of Opus Dei, according to that year's *Anuario Pontificio*), Severino Monzó (who would later become officer of the Sacred Congregation of Religious, taking on this job almost as soon as Salvador Canals left it), and three members of the women's branch, Señoritas Encarnación Ortega Pardo, Patrocinio Sind, and Catherine Bardinet.

In order of appearance, the first paper is one by Salvador Canals (*Actas,* 300–302), in which he asserts that the apostolic constitution *Provida Mater Ecclesia,* regulator of the juridical format of the secular institutes, "envisioned the existence of secular institutes having priests as members who, without losing their secular condition, could live in a juridical state of perfection" (ibid., 302). In a footnote Canals quotes on this point a text by Escrivá with which we are familiar, *La Constitución Apostólica Provida Mater Ecclesia y el Opus Dei* (Escrivá, 1949). Opus Dei being one of the institutes which embodies these characteristics, it seems that neither Canals in 1956 nor Escrivá himself in 1949 would have refused to admit into the bosom of their organization the kind of people whom the Founder would later designate, in his letter of 1958, as "those who say they are in a state of perfection."

The second paper is by Msgr. Álvaro del Portillo (*Actas,* 344–48). After asserting that *Provida Mater Ecclesia* recognizes in the secular institutes "a new complete state of perfection" (ibid., 345), Portillo goes on to say that the secular institutes are "societies, clerical and laical, whose members, in order to acquire Christian perfection and to exercise the apostolate fully, profess the evangelical counsels in secular life." The secular institutes thus signify "the union in one juridical format of the complete state of perfection and the 'conditio saecularis' of the person" and "the participation of each institute in the apostolate of the Church is manifested in very different ways, according to the means or the manner of exercising the apostolate the Church approves for each institute" (ibid., 347). Portillo concludes that the requirement of complete dedication to the apostolate is indispensable for the approval of a secular institute (ibid., 348), and that this is precisely the difference between the secular institutes, on the one hand, and the religious orders and "societies of common life without public vows," on the other (which is what Opus Dei had been from 1943 to 1947). In all three cases, the obligation to practice the evangelical counsels is the same. But in the secular institutes the members "remain in the 'saeculo' and in it they must sanctify themselves and exercise the apostolate" (ibid.).

Obligatory practice of the evangelical counsels and complete dedication to the apostolate, then, are requisites of every secular institute. At first glance, there seems to be *such a flagrant contradiction* between these statements by Portillo and the position Escrivá would adopt two years later that we are prompted to ask if Portillo might here be referring to the secular institutes in general, but not to Opus Dei in particular. But this interpretation runs into serious problems, because the rest of the contributions to the congress by Opus members all point in the same direction.

Thus the brief paper by Antonio Pérez (*Actas,* 399–401), echoes this same note of exigency. "The fundamental element of the theological state *perfectionis adquirendae* consists of total consecration to God; that is, through the free acceptance of a stable obligation in order to acquire the perfection of charity through the indispensable practice of the three general evangelical counsels" (ibid., 400). "There is, then, a requisite juridical obligation, even coercively, to tend to perfection through specific means: the practice of the evangelical counsels" (ibid., 401).

Likewise, Salvador Canals in his second essay (*Actas,* 414–18) writes that

> although every Christian is surely called to holiness [as the members of Opus always say that Opus has always said], nevertheless those souls who profess the state of complete perfection in the societies officially approved by the Church, pursue holiness through their own path, using means of higher nature ("celsiores naturae"). These means, which are the evangelical counsels, accepted voluntarily and made stable through a link which obliges in conscience, give rise in the persons who profess them a new obligation to tend toward perfection, which, on being sanctioned and regulated by the Church, in the external domain, obliges not only in conscience, but also juridically. Therefore, the cleric or the layperson who desires to live in a state of perfection *must join one of the societies approved by the ecclesiastical authority as "status iuridici perfectionis,"* whether it be one proper to the religious life or one proper to the secular institutes. [ibid., 415; emphasis mine]

Severino Monzó, also an Opus Dei member, concluded his paper (*Actas,* 472) by saying that "Spain reaped the honor of being the cradle where the first secular institute that obtained pontifical approval was born" (and this was none other than Opus Dei). We must conclude that although it is not mentioned expressly, all these texts in fact concur in pointing to Opus Dei—as Canals wrote in 1954 (2d ed., 1961, 82)—as the embodiment of the "perfect type of secular institute." This strikes me as the most plausible interpretation. It coincides, furthermore, with observations I gathered in two interviews during the course of my research. Both interviews were with priests who attended the congress of 1956 but who did not know each other and were interviewed at different times and in different places. According to both of them, at the 1956 congress the members of Opus Dei, far from appearing at odds with the format of the secular institutes, acted as their best representatives, and advocated very high standards before the Church could recognize any association as a secular institute.

This is perhaps what Bishop Laureano Castán had in mind in his contribution to the congress, when, without explicitly mentioning Opus Dei, he referred nevertheless to the existence of a "historical constant": initially, those in the state of perfection already officially recognized (the religious) "look with misgivings at the efforts to broaden this notion of the state, and struggle against the inclusion of those who have accepted more contact with the world so as to better exercise the apostolate" (Opus Dei, for example?). But as soon as these new forms have been admitted (with the creation of the format of the secular institutes), "there is a kind of resistance by these latter forms to accepting new broadenings and admissions." Castán observes, among other things, in those who have already obtained recognition, "a kind of superiority complex" (*Actas,* 372). This "kind of superiority complex" was precisely what our two interviewees found reproachable in the attitude of Opus Dei members during the congress.

(*b*) The third session of the congress was dedicated to "The Secular Institutes and the Organization of Perfection and Apostolate in the Church," and the papers take up more than a hundred pages of the volume (*Actas,* 419–534). Contrary to what one might expect after a hasty reading of the Padre's 1958 letter, Opus Dei members played a highly visible role: the session was moderated by Salvador Canals, and after a "lecture" by the Claretian Father Anastasio Gutiérrez three papers were presented "on the topic as a whole," all by Opus Dei members. These were followed by twelve communications, three of which were also given by Opus members.

In his lecture Father Gutiérrez, officer of the Sacred Congregation of Religious (*Actas,* 421–45), gave a concise summary of the nature of the secular institutes from the theological, juridical, and apostolic points of view, as compared with the "religious" (in the sense of the societies of religious) and the societies "of common life."

Then papers were presented by Álvaro del Portillo (*Actas,* 445–50), Encarnación Ortega (ibid., 450–53), and Patrocinio Sind (ibid., 453–56). Monsignor del Portillo asserted that the Church "has acted positively to organize and channel the desires of so many souls to live in perfection and to realize an effective apostolic labor *in* and *from* the very bosom of civil society" (ibid., 446); he thinks that with the secular institutes "a new state of perfection is recognized, different from those which have until now existed juridically" (in a footnote he quotes again Msgr. Escrivá's 1949 text), and he upholds the thesis that in the case of the secular institutes, their secular condition is what differentiates them from the religious (and from the societies of common life), while their nature of "juridical state of perfection" is what distinguishes them from the simple "associations of faithful" (ibid., 446f.).

Further on, Portillo warns against the danger of "giving juridical status as secular institutes to associations or societies that pretend to fit the framework of the type of institutes sanctioned by the *Provida Mater Ecclesia*" (*Actas,* 447). Before they can be recognized it must be verified that they conform fully to the norms fixed both by the Apostolic Constitution and the instruction *Cum Sanctissimus* of the Sacred Congregation of Religious (1948, which can be found, in Spanish, in the appendix of Canals, 1954, 189–98). According to Portillo, it is a question of preventing "possible adulteration of the nature of these institutes, insisting on the essential importance of their specific character: a state of complete consecration to God *in saeculo*" (*Actas,* 449).

The text seems to confirm, then, the defense of a restrictive or at least a strict policy regarding the admission of new secular institutes, as we suggested above.

Señorita Patrocinio Sind's communication begins: "The Secular Institutes, which constitute a true juridical state of perfection, approved and regulated by the Church, have as the primary fundamental of their spirit the strict observance of the three evangelical counsels" (*Actas,* 453). This assertion really seems very difficult to reconcile with the Founder's categorical declarations to the effect that "Opus Dei is not nor can it be considered a reality linked to the evolutionary process of the state of perfection in the Church," and that "Opus Dei is not interested in vows, or promises, or any form of consecration for its members, other than the consecration they all received with baptism," and that "our association does not pretend in any way that its members should change state, that they should cease to be simple faithful, just like everyone else, in order to acquire the peculiar *status perfectionis"* (*Conversations,* no. 20; the quotations are from an interview in 1967).

I wish to emphasize, apropos of Patrocinio Sind's remarks, a second element that has not appeared until now: the assertion that, still according to the *Provida Mater Ecclesia* (art. 3.2), the members *strictu sensu* of the secular institutes "must offer themselves totally to God, in a life consecrated to him through a complete surrender to the Institute. In this way, the Institute can *make absolute and complete use of them"* (*Actas,* 454; emphasis mine). In a footnote Sind quotes a work by Msgr. del Portillo, *Un nuevo estado jurídico de perfección: los institutos seculares* (Rome, 1952).

For Encarnación Ortega, who was one of the first members of the women's branch from 1941, when the Padre "had to take up the work again starting from scratch" (Helming, 45), the secular institutes assume the union of perfection and the apostolate "as a necessary and essential union, so intimate and complete that it penetrates and suffuses the whole life of the institutes" (*Actas,* 451). She concludes by saying that "this marvelous unity of life—and we feel joy in writing it—is not just wishful thinking. It is a splendid reality in some institutes, and it is to be desired in all of them" (ibid., 453). I do not believe it would be too bold to suggest that one institute where this "splendid reality" existed was, specifically, Opus Dei.

The rest of the papers by members of Opus Dei in this session of the Madrid congress do not contribute anything new, but neither do they break at any time with the unanimity we have observed till now.

(*c*) Let us finish this long review of the first volume of the *Acts of the National Congress of Perfection and Apostolate* with a summary of the important lecture by Amadeo de Fuenmayor which opened the ninth session on the secular apostolate. The text holds special interest for us, both for its content and because it was written by one of the authors of a basic reference work, *El itinerario jurídico del Opus Dei.*

Fuenmayor's entire lecture (*Actas,* 1199–1207) can be read as an exposition of the Opus Dei point of view, at least at that time, on the question of the apostolate. In his first sentence he defines the apostolate as "the great divine mission of restoring and reordering all things in Christ, of including all men in the order of grace and, thus, of reestablishing, in the fullness of its integrity and its vigor, the kingdom of God on earth" (ibid., 1199). He then distinguishes three major types of apostolate: the contemplative life, the active life "dedicated to remedying the world's evils from outside it" (ibid., 1199—a phrase that repeats verbatim Escrivá's words in *La Constitución Apostólica Provida Mater Ecclesia y el Opus Dei,* 1949, 16, quoted in Fuenmayor et al., 218)—and, third, the "labor of sanctifying the everyday life and activities of men" (*Actas,* 1200). This last, says Fuenmayor, is the secular apostolate; it is also, obviously, the one Opus considers typically its own.

He goes on to enumerate a series of characteristics of the apostolates proper to the secular institutes:

1. "The practice of the evangelical counsels, in order to fully exercise the apostolate, is without doubt an essential and necessary requisite" (*Actas,* 1202). According to later texts (Fuenmayor et al., 331) it appears that after 1961 the Padre would speak of "the need to eliminate all use of the expression 'evangelical counsel.'"
2. The requirement of explaining the specific and general goal of the secular institute. This requisite too would be criticized in later writings by Opus Dei authors.
3. It is an apostolate "fully lived and wholly secular," which makes the secular institutes "providential instruments of social penetration" (*Actas,* 1203). The Padre would say in 1968 that he hoped, "the day will come when the phrase 'the Catholics penetrate the social environment' will no longer be uttered, and that everyone will realize that it is a clerical expression. In any case, it does not apply at all to the apostolate of Opus Dei" (*Conversations,* no. 66).
4. "With the goal of encouraging the maximum efficacy of this apostolate . . . it is suitable and laudable that in nearly all the institutes a certain discretion be maintained with respect to the members, the works, and the houses in which the members live in common" (*Actas,* 1203); it is a discretion "which on the ascetic plane favors humility and on the apostolic plane, efficacy." Let us recall that although in 1968 Msgr. Escrivá would say that "we, from the first moment, have acted always in the light of day" (*Conversations,* no. 65), article 189 of the Constitutions of 1950 said that "the Institute, as such, wants to remain concealed," while article 191 seems to have directly inspired Fuenmayor's statement here in affirming that "collective humility" must make the members of Opus live their consecrated life with a "discretion" that is appropriate for "the efficacy of the apostolate."

Finally, after having insisted that the first of the "principal apostolic ends of the secular institutes" is an apostolate "of social, intellectual, etc., penetration into civil society, which seeks to bring the Christian spirit to all public activities, governmental or not" (*Actas,* 1204), Fuenmayor devotes a paragraph of his talk to the "dangers and difficulties of the apostolate in the world."

The first difficulty lies in the need to find a balance between outside activity, on the one hand, and prayer and mortification, on the other. Second, "another possible danger is that of an individualistic apostolate, which does not let itself be easily led and directed" (*Actas,* 1205). In fact, the spirit of initiative, "which is good when it allows itself to be informed by obedience, can be prejudicial if it leads to an apostolic labor performed with too much personal judgment." In the case of the secular institutes, however, these dangers are not so serious—the first because the life "is active and contemplative at the same time," and the second because of "the close dependence that necessarily exists, through the link of obedience, between the members and the internal superiors" (ibid.). Earlier (ibid., 1203) Fuenmayor had already said, in the same vein, that in the secular institutes "each member, closely united to his internal superiors—because he is sanctifying himself in the world through the practice of the evangelical counsels, specifically, obedience—always pursues, in his personal behavior, the general and specific goals of his institute."

It is surprising to find, in this last passage, the use of the phrase "sanctification in the world through the practice of the evangelical counsels," when we have become accustomed to hearing the phrase "sanctification through work" as the only and permanent expression proper to the "message" of Opus Dei. It is also surprising to see, in light of the Founder's oft-repeated assurance to the members of Opus Dei—"you are free, my children"—this insistence on the links of "dependence" and "obedience." Indeed Amadeo Fuenmayor himself, on pages 356 and 357 of *El itinerario* (of which he is coauthor), would include quotes from Escrivá like these (from *Conversations,* nos. 19, 27, 53, 27, and 49):

1. "We give primary and fundamental importance to the apostolic spontaneity of the person, to his freedom and responsible initiative, guided by the action of the Spirit; and not to organizational structures, mandates, tactics and plans imposed from above, from the seat of government." In 1956, by contrast, Fuenmayor was warning of the danger of "too much personal judgment" when faced with the idea of "the apostolic spontaneity of the person."
2. "The principal activity of Opus Dei consists of giving its members, and other persons who desire it, the necessary spiritual means to live like good Christians in the midst of the world." In 1956, however, we were told that "another danger is that of an individualistic apostolate, which does not let itself be easily led and directed."
3. "This is the fundamental mission of the directors of our Work: to facilitate in all members the knowledge and practice of Christian faith, so they may make it a reality in their lives, each with complete autonomy." In 1956, Fuenmayor was stressing "the dependence that necessarily exists—through the link of obedience—between the members and the internal superiors."
4. "All the behavior of the Directors of Opus Dei is based on an exquisite respect for the professional freedom of the members." Let us recall that in 1956 we were told that "each member is closely united to his internal superiors, because he is sanctifying himself in the world through the practice of the evangelical counsels, specifically obedience."

5. "Opus Dei does not give any orders to its members about how to conduct their work; it does not try to coordinate their activities." Fuenmayor asserted in 1956, however, that "each member . . . always pursues in his personal activities the general and specific goals of his Institute."

Here we reach the end of our summary of the *Acts of the National Congress* held in Madrid in 1956. To conclude this chapter we shall now return to the text of Msgr. Escrivá's 1958 letter, which we took as our point of departure (Fuenmayor et al., 563–65), and look at some of the principal questions it poses.

Some Questions

(*a*) In the first place, to begin with the less important questions, we ought to clarify the meaning of the Padre's assertion at the end of his letter (Fuenmayor et al., 565) that "we never wanted to attend the congresses or assemblies of those who say they are in a state of perfection." Could it be possible that he was referring to some other type of congress, and that in spite of its title, the "National Congress of Perfection and Apostolate" of 1956 was not a congress "of those who say they are in a state of perfection"? Since the members of Opus Dei who participated in it kept insisting on the fact that their own institution had been officially recognized as a "state of perfection," this hardly seems plausible.

Nor does it seem logical to assume that Msgr. Escrivá was using the "royal we," and that he meant us to understand that he personally did not want to attend such meetings, even though several members of Opus Dei did, and those persons in general were very important in the Work. And it seems even less likely that there was a disagreement between the Founder and some of his principal colleagues on a question that affected the very conception of Opus Dei. Therefore, rather than go on formulating additional hypotheses, which would necessarily become more and more absurd, the best thing is to let the matter drop, and wait for the day when more knowledge will make it possible to resolve this apparent discrepancy.

(*b*) Turning to more important matters, because of their bearing on questions of content, we must pause to consider two of the most forcible statements in Msgr. Escrivá's letter: "We do not desire the state of perfection for ourselves" and "nor can Opus Dei be confused with the so-called movements of apostolate" (both in Fuenmayor et al., 564). Discarding the hypothesis of any frontal disagreement between Msgr. Escrivá and the participants in the 1956 congress (according to the data in the *Anuario Pontificio* of that year, three of them occupied posts of maximum responsibility in the Work immediately under Msgr. Escrivá, who was, of course, its president general: Álvaro del Portillo, general counsel; Amadeo de Fuenmayor, one of four consultants; and Antonio Pérez, general administrator), but at the same time taking account of the obvious discrepancy between these two statements and the materials from the congress which we just finished analyzing, we find ourselves facing a new set of questions.

First of all, it seems clear that the materials from the congress are not consistent with the general thesis presented by Fuenmayor, Gómez-Iglesias, and Illanes, according to which the format of the secular institute, from the very moment of its creation, had been viewed by Opus Dei as unsatisfactory, partly unsuitable, or insufficient, even though it offered some advantages. It is necessary to recognize that this thesis has been supported by three other Opus Dei authors who have written on the topic recently. Let us confine ourselves to two brief examples, from books that are equally brief and concise: Helming states that "the Holy See granted (1950) Opus Dei the necessary approvals, although the juridical framework given to it was not what the Founder had envisioned" (Helming, 52), while Dominique Le Tourneau, in a little volume from the *Que sais-je?* series, writes that the solution "did not respond completely to the exigencies of the proper nature of Opus Dei" and that Msgr. Escrivá "accepted these partial solutions, which did not respond wholly to the character of Opus Dei, without in so doing ceding what constituted the charisma of the foundation" (Le Tourneau, 59). Le Tourneau's book was published in 1984 (2d ed., 1985), but this formulation in fact offers a good synthesis of the thesis that would later be argued—at much greater length and in much greater depth—by Fuenmayor, Gómez-Iglesias, and Illanes in 1989.

Other authors tend to date the beginnings of dissatisfaction with the format of the secular institutes somewhat later, considering it furthermore a gradual development, due less to the formula arbitrated in the apostolic constitution *Provida Mater Ecclesia* of 1947 than to its subsequent application, through which the Sacred Congregation of Religious had progressively assimilated the secular institutes to the religious, to the consequent displeasure of Opus Dei.

This interpretation presents clear advantages. It permits us to understand why, after the Decree of February 24, 1947, which bestowed on it the title Primum Institutum Saeculare, Opus Dei made haste to obtain final approval as a secular institute, an objective it in fact achieved before anyone else, with the Decree of June 16, 1950, which begins with the words "Primum inter Instituta Saecularia." Furthermore, it also helps us to fit in what Salvador Canals wrote in 1954 (*Institutos seculares y estado de perfección*): "the content of this constitution (*Provida Mater Ecclesia*) holds special interest for Spanish readers by virtue of the fact that it was specifically an institute of Spanish origin, the 'Priestly Society of the Holy Cross and Opus Dei,' that was the first to enter the new juridical molds created by the Church" (Canals, 1954, 14). Canals states further on that the fourth article of this same papal document distinguishes the secular institutes, dependents of the Sacred Congregation of Religious, from other societies which, "because they do not have a complete state of perfection," depend on another sacred congregation; with this he even seems to be defending this ascription to the Sacred Congregation of Religious, which will later be represented, on the contrary, as one of the greatest inconveniences Opus Dei had to face.

Canals says, in fact, that "concretely, we can call complete that state of perfection in which the three general counsels are professed—poverty, chastity, and obedience—which are formally known as the evangelical counsels" (Canals, 1954, 47), whereas otherwise one would have to speak of an incomplete state of perfection. Finally he describes (as we said earlier) Opus Dei as the "Spanish Institute which enjoyed the glory of being the first secular institute of pontifical law" (ibid., 81), because "it seemed clear that 'Opus Dei' embodied the perfect type of secular institute" (ibid., 82).

The problem is that while this interpretation postpones Opus Dei's dissatisfaction with the juridical framework of the secular institutes to a date after 1954, the documents of the Madrid congress oblige us to postpone it to an even later date, thus leaving as the only possible time frame for its development the two-year period between the end of the congress and the date of Msgr. Escrivá's letter.

In the previous chapter we were able to verify that none of the events presented as particularly important by the official literature seems susceptible to being interpreted in such a way. As for the rest, on only one occasion have we come across an article (by a non-Opus writer) sustaining the thesis that one reason Opus Dei may have wanted to dissociate itself from the secular institutes was that, from a juridical point of view, the activities of the members of the Work compromised the institution as such (Ortuño, 1963). Obviously this assertion would have to be understood in connection with the participation of some members of Opus Dei in certain public jobs in the Franco regime, a phenomenon that became especially visible after 1957.

Working against this thesis is the fact that it has hardly ever been argued. Moreover, Opus Dei has consistently—and I would say reasonably—insisted that the activities of the members of the Work in various areas of public life neither involve nor compromise the institution as such. In any case, the 1950 Constitutions themselves expressly envision that lay members will take on jobs in public administration and university teaching, practice professions in law and medicine, and devote themselves to business and finance (art. 15). Furthermore, a member who held one of these positions, López Rodó, reported in his memoirs that at a meeting in Rome on June 15, 1960, Msgr. Escrivá said to him, in so many words: "The presence of some members of Opus Dei in public life is a natural consequence of the growth and expansion of the Work. Today it no longer shocks anyone, everyone admits it as normal for members of Opus Dei to hold public positions, since they are citizens with the same rights as anyone else" (López Rodó, 1990, 226).

(*c*) Before abandoning this search for clues that so far do not seem to lead anywhere, perhaps we ought to pay attention to another coincidence of dates, accidental or not. Monsignor Escrivá's letter is dated October 2, 1958; this was exactly the moment when Pius XII, gravely ill, began what would be the last week of his life (he died on October 9).

We might ask if the letter, dated October 2 to commemorate the "myth of the origins" of the foundation of Opus (October 2, 1928), could have been written at this time in anticipation of the possible results of the conclave that would take place to elect Pius XII's successor. Depending on who was elected, the letter might perhaps have been used to try to speed up a possible juridical transformation. But surprise at the election of Cardinal Roncalli, and above all the accumulation of surprises during the early part of John XXIII's papacy, seems to have suggested waiting for a prudent length of time (specifically, until 1962) before taking the first step, in the form of a letter from Msgr. Escrivá to John XXIII and the secretary of state (both dated January 7, 1962; reproduced in Fuenmayor et al., 568, 569); it was a frankly timid first step, furthermore, since the tone adopted is much more moderate than that of the 1958 letter which, by the way, *is not mentioned at all.*

We must add that there is at least one text that points to some relationship between Opus Dei's juridical problem and the death of Pius XII. This is Albert Moncada's book *Historia oral del Opus Dei.* Curiously, the reference appears in a place in Moncada's book where he commits a serious error. Concerning Opus's efforts to obtain Vatican approval, Moncada states: "The end of the procedure took place during the last months of the papacy of Pius XII, and by then Escrivá and his men were quite familiar with the corridors of the Vatican and boasted of having made some clever bureaucratic moves to further their plans. Pius' final signature was obtained literally on his deathbed. It seems that the original document even contains traces of this circumstance" (Moncada, 1987, 24). The next sentence in the book begins with the words "the final approval bears the date June 16, 1950. . . ." This date is correct, but it implies an error of eight years in dating the death of Pius XII, who died in October 1958.

Even assuming that Alberto Moncada—who knew Opus Dei inside out, since he belonged to it for many years—might have heard of an important event at the moment when Pius XII lay "on his deathbed," and that later he had confused this with "the approval as secular institute" when in fact it had to do with Msgr. Escrivá's wish that "in the future this name not be applied to us," we are left only with the fragile hypothesis that the letter of October 2, 1958, was written in anticipation of what might occur in the Vatican after the death of Pius XII.

It is clear, then, that the search for important clues between 1956 and 1958 that might lead us to the reasons for the radical divergence between the position held by Opus Dei at the Madrid congress and the one adopted by Msgr. Escrivá in his letter only leads us further into the morass.

Is the problem our inability to find the key event that provoked the change? Perhaps, although it would seem logical to think that the "official" literature could hardly have omitted such an important event. Everything seems to lead, therefore, to the conclusion that to the symbolism of the letter's date—October 2—must be added the irony of its title: *Non ignoratis*. "Don't be ignorant," the Padre began his letter; and we must end by saying that we are ignorant of everything concerning the reasons why he wrote the letter.

(*d*) But suppose the problem is something else. What if the problem does not hinge on finding or failing to find a decisive event? Suppose it is rather a problem of method. In this case it would be necessary to try using the method of Chesterton and his Father Brown again, since the attempt to reconstruct the facts from the clues offered by the "official" literature does not seem to lead anywhere.

How would we apply Father Brown's method in this case? It would mean looking for something that is obvious or, rather, something that has been taken for granted all this time. That is, the letter: the letter of October 2, 1958.

The first alternative hypothesis, then, would be: could Fuenmayor and his colleagues be mistaken in dating the letter? If it was actually written later than 1958, wouldn't all the problems of interpretation vanish?

But this raises another, practically insurmountable, problem: is a mistake like this conceivable in the book by Fuenmayor and his colleagues? It is true that we just saw a similar error, an error of eight years, in Moncada's book, and in that case it was undeniably an error. . . .

The way the letter is presented in Fuenmayor's book (pp. 321ff.), there are no data that would help to certify the authenticity of the date; but neither is there anything concrete that would cast doubt on its correctness. As for the reproduction of the document (ibid., 563), it is headed by a simple reference: "RHF-EF-581002-1." RHF means "Historical Register of the Founder" (Fuenmayor et al., 17), and the figures clearly indicate the year, the month, and the day, confirming that it is from the year 1958.

It is Giancarlo Rocca's book that permits us to establish that the error, assuming that there is one, is not attributable to Fuenmayor and his colleagues but goes further back and higher up. In fact, Rocca also publishes Msgr. Escrivá's letter, but in Latin, not in Spanish. And he repeats the date: October 2, 1958. However, Rocca had to get it from another source, since he did not have access to Msgr. Escrivá's personal archives; he took it from an edition of the Statutes of the Prelature of the Holy Cross and the Work of God (*Codex iuris particularis*), published in Rome in 1983, where the letter appears as an introductory text (pp. xxiii–xxx).

The document reproduced by Rocca includes a note (absent in the *Itinerario*) right after the date, which says—in Latin— that on February 14, 1964, the Founder sent Paul VI a copy of the letter, together with a copy of the Constitutions or Statutes then in force, and that with the kind of juridical recognition the Padre longed for now (in 1983) obtained, the document is included in the current edition of the Statutes of the Prelature (Rocca, 1985, 187).

This clarification, which, we repeat, does not appear in the document reproduced in *Itinerario,* nevertheless appears in the text of the volume. Let us see what the authors have to say (Fuenmayor et al., 350): after the first audience granted by Paul VI to the Founder on January 24, 1964, in which "he was interested in the institutional problem of the Work," Msgr. Escrivá sent the Holy Father a letter, to which he appended, "complying with a desire expressed by the Pope," a copy of the Constitutions "in their 1963 edition" (which were the ones approved in 1950, with a series of modifications introduced in 1963, basically the elimination of chapters 3 and 4 in the second part). He also included "by way of introduction the letter of October 2, 1958" (ibid.), in addition to "another little volume relative to the spirit of Opus Dei, and a long note—*Appunto riservato all'Augusta Persona del Santo Padre*—in which, by way of an account of conscience, he expounds and comments on some questions that fill his spirit."

Since the documentary appendix includes the letter from Msgr. Escrivá to

Paul VI (Fuenmayor et al., 574f.), let us look at the exact words of the Founder (in Italian). After a brief preamble in which he expresses how moved he is for having been able to "avvicinare e ascoltare" the Holy Father, and after saying that this fact alone, as well as his blessing, "mi hanno ripagato dei 36 anni di servizio alla santa Madre Chiesa attraverso la mia vocazione all'Opus Dei," he adds that he is sending him a "volumetto" with Opus's Constitutions, and another on the spirit of the Work, as well as an "appunto" and a photocopy of some pages from a book by Cardinal Suenens. Curiously, however, Escrivá makes no explicit reference to the inclusion of the letter of October 2, 1958, to his "dearest daughters and sons."

Once again we find ourselves facing problems and apparent contradictions. How should we approach them using the "methodology of Father Brown"?

The initial evidence says that Msgr. Escrivá sent a letter to the members of the Work on October 2, 1958 (the anniversary of the founding of Opus Dei), that a copy of this letter was sent to Paul VI on February 14, 1964 (the anniversary of the founding of the Priestly Society of the Holy Cross), and that it was published in 1983 (together with the Statutes of the Prelature). Not to accept the initial evidence would imply the formulation of questions like the following: Did Msgr. Escrivá really write the letter *Non ignoratis* on October 2, 1958? If he wrote it, did he send it to his "daughters and sons"? Were there any conditions, such as a prohibition on using it? Could the letter have been written later than 1958, and before 1964? Was it included in the documents Msgr. Escrivá sent to Paul VI in 1964? Could it even have been written between 1964 and 1975, the year of the Founder's death? Finally, so as not to leave out any possible question—likely or unlikely—could it have been written after 1975 and not, consequently, by Msgr. Escrivá at all?

The hypothesis of its having been written after 1964 is possible, but not necessary, the only condition being that sufficient emphasis must be placed on the fact that the letter was not made public until 1983, and that before the approval of Opus as a prelature it is never cited as a source in any study.

The hypothesis that it was written on the official date of October 2, 1958, is also possible. But it seems to require a set of complementary explanations, which up to now have never been offered. There are basically two questions here: What kind of diffusion did it have, and what ability did those who received it have to make public reference to it? This hypothesis has the immense advantage of being the only one that meshes with the official assertions of Opus Dei, but as long as the apparent contradictions pointed out on these pages cannot be satisfactorily resolved, it does not seem to me that one can entirely ignore the third hypothesis, that it was written between 1958 and 1964; or, alternatively, that it was written in 1958 but remained unknown for a period of time, even to some of Msgr. Escrivá's closest colleagues.

How else can we explain the fact that in July 1960, Ediciones Rialp published a book by Salvador Canals, *Los institutos seculares,* whose contents are comprehensible only if we assume that the author did not know the text of the Padre's letter?

Canals wrote (1960, 19): "Those who confuse a secular institute with a simple association of faithful are mistaken." We read in *Itinerario,* however (p.

354 n. 162): "It must be noted that in order to avoid the use of the term 'secular institute,' Msgr. Escrivá substituted, during this period, the term 'association'; he thought it was legitimate to do so, in accordance with his assertion—which is also that of the Apostolic Constitution *Provida Mater Ecclesia*—that the secular institutes are associations of faithful."

Canals wrote (1960, 35): "Historical evolution shows us that today the state of perfection has diversified into two great branches: the religious state and the complete state of perfection professed in the secular institutes." Theoretically, Msgr. Escrivá had asserted *two years earlier* that "we do not desire the state of perfection for ourselves."

In his bibliography Canals lists the titles of two contributions, by Msgr. Escrivá and by Álvaro del Portillo, at the Congressus generalis de statibus perfectionis, held in Rome in 1950. Yet, "We never wanted," said the Padre in his letter, "to attend the congresses or assemblies of those who say they are in a state of perfection."

Canals's book is hardly ever cited in studies by members of the Work, and today it is practically impossible to locate a copy. Of the first hundred volumes of the Patmos series of Ediciones Rialp, this book is the *only* one by a Spanish author that has not been reprinted. In the 1961 *Anuario Pontificio,* for the first time Canals's name does not appear among those who worked in the Sacred Congregation of Religious, where other members of Opus Dei continued to work for many years. Beginning in 1962, Canals seems to have specialized in the Vatican organization on matters relating to cinematography.

Monsignor Julián Herranz has been bishop of the Prelature of Opus Dei since 1991. From 1984 he was secretary of the Papal Council for the Interpretation of Legislative Texts (to which Amadeo de Fuenmayor also belongs), formerly called the Papal Committee for the Authentic Interpretation of the Code of Canon Law. Before that he was an officer of the Committee for the Revision of Canon Law, for which Álvaro del Portillo was a consultant. He took his first job in the Curia in 1961, in the "section of pastoral and catechistic activity" of what was called the Sacred Congregation of the Council before the Second Vatican Council, and was basically concerned with "De disciplina cleri et populi christiani," working beside Álvaro del Portillo and under its president Cardinal Ciriaci, named in 1962 as "protettore" of Opus Dei (data extracted from the appropriate pontifical yearbooks).

In 1964, Julián Herranz published an article in the magazine *Ius Canonicum* (published by the University of Navarre; 1964, 303–33), entitled "La evolución de los institutos seculares." This article has been considered by many authors as the first public exposition of the new attitude adopted by Opus Dei. To put it, albeit very simply, in a single sentence, Herranz's thesis is that after the approval of the first secular institute (Opus Dei), the practice of the Sacred Congregation of Religious made it evolve progressively toward the typical forms of the religious; that the only one to remain faithful to the original form was Opus Dei; and that in view of the situation, "Opus Dei can no longer be considered, because in fact it is not, a secular institute" (ibid., 331).

The expression is identical to that of Msgr. Escrivá's letter ("in fact we are not a secular institute"), and it is published here for the first time.

Rocca assigns considerable importance to Herranz's article, which he summarizes and comments on in detail (Rocca, 1985, 93–97). Fuenmayor, Gómez-Iglesias, and Illanes, on the contrary, do not seem to want to accord it this degree of importance, since throughout their book they refer to it only to say that this phenomenon "of the secular institutes approaching the religious, which implied a modification in practice from what was established in papal documents," "was strongly emphasized, although alluding to events that dated from earlier years, by J. Herranz" (Fuenmayor et al., 313, 313 n. 40). Monsignor Escrivá, too, when questioned in 1967 about Herranz's article, answered only that "Dr. Herranz expresses, under his own personal responsibility, a well-documented thesis; on the conclusions of that work, I prefer not to speak" (*Conversations,* no. 25). According to Herranz, Msgr. Escrivá had raised "respectful protests" (Herranz, 1964, 331) about this evolution of the secular institutes, dating back to 1948; the documentary appendix of *Itinerario* does not include any of these supposed protests: the first one is from 1960 (Fuenmayor et al., 566).

In any case, it seems difficult not to draw a relationship between what today is presented as a letter by the Padre from 1958 and Herranz's conclusion: Opus Dei is not, de facto, a secular institute. But Herranz goes a little further—perhaps too far—when he finishes the thought as follows: ". . . and in reality by now no one thinks that it is" (Herranz, 1964, 332). The corresponding footnote bases this assertion on an article that appeared in a León magazine in the same year, 1964, and on the entries under "Opus Dei" in an *Encyclopedia of Spanish Culture,* a *Dictionary of Spanish Ecclesiastical History,* and the *New Catholic Encyclopedia.* Herranz, going against his custom in the rest of his footnotes, does not give the year of the edition of these encyclopedias, but we can surmise that none of them can be much before 1964. On the other hand, Herranz seems to forget that the official Vatican documents still "considered that it was one" in 1964, and that they continued to consider it one until the 1982 edition of *Anuario Pontificio,* eighteen years later.

Regardless, then, of the exact date of the letter which is officially given as 1958, 1964 seems to have been the year of the public espousal of the rejection of the format of the secular institutes, following an earlier, timid, and frustrated attempt before John XXIII in 1962 (Fuenmayor et al., 568–72). This leads one to think that possibly the old book—much criticized by Opus—by Daniel Artigues (pp. 134–37) was right in attributing the origin of this whole problem to the "crisis" Opus went through in 1964, a crisis whose nature has not been clarified to the present day. We shall have to discuss the question of this crisis at least briefly in the next chapter, but not before concluding that we are still ignorant of many things about a letter ironically titled *Non ignoratis.*

Perhaps this maxim of Escrivá's is applicable in this case: "It is worth the trouble to clarify inconsistencies, and to clarify them with as much patience as necessary" (*Surco,* no. 574). To my thinking, yes, it is. Furthermore, to my thinking, as long as those patient efforts at clarification are not made, we ought to apply another of Msgr. Escrivá's maxims to the case of this letter of October 2, 1958: "You tell a half truth, with so many possible interpretations, that it can be considered as . . . a lie" (ibid., no. 602).

11

The Consolidation of Opus Dei

A New Epoch

Even assuming it had merely symbolic value, the truth is that the date of Msgr. Escrivá's letter, which we dealt with in the last chapter, could not have been more shrewdly chosen, at least for our particular purposes. In fact, the year 1958 marks the end of the first period analyzed up to this point and signals the beginning of the next period, where we will be able to focus our attention on the specific themes of the second part of this study.

The month of October 1958 effectively satisfies all the requirements of such a dividing line. In the history of the Catholic Church it signifies the end of an entire epoch and the beginning of a new phase on may fronts. It was the month that saw the death of Pius XII (on October 9) and the election of John XXIII (on October 28). The announcement of the convocation of the Second Vatican Council, coming less than three months later, signified an upheaval with extremely important consequences in the bosom of Catholicism, both *ad intra* and *ad extra*. In this sense, the change of pope symbolized a transformation that went far beyond the mere (albeit significant) differences in appearance and personal style between the two radically different figures of Pius XII and John XXIII. The transformation affected important aspects of the Church's own definition of itself, especially as regards the ways in which it exercised authority and the type of relationship existing between the laity and the clergy and between the clergy and the hierarchy. The transformation also affected relations between Catholicism and other religious groups, both Christian and non-Christian. And, finally, it supported the adoption of a new model of the "presence of the Church in the world," to put it in the language of the epoch, or, in the more usual sociological terminology, a new model of insertion of the ecclesiastical institution into global society.

This new model of the insertion of the Church in society had serious consequences for Franco's Spain, specifically because the legitimation of Franco's political regime had been supported, in good part, by another, different model of insertion, from which the Catholic Church itself was now, in a way, withdrawing its authorization. As an officially Catholic state, the Spanish state was able to justify the absence of any kind of "pluralism" in the name of the need to control the pernicious effects of the "process of secularization," which in other countries—according to the official line—was contributing decisively to the inherent decadence of the liberal democracies. Let us recall again the formulation of Rafael Calvo Serer, a member of Opus Dei, which appeared in a 1957 newspaper article (*ABC*, May 23): "Freedom of conscience leads to the loss of faith, freedom of expression to demagogy, mental confusion, and pornography, and freedom of association to anarchism and totalitarianism."

From the moment the Catholic Chruch officially accepted "pluralism"—and gave its blessing to freedom of conscience, freedom of expression, and freedom of association—Franco's regime was suddenly delegitimized and progressively began to look like what it really was: a genuine anachronism in the western Europe of the second half of the twentieth century. (This is why, of all the documents of the Second Vatican Council, the one relating to "religious freedom" provoked the greatest uproar on the political level and met with the greatest resistance among the Spanish bishops closest to the regime.)

In this sense, then, the month of October 1958 symbolizes a crucial moment of change that had enormous repercussions, a change that generated uneasiness and reticence in Francoist Spain as well as in the conservative—by definition—bureaucracy of the Church.

Even though with the start of its international expansion Opus Dei was beginning to have many arms (it wasn't long before jokes about "Octopus Dei" began to circulate), its two basic supports continued to be found in Franco's Madrid and the Vatican Curia. This is why the whole set of changes resulting from the election of John XXIII caught them so completely by surprise, so much so that the whole history of Opus during the sixties and seventies can be read as a history of the tension between resistance and adaptation to those changes.

But this is not the end of the symbolic significance of October 1958. In fact, this was precisely the moment when the business school established in Barcelona by Opus Dei, but linked to Opus's University of Navarre, began its operations. Simultaneously the Society of Jesus was starting a similar school, also in Barcelona. The two institutions, both of which became quite important, are today located only a few hundred yards from one another. Finally, in the same month—October 1958—the School of Public Administration, created by Opus member Laureano López Rodó, opened its doors in the outskirts of Madrid.

Thus it is also possible to designate October 1958, even if only symbolically, as the date of the institutionalization of two conceptually distinct processes—economic development and modernization (Berger, 1986)—which in Spain of the seventies would be concomitant and strictly parallel. The business schools symbolized the incorporation of the Spanish economy into the modern capitalist system, while the school of public administration played a role of the first magnitude in the modernization of the state apparatus through the training of technocrats, fostered by the team responsible for the Plans for Economic Development.

In light of this whole set of factors, in all of which Opus Dei was implicated in one way or another, the fact that the month of October 1958 began with Escrivá's letter *Non ignoratis* and the manifestation of his wish to detach himself from the juridical format of the secular institutes and the forms of the so-called state of perfection has unquestionable symbolic value.

Written in commemoration of the thirty years since the no less symbolic date of Opus Dei's foundation, the document we analyzed in the preceding chapter symbolizes the position taken by the movement in the face of all the changes confronting it. In this sense one must say of the letter of October 2, 1958, and above all of its date, *se non è vero è ben trovato:* welcome even if not true.

From this moment, therefore, we could abandon the puzzle of Opus Dei and concentrate exclusively on the relationship between its economic ethic and its professional asceticism, on one side, and modernization and economic development, on the other.

At the same time, however, we do not wish to suddenly interrupt our summary of the historical evolution of Opus Dei from this point until the present day. We will take up the thread of the different aspects of the life and actions of Opus where we left off earlier, and look at the main events that took place between then and now.

The evolution of Opus Dei throughout all these years is characterized by the progressive widening of its base of recruitment, the consolidation and extension of its international implantation, and the growing diversification of the activities it directly or indirectly sponsors. As a consequence of its growth and diversification, there was naturally an increase in the separation between "the simple members," who grew ever more numerous, and "the leaders," who became more and more specialized in the tasks of direction. This separation did not necessarily lead to internal conflicts, such as are to some extent inevitable in any large organization, but which Opus Dei has been largely able to avoid (except perhaps at particularly critical junctures, such as the late 1960s, when a number of young priests left Opus over disagreement with its "political commitments" and "nine important officials asked to be relieved of their duties" (Hermet, 1981, 2:470). But it is a fact that, while among the "group of friends" of the forties and most of the fifties a fairly close reciprocal knowledge and cooperation was possible, in the complex international organization of today's Opus Dei such relations are necessarily much weaker. Two consequences of this fact must be kept firmly in mind.

First, it explains Opus Dei's ideological evolution: the "specific objective" of the Institute, as defined in the statutory texts, became increasingly inconvenient as the group became too numerous and too diverse to be able to feel fully identified with the label of "intellectual" and the "directing part of society." This led to an increasing insistence on "the sanctification of work" (without further qualification) as the only common denominator of all members of the Work. It also led to more and more insistence on "freedom" of action for every individual member, another area where Opus had to face the fact that its members were too diverse and too numerous to have homogeneous standards imposed on them.

The second consequence was the emergence of "experts" in every one of Opus's areas of action, among whom there was often little communication. Between an Opus Dei priest in Peru, a regular member working in a university bookstore in Pamplona, the director of a residence or a club in Chicago or an African country, and an economist working in the training of entrepreneurs in Manila, relations were, as we said, much weaker than in the prior period of the organization's history.

As a result of all this the separation we mentioned took place, and this meant that most members of the Work were not necessarily kept up to date on the actions of their leaders, especially those in the Roman "cupola." This is underscored by the fact that the "official" literature is concerned with publicizing certain kinds of activities while letting others remain discreetly in the shadows.

Expansion and Diversification

Let us take a brief look at the ongoing progress of Opus Dei's internationalization from 1958 on. We will focus both on its expansion and on the growing diversification of its activities, and we will distinguish three main types or models of implantation: its most recent ventures in non-European countries (without any pretense at being exhaustive, we will choose four of these as examples); in South America, which presents some clearly different characteristics; and in Europe (except Spain and the special case of Rome and the Vatican, which we will treat later in separate sections).

Recent Implantation in Four Countries: Australia, the Philippines, Japan, and Kenya

Australia. Opus has been present since 1963; the first members were Spanish and North American. In 1965, the first women's group was formed by three Spanish women (auxiliary members).

ACTIVITIES. Inauguration of the first cultural center in 1965, with the objective of preparing students to enter the university. Inauguration in 1971 of Warrane College, a residence for two hundred university students near Sydney, which is also a teaching center. Later a women's residence was opened, a center for training in domestic labor, and, for the young men, two centers for study and a club for students. There was also a girls' school and a boys' school, promoted by Opus members, but they do not count as official Opus centers (Sastre, 495ff.; West, 161ff.).

CONFLICTS. Three years after Warrane College was inaugurated, the university opened an official investigation after a series of student protests: because women were forbidden to enter the rooms, according to West; because of censorship of newspapers and television; and because of pressures put on students to join Opus, according to Walsh (Walsh, 72). There is no reference to any conflict in Sastre's book.

STATISTICS. In 1987 there were three hundred members including nine priests. The members "practice the most varied professions, some scientific, academic, and highly intellectual, others manual" (West, 173).

Philippines. Opus has been present from 1964; the first women arrived in 1965.

ACTIVITIES. A cultural center, a center for lectures, a school for professional training for future maids and waitresses; and most important, the Center of Research and Communication, founded in 1967, which is an institute for economic studies and business training that has become practically "a specialized University for Economics, Business Sciences, Pedagogy, and Humanities" (West, 151). Subordinate to this institution are two centers for training in the trades (mechanics, electricity, and electronics), one of which is subsidized by a German foundation closely linked to Opus Dei (Hertel, 65f.).

Japan. In 1958 one of the first disciples of Escrivá was sent to Japan; soon he was followed by two other Spaniards, also priests. The first women arrived in 1960.

ACTIVITIES. The most important work is a language institute, with places for about fifteen hundred students, and a residence, in Ashiya. In Nagasaki they opened a school with two sections (male and female). Intended for university students and professionals, the language school "brings the Japanese in contact with Western languages and civilizations," while at the same time it is a "center for evangelization among the persons who come to study languages" (Sastre, 466f.). In this case too, then, priority is given to the "apostolate of the intelligentsia" because, as Msgr. Escrivá said, "men, like fish, have to be caught by the head" (Vázquez, 354).

Kenya. Identical pattern as in the previous case: initially the Padre sent one of his first disciples (1958), Pedro Casciaro, a Spanish priest. After the preliminary arrangements were made, the first stable group arrived to take charge of the activities; later came a small group of women.

ACTIVITIES. Creation of a university college and residence, later expanded to include a business school (1966), a secondary school (1978), and a primary school (1987). For the girls, in Nairobi, there was a secretarial school, a residence, a college, and in another building a women's preparatory school. Here too the pattern was the same: training of leaders and evangelization. In his book West mentions his interviews with former pupils at the university college: an engineer who works for the government and is an Opus associate; a professor of accounting in the college itself; an oil company director; a professor of physics and mathematics. Among the women, a jurist, member of the Supreme Court, converted to Catholicism and became a member of Opus; a social worker who worked in another field "because she could not accept the contraception policy of the government" (West, 71); and a married couple with six children, one of them named *Josemaría* (ibid., 72).

CONFLICTS. Initially in Kenya there was an interesting conflict when the local ecclesiastical authorities decided the Opus group was a "religious missionary group," and therefore the government subsidy for compensation of their teaching personnel was cut in half. In 1960 Opus protested to the corresponding Sacred Congregation that the "members of the *real* secular institutes should not be treated as if they were religious" (Fuenmayor et al., 566f.; emphasis in the original). The document is extremely interesting, partly because it is one of the few concrete proofs we have of the confusion between secular institutes and religious orders, which the directors of Opus Dei complained about so much; but also partly because *a year and a half after* the letter in which Msgr. Escrivá said "we are not a secular institute, nor can that name be applied to us," the general counsel presented his institute as a "*real* secular institute."

These four countries give a fairly good idea of the first model of Opus Dei implantation by means of the creation of corporative works of a basically educational nature.

It should be added that the "official" literature is enormously selective, in the sense that it provides information only about certain countries, explaining absolutely nothing about others. For example, the authors all talk about Kenya in some detail (Sastre, 460–63; West, 57–74), while all they say about another African country, Nigeria, is that Opus Dei is present there, without offering any more information. Likewise, they talk about the United States, albeit vaguely (Sastre, 388–92; West, 125–42), but they merely mention the presence of Opus Dei in Canada, without any further explanation.

Latin America

(*a*) The official literature gives a picture of the implantation in Latin America that seems to reflect a very different kind of model from the one we have just discussed. In addition to the usual residences and colleges in all the countries, other kinds of initiatives are undertaken in Latin America.

In Mexico, for example (West, 111–23), the Montefalco complex (in the state of Morelos) comprises a center for peasant promotion, a retreat house, two colleges, a school of agriculture, and a women's school of "domestic sciences." In addition to several other centers, Mexico City also has an ophthalmological clinic linked to the Panamerican University, a university "founded by members and friends of Opus Dei" (West, 121).

In Peru Opus created, in addition to a rural institute (with its own radio station), the University of Piura in the north of the country, which offers courses in engineering, business administration, and journalism (Sastre, 450). Another University is that of La Sabana, in Colombia (Seco, 165).

(*b*) In various Latin American countries Opus Dei has been accused of involving itself directly in politics, as in Spain. And as in Spain, it is officially said that Opus Dei as such never participates in politics, while its members enjoy complete freedom of action in this field. Harsh accusations—which no one has been able to substantiate—have been made in the case of Chile: links with the CIA and with Pinochet's dictatorship have been alleged. Two German authors drew connections between Opus Dei and the traffic in arms and the "death squads" and the so-called Argentinian Triple A (Roth-Ender, 1984). These authors were sued by Opus Dei. The judge in the case ruled in Opus's favor, saying that the authors lacked proof of their allegations. Similar accusations were made against Opus Dei in Colombia.

I believe that it would not be incorrect or excessive to state that in various countries of Latin America some members of Opus Dei have made political commitments—as they have in Spain but tend *not* to do in most European countries—and, furthermore, that Opus Dei as such has clearly been a belligerent in the conflict that has been going on for years between the Roman authorities and the representatives of the so-called theology of liberation.

(*c*) A third characteristic feature of Opus Dei's implantation in Latin America is its institutional presence in assuming pastoral responsibilities. In 1957 Pius XII granted Opus Dei a "territorial prelature" in Peru (this means a territory which, without juridically constituting a diocese, has its own clergy, with a "prelate" who exercises the functions of a bishop). By the end of 1988, according to data in the *Anuario Pontificio,* there was a total of eleven bishops who were members of Opus Dei, all in Latin America. Five were in Peru: one a Basque, Ignacio María de Orbegozo, who had in 1957 been the first prelate of Yauyos, was currently bishop of Chiclayo. Luis Sánchez Moreno, a Peruvian, had been a bishop since 1961 and was presently in the prelature of Yauyos. (These two were the only members of Opus who participated in the sessions of the Second Vatican Council as "council fathers," something the members of the Roman "cupola" doubtless found hard to swallow, especially Msgr. Escrivá; see Moncada, 1987, 29). The other three are a Catalan, Enric Pèlach, bishop of Abancay since 1968, and two Peruvians, Juan Antonio Ugarte, assistant bishop of Cuzco, and Juan Luis Cipriani, assistant bishop of Ayacucho.

In Peru, then, we find nearly half the Opus Dei bishops. The other six were, in 1988, in six different countries: Francisco de Guruceaga, bishop of Laguaira (Venezuela); Juan Ignacio Larrea, a native of Argentina, bishop of Guayaquil (Ecuador); Hugo Puccini, bishop of Santa Marta (Colombia); Fernando Sáenz, a Spanish native, assistant bishop of Santa Ana (El Salvador); Alfonso Delgado, bishop of Santo Tomé (Argentina); and Adolfo Rodríguez, a Spanish native, bishop of Los Angeles (Chile).

This Spanish-American monopoly was broken for the first time in 1989, when a European Opus Dei priest was consecrated bishop after a conflict that lasted more than two years (Hertel, 184ff.): Klaus Küng, bishop of Feldkirch (Austria). Later another bishop was consecrated in Brazil. The prelate of Opus, Álvaro del Portillo, and Julián Herranz, another resident of Rome with an important post in the Curia, also became bishops.

(*d*) Finally, one last feature also differentiates the presence of Opus Dei in Latin America from its presence in other countries (except Spain): the huge mass gatherings surrounding Msgr. Escrivá's visits. Such demonstrations never occurred in other European countries visited by the Padre, but they were characteristic of his three trips to America (Mexico in 1970; Brazil, Argentina, Chile, Peru, Ecuador, and Venezuela in 1974; Venezuela and Guatemala in 1975). Aside from the obvious fact that he spoke the language in these countries, these responses are probably a reflection of the type of implantation we have just examined. Opus Dei has a more visible presence in the life of these countries, and consequently there is a higher objective likelihood for thousands to congregate. Although Opus Dei began in Spain in the forties as a work "not of the multitudes, but of the select," and while in most countries it continues to be that today, in Latin America it was gradually transformed into a work both of "selection" *and* of the multitudes.

Europe

This distinction between selection and multitudes is probably the best clue to an understanding of the diversity of the initiatives promoted by Opus Dei in different European countries. Residences, cultural centers, and youth recreation centers are always the common denominator, and the favorite places from which Opus spreads its influence and makes its selection of candidates for membership. As Luis Ignacio Seco, an Opus member, recognizes, "I have not found any exhaustive publication of these works, which seems natural to me, given the scant enthusiasm of Opus Dei members for statistics and growth charts" (Seco, 151).

The next level in the expansion process is the teaching centers proper: primary and secondary schools, trade and agricultural schools, and women's schools. Then comes the next level, activities related to the communications media: presses, bookstores, magazines, and the like.

(*a*) The implantation of Opus Dei in Great Britain and Ireland is briefly documented thanks to West (pp. 97–109), Seco (p. 152f.), and, from an "unofficial" point of view, Walsh (pp. 159–65). One outstanding item not explained by the official literature is a document published in 1981 by Cardinal Basil Hume, bishop of the diocese of Westminster. After having "carefully studied certain public critiques of the activities of Opus Dei in Great Britain, as well as correspondence I have received on the topic," including both critical and laudatory letters, and after having "held meetings with the official representatives of Opus Dei," the cardinal decided to make public some recommendations that had previously been presented to those representatives. The underlying basis for them, according to the English Benedictine bishop, is that any international movement must respect the traditional character and normal manner of acting of the specific societies in which they are working. The four recommendations are as follows: that no vows or promises be demanded of minors; that parents and teachers know about the youths' relations with Opus Dei; that individual freedom of association and freedom to leave the organization without coercion be guaranteed, as well as freedom to choose a spiritual director without his necessarily having to be a member of Opus Dei; and finally, that in all its initiatives and activities, Opus Dei must clearly indicate who is directing them and who is sponsoring them (the full text of the document may be read in Rocca, 1985, 205f.).

Minors, coercion, and secrecy: Cardinal Hume's document is especially interesting because it allows one to specify fairly exactly the kind of accusations to which Opus Dei has been subject. In the case of Great Britain specifically, given "the obstinacy of the English on personal questions" and their resistance "to any conversation touching on the private domain of each person, especially the topic of religion," Msgr. Escrivá had instructed his British children: "you have to go into the lives of others as Jesus Christ came into mine, without asking permission" (in Berglar, 282).

(*b*) One locus of implantation that has become very important recently, and to which the "official" literature hardly alludes, is the German countries, specifically, the north of Switzerland, Austria, and the southwest part of Germany. The reason for interest in this area, like the reason for this very discreet silence, might be twofold: in the first place, the until recently strategic situation of Vienna as "a frontier city between the two Europes" (Sastre, 442), which made Msgr. Escrivá exclaim, "my Austrian children, it will be you who will give a good hard push, from your land, to the work in all of Eastern Europe."

Undoubtedly this "labor" has already begun, and it is not impossible that it will become in the next few years one of the primary objectives of Opus Dei. It might even give rise to a new episode in its historical confrontation with the Society of Jesus. In an interview published in 1990, the French Jesuit Pierre de Charentenay, after mentioning the Society's vocation for "the training of elites," which translates into a "will to be present where important questions can be decided" (Charentenay, 71), comments that in 1990 there was not one Jesuit in Russia, but that a team might go there soon; that in the Baltic countries there were already a few Jesuits, who "doubtless will be receiving reinforcements," and that the same was going to happen in Czechoslovakia and Hungary (ibid., 72).

Second, it must be said that the implantation in this Germanic zone was accomplished, at least in part, through a penetration of the diocesan ecclesiastical structures of the practically contiguous bishoprics of Feldkirch (Vorarlberg, Austria), Augsburg (Germany), and Chur (the Swiss Canton of Grisons and Liechtenstein). The author who offers the most abundant documentation on this area tells about the formation of an important financial center linked to Opus Dei, which operates through two large foundations (Hertel, 168–87).

(*c*) In the case of Italy—leaving the Vatican aside for the moment—in addition to the usual initiatives in such places, special mention should be made of the ELIS center (*Educazione, lavoro, instruzione, sport*), inaugurated in 1965, if only because all the Opus Dei authors speak expressly and at length about it. Built on the outskirts of Rome, the center includes a trade school, a residence, and a boys' sports club, as well as parallel facilities for girls. What is original about ELIS is not so much the nature of its activities, which are similar to so many others, but two events that are symbolically significant for Opus Dei.

The first of these has to do with the way it began. On the occasion of Pius XII's eightieth birthday, a worldwide collection was taken, and the sum was offered to him as a birthday gift. The pope died without having used the funds, and through the grapevine the directors of Opus Dei found out that John XXIII wanted to designate the money for a specific project. After drawing up and presenting a highly detailed plan, the directors of Opus were awarded these funds for the creation of the ELIS center. (One ecclesiastical dignitary has mentioned on several occasions that one day when he was received by John XXIII, the pope exclaimed to him, "Those people from Opus just left a minute ago; the whole time they were talking about money, so much that my head is still spinning.") The second event, no less significant for Opus Dei, is that the official act of inaugurating the center was presided over personally by Paul VI. Not one of the authors of the "official" literature omits this detail, probably because it gives them a rare opportunity to quote some words—obviously of praise in this case—from Paul VI about Opus Dei.

Another event that also took place in Italy but is much less publicized by the "official literature" was an inquiry ordered in 1986 by the minister of the interior, Scalfaro, after a parliamentary interpellation on the topic—once again—of secrecy and the supposedly "secret society" nature of Opus Dei. The matter was broadly covered in the press, and with greater or lesser doses of sensationalism, connections were drawn between controversial and unresolved questions such as that of Bishop Marcinkus and the Vatican financial investments, Calvi and the Ambrosian Bank, Ruiz-Mateos and Rumasa, Gelli and the Masonic Lodge P2, and Opus's relations with foundations (see, for example, Walsh, 167ff., and especially, from Italy itself, Magister, 1986). One of the rare Opus authors who mentions the topic (West, 24f.) concludes that the whole thing was in the interests of a few people who wanted to foment a scandal and smear Opus Dei, but that the investigation ordered by Minister Scalfaro clearly showed that "Opus Dei was not a secret society." However, he chooses to remain silent on one fact that Giancarlo Rocca, on the other hand, presents: the Italian minister based his argument on a memorandum of the Holy See to the effect that "anyone who belongs to Opus Dei (priests and laypersons) is obligated to shun secrecy and the clandestine," and that "they have the duty to reveal their affiliation whenever they are legitimately asked to do so" (Rocca, 1989, 389). The same author comments that this assumes an important change, not just with respect to the former Constitutions of 1950, but also with respect to the Statutes of the Prelature in force today, which specify that only "on the request of the bishop are the names of the priests and the Directors of the Centers of the corresponding diocese given out" (*Statutes,* 1982, art. 89.2).

Spain

For obvious reasons, the Spanish case has always been, and is today, a special case in the history of Opus Dei. All the different characteristics we have just observed in Europe, America, Asia, and so on, relating to the expansion and diversification of its activities in the last few decades, can be found in the case of Spain. There is only *one exception* to this: in spite of the fact that some Spanish members of Opus are bishops, there has never been an Opus bishop in a

Spanish diocese. On the other hand, even without going so far as the Australian William West—"They say that in some Spanish cities there is an Opus Dei center on every corner" (West, 95)—we can agree with this author that in Spain Opus "like the evangelical grain of mustard, has grown into a leafy tree where *all kinds of birds* make their nests" (ibid., 96; emphasis mine).

(*a*) In accordance with the official data in the *Anuario Pontificio,* at the beginning of the nineties the Prelature of Opus Dei had 1,400 priests, nearly 350 seminarians, and 75,000 lay members in the whole world. (The *Anuario* began publishing these statistics in 1987; in the five-year period between 1987 and 1991, the total growth of the number of Opus priests was 9.5 percent, and of the number of lay members, 0.5 percent.) These are the only official figures available to the public, but it is believed that four out of every ten members are Spanish. Furthermore, if we keep in mind that the regular members represent approximately 20 percent of the total (Hertel, 158), it could well be that the percentage of Spaniards among them is even higher. But in this area it seems as if the old 1950 Constitutions are still in force: they stated that "to outsiders, the number of members is concealed" (art. 190). Our hypothesis is that the percentage of Spanish members is higher among the regular members than in the membership at large, and even higher among the members who are priests. In other words, the closer we came to the "cupola," the more obvious it becomes that Opus is still, even today, a fundamentally Spanish movement, or at least one that is fundamentally directed and controlled by Spaniards.

The meager official information that can be extracted from the *Anuario Pontificio* at least suggests the plausibility of this hypothesis: not only is the prelate Spanish, but so is the vicar general of the Prelature. In the Roman Atheneum of the Holy Cross (an institute of higher studies), the rector, the vice-rector, the two deans (Philosophy and Theology), and the director of the library are all Spanish; the secretary and the administrator are Italian (*Anuario Pontificio,* 1991, 1612). Among those who carry some honorific title (honorary prelate, etc.), there is a clear majority of Spaniards (ibid., 1909–2276). Spaniards also clearly predominate among the Opus members who hold some post in the Vatican Curia.

(*b*) Among the "corporate" works or works acknowledged as their own, and those promoted by the free initiative of members or friends of Opus Dei, who no less freely entrust their spiritual direction to the Prelature, we find all kinds of teaching institutions in Spain: primary schools, secondary schools, trade schools, special education schools, agricultural schools (of these last, West says there are "about 36"; West, 82). To these we must add residences, cultural centers, language schools, domestic training schools for women, retreat houses, clubs, recreation centers, and athletic schools which "together with the training and enjoyment of sports offer spiritual and cultural education" (ibid., 83). There are several publishing houses, a chain of bookstores, magazines (in spite of the fact that in the name of "collective humility" "Opus Dei does not publish under the name of the Work," according to article 89.3 of the 1982 Statutes), and even a few stores where "morally acceptable" videotapes can be rented.

All this—and it is well not to forget it—applies to an institution which at the moment of its papal approval (1950), without excluding the creation of "its own schools," asserted that "it would prefer, to the extent possible, to lend its anonymous collaboration in public schools" (Fuenmayor et al., 548).

(*c*) Last but not least, of course, on this list of initiatives, we must add the University of Navarre, which began operations in 1952 as a center subordinate to the civil university of Zaragoza with the name *Estudio General,* but which starting in 1960 "added to its first name the prestigious title of Catholic University with which the Church had honored it" (Fontán, 1961, 148). The accreditation which made a degree from Opus's university equivalent to one from any Spanish university—and was attained after a laborious process ending with the signing of a covenant between the Holy See and the Spanish government—was from the outset a source of debate and protest in the academic world. But the University of Navarre had been gradually consolidating itself, and some of its centers (medicine and journalism, for example) had come to enjoy a very good reputation. The cost of tuition is certainly higher than in the state universities: five or six times higher, for example, in the specific case of studies in the Faculty of Economics for the school year 1989–1990. But this fact has to be seen in relation to the system of the distribution of scholarships. Furthermore, the design of the curriculum is strictly parallel to that of all the other universities, the only exception being the inclusion of an annual supplementary course in theology. Again taking the example of the Faculty of Economics, a third-year student in the business management section would have to register (in 1989–1990) for Economic Theory II, Statistics, Public Finance and Fiscal Systems, Business Organization and Administration, Cost Accounting, and Theology II.

The argument has frequently been made that students at the University of Navarre are not authorized to freely consult all kinds of books, which could pose notable difficulties in the case of studies in the humanities and social sciences, for example. This seems to be a kind of perpetuation of the now extinct "Index of Prohibited Books" formerly in force in the Catholic Church. The argument is given the silent treatment but not denied by the "official" literature; it is indirectly confirmed, however, by a manual of sociology published by Ediciones Rialp, which includes a bibliography describing twenty selected titles, with observations like the following: "In its exposition it includes theories not in accordance with the Christian view of man," for which reason the work in question is recommended to "specialized readers with solid doctrinal standards"; and "a work based on a conception which argues for the return of man to the values of the spirit; nevertheless, in some paragraphs there are ambiguous statements" (Carreño, 333, 334).

One of the most spectacular conflicts over the University of Navarre took place in 1968, arising from a debate in the Spanish Cortes (which replaced Parliament during the Franco era) over the Budget Law. Fernando Suárez, a Parliament member, denounced the fact that the state gave a subsidy of a million pesetas to the Pontifical University of Salamanca, two million to the Jesuits' University of Deusto, and one hundred million to Opus Dei's university in Navarre. López Rodó has recently recounted that a few hours later the situation was explained to Franco (López Rodó, 1991, 288). According to him, the figure was only thirty million; he says nothing, however about the inequality with respect to the other two nonstate universities. In any case, he confirms the existence of the state subsidy, although less than six months earlier Msgr. Escrivá de Balaguer had asserted that "the Spanish government gives no help for the operating expenses of the University of Navarre" (*Conversations,* no. 83). Earlier in the same year (1967), in an interview with *Time Magazine,* Escrivá had declared that "in few places have we encountered less financial help than in Spain," and that "the governments of countries where the majority of the citizens are not Catholics have been much more generous than the Spanish state in helping finance the teaching activities and good works promoted by members of the Work" (*Conversations,* no. 33).

(*d*) Although Msgr. Escrivá lived in Rome, he traveled frequently to Spain. And as in Latin America—perhaps even more so—some of these trips were prepared by his "children" in such a way that they became occasions for mass assemblies and acts of homage to the Opus's founder.

The establishment of the University of Navarre in 1960 was in fact the first occasion when the Padre found himself "with the real masses" (Fuenmayor et al., 307). On a ten-day journey during the month of October, he visited Madrid, Zaragoza, and Pamplona. In Madrid "the nave of the church [where he celebrated mass] was packed with members of Opus Dei" (Sastre, 473). In Zaragoza he was named doctor *honoris causa:* "The main hall was filled to the rafters with people" (ibid., 474). In Pamplona, the grand chancellor of the university (later he would also be chancellor of the university of Piura in Peru) was named "adopted son" of the city: "The crowd gave him a warm ovation" (Gondrand, 222). Also in 1960 he was decorated with the Grand Cross of Charles III (Vázquez, 550).

During the summer of 1964 he was named adopted son of Barcelona (Vázquez, 356). But although around the same time Msgr. Escrivá was in the Basque Country (López Rodó, 1990, 473), and although in November he went to Pamplona again (Sastre, 488), the official act of awarding the title was not celebrated until the fall of 1966. It was an "intimate act" that time (López Rodó, 1991, 86), which represented "something more than a gesture of thanks on the part of the city," but which Msgr. Escrivá did not want to be interpreted "as a retaliation" twenty-five years after a conflict (see Chapter 7) which in Barcelona had had particularly serious effects (ibid., 88).

In 1967, again at his university, "he spoke on campus before a crowd of more than twenty thousand people" (Sastre, 503). The text of his sermon, quoted in its entirety in *Conversations* (pp. 233–47), is like a programmatic document of this new stage in the history of Opus Dei after the Second Vatican Council. Escrivá spoke in his homily of the "sanctification of ordinary life," of a "lay mentality" opposed to an unacceptable "official Catholicism," of freedom in contrast to "intolerance and fanaticism," and of the rejection of the "mentality of the select." In the same year that Msgr. Escrivá made these affirmations, the *Anuario Pontificio* (1967, 905) was still presenting Opus Dei as a secular institute whose objective was "to promote among all classes of civil society, and especially among the intellectuals, the search for Christian perfection in the midst of the world."

In his homily the Padre added that "also the works which Opus Dei—as an association—sponsors have these eminently secular characteristics: they are not ecclesiastical works" (*Conversations,* no. 119). In the middle of the campus of the University of the Church, so spoke the man who had been named grand chancellor of that institution by a decree of the Sacred Congregation of Seminaries and Universities (Fuenmayor et al., 307).

(*e*) In 1972 the Padre made his most spectacular visit to the Iberian Peninsula, a journey presented as a "heroic decision" by Sastre (p. 528) and as an "apostolic run" by Vázquez de Prada (p. 387). It was a journey of two months' duration ("two months of catechism"; Gondrand, 264), justified by the "crisis which God is permitting in the Church" (Sastre, 529), a Church which "was so sick inside" that "the Padre felt that he needed to fight" (Vázquez, 387).

Thus, given the post–Vatican II crisis of the last years of the papacy of Paul VI, "he fought." In Pamplona, in "his" university, he asserted that "all the apostolates of Opus Dei are reduced to a single one: to impart doctrine" (Gondrand, 264). And that is what he did at each and every one of the stops on his trip: Bilbao, Madrid, Oporto, Lisbon, Jerez, Valencia, and Barcelona.

- "Doctrine" pertaining to priests: "There are priests who, instead of talking about God, which is the only thing they are obliged to do, talk about politics, sociology, anthropology. Since they don't know a word, they make mistakes; and, in addition, the Lord is not happy" (ibid., 268). Consequently, he states, "we are going to rescue priests" (Vázquez, 390).
- "Doctrine" pertaining to marriage (and divorce): "If you're very lenient, and you let them, at the first squabble, goodbye! This way, they're on bad terms for eight days, and then they love each other again, and they live happily ever after" (Vázquez, 393).
- "Doctrine" on confession, a practice that some Catholics are abandoning, and on which Opus Dei strongly insisted during this period.
- "Doctrine" on liturgy, which has to be properly done, without rushing, not "in a slapdash manner, as one sometimes sees" (Gondrand, 269).
- And "doctrine" about children, at a time when there are people who "are committing horrible crimes, true infanticides" (Vázquez, 394).

The Spain Escrivá had wanted to "re-Christianize" thirty years earlier was now a Spain thoroughly caught up in a "process of secularization," and it was turning its gaze more and more toward the Europe of the Common Market, and less and less toward the Europe of Fátima and Lourdes.

"Since the close of the sessions of the Second Vatican Council, the Church has suffered violent upheavals" (Sastre, 528). As Berglar says, "the inspiration of the Holy Spirit was not absent at the council. But this does not mean that every one of the council fathers was, at every moment of the four sessions, a receptacle and spokesman for the Holy Spirit without need of interpretation" (Berglar, 296). Seeing the results of the Vatican Council, Msgr. Escrivá had in 1970 been able to assert before the General Council of Opus: "I suffer greatly, my children. We are living in a time of madness. Souls by the millions feel confused" (ibid., 305).

Let us move on now to Rome, the other great center of Opus Dei's implantation and activity, and the focus of this "time of madness" that the Church, according to Escrivá, was going through.

Rome

The Second Vatican Council

We have already mentioned the topic of the council on a number of occasions: we have pointed out the lack of enthusiasm with which Opus Dei embraced its announcement and the preparations for it, and the not very positive or serious image the "official" literature presents of John XXIII (an image and a lack of enthusiasm also rather widespread, in fact, in certain circles of the Vatican Curia). In addition, and in spite of the fact that he tried retroactively to present Escrivá as a precursor of the council, and his message as an anticipation of some council documents, Berglar's overall judgment of the council is not very favorable. The same can be said of Vázquez de Prada, who begins his Chapter 10 by recalling the maxim that all councils pass through three phases—the first is the devil's; the second is man's; and the third is God's; he adds that in the case of the Second Vatican Council "the first sequence lasted too long" (Vázquez, 361).

While other institutions suffer "a kind of religious cataclysm," by contrast "the spirit of Opus Dei perseveres, of course, in its pristine identity." Even so, in the face of "doctrinal upheaval and ecclesiastical disorder," the Padre's "temperament rebelled" because "passivity and vacillation seemed unacceptable to him, and no one dared to put the situation to rights. He was even more upset by the tendency to make anything into a problem, letting it lead to resignation instead of looking for solutions" (Vázquez, 363). Although no names are mentioned, his judgment of the doubting and "Hamlet-like" Pope Montini is almost as explicit as it is harsh.

In the three pages that follow, Vázquez de Prada does not mince words. Some "sectors of the clergy" demand freedom, equivalent to the power "to impose their own standards." These are priests "contaminated by social propaganda of a Marxist bent," who talk of "the Church of the poor, wrongly hanging their cassock on any hook," if they do not go to the extreme of "enlisting with the guerrillas." The author then sequentially criticizes "the Sunday homilies," the "senseless liturgical experiments," and the abandonment of the "traditional garb" of the priest. He underscores the "defections in the religious communities," the emptying of the seminaries, "the arbitrary practices" in the administration of the sacraments. By fault of the "negligence" of many bishops, the faithful are "like sheep without a shepherd" (Vázquez, 364–66).

"In this period of ecclesiastical overthrow" (Vázquez, 366), Opus Dei tends to conceive of itself as "the little that remains of Israel," the faithful nucleus which must rescue the Church from the shipwreck provoked by the Second Vatican Council.

How, specifically, did the members of the "cupola" of Opus Dei participate in the tasks of the council? In their peculiar oblique style, Escrivá's biographers convey the idea that the Founder felt very bad not to be able to participate directly in it. In fact, Gondrand asserts that "in agreement with the Holy See, he did not participate as a council father in the tasks of the Vatican Council, but he followed them closely" (Gondrand, 231)—as if the decision to attend the council, as "council father," were the result of agreements reached between the Holy See and the interested parties! The council fathers were the bishops "et alii a Summo Pontifice ad Concilium vocati" (*Actae Apostolicae Sedis,* 1962, 612). Escrivá was not a bishop, nor was he summoned by the Holy Father. Berglar is a little more cautious, but even so he cannot avoid the temptation to assert—without offering any proof, of course—that "in reality John XXIII would have liked to name both the President General and the Secretary General of Opus Dei as consultants to the Council, but this would have presumed an enormous overload of work and expenditure of time on the part of the Founder, who had almost superhuman demands placed on him as it was" (Berglar, 299). In a word: the Padre was not considered, and his name never came up in direct connection with the council.

But that is not the end of the muddle in the "official" literature. Álvaro del Portillo did in fact participate directly in some of the preparations for the council, "as president of one of the preparatory committees, the lay committee," in addition to acting, during the council, as secretary of the committee "on discipline in the clergy" and as "an expert, also, on other committees" (Gondrand, 231; see, along the same lines, the curriculum of Álvaro del Portillo in Rodríguez et al., 34; Portillo, 1981, 11; etc.). On this point also Berglar is more ambiguous: "Don Álvaro del Portillo participated from the very beginning in preparations for the Council, especially in matters relating to the topic 'lay persons in the Church'; later he was named secretary of the council committee *'De disciplina cleri et populi christiani'* and an expert on several other committees" (Berglar, 299). In fact, the council committee on discipline in the clergy was presided over by Cardinal Ciriaci, who was also the "cardinal protector" of Opus Dei; the secretary was Portillo, and the secretary of minutes, Julián Herranz, also of Opus. Furthermore, in the volumes of the *Actae Apostolicae Sedis* for those years, Álvaro del Portillo appears at the end of a list of experts (1962, 784), and also figures as a member of the "Pontificia Comissione *dei Religiosi* per la preparazione del Concilio Vaticano II" (*Actae Apostolicae Sedis,* 1960, 844; emphasis mine); Amadeo de Fuenmayor was a consultant on this same Committee of Religious (ibid., 845). On the other hand, in the list of members and consultants of the "Comissione dell'Apostolato *dei laici*" (ibid., 852f.), no member of Opus Dei appears.

The other source that reflects these appointments and jobs, the *Anuario Pontificio,* offers the following information on the participation of Opus members in the tasks of the council: in the 1961 edition it confirms that Álvaro del Portillo was a member of the Committee of Religious, and that Amadeo de Fuenmayor was consultant to that same preparatory committee, while José María Albareda, then rector of the Catholic University of Navarre, was consultant to the committee on studies and seminaries. The *Anuario* of 1962 presents Salvador Canals as consultant to the "Secretary of the Press and Publicity" in preparation for the council. But in the list of names of the members of the preparatory Committee of Apostolate of Laypersons, no member of Opus Dei appears.

There seems to be some discrepancy here. Either the official sources of the Holy See did not give sufficient documentation to the presence of Opus on the council committees, or else it is difficult to understand what form their participation took, "especially in reference to the topic of laypersons in the Church" (Berglar, 299). It is quite clear that it was fundamentally on this topic that Opus would later want to present itself as a movement which had anticipated the council, along a line of thinking that the council had finally recognized and made official (Benito, 1967, 566). In light of this it seems curious that in the thirteen hundred pages of the *Journal du Concile* by Henri Fesquet, Opus Dei is mentioned only once in relation to the theme of laypersons. And Fesquet only cites Opus to say that if recognition of the maturity of the laity is a condition for the authentic presence of the Church in the modern world, certainly "until now sufficient measures have not yet been taken to assure this except through such suspect secular institutes as Opus Dei" (Fesquet, 610).

Opus Dei and the Vatican Curia

But if the level of direct participation by the Opus cupola in the tasks of the Second Vatican Council was definitely minuscule—much less than the "official" literature would like to make it appear, and possibly less than the directors of Opus Dei desired—one cannot say the same of its progressive penetration into the organisms of the Vatican Curia.

Throughout the years of preparation for and celebration of the council, the bureaucratic apparatus of the Church continued to function, partly through the inertia inherent in any bureaucracy, partly by placing itself at the service of the council, but also partly by acting in the service of diverse attempts to boycott the council. The reorganization and restructuring of the Curia was not completed until the end of the sixties, and during the time this reform of the Vatican bureaucracy remained incomplete, Opus Dei increased its presence there significantly.

At the beginning of John XXIII's papacy, Álvaro del Portillo was consultant to the Congregation of Religious and the Congregation of the Council. (This curial congregation had nothing to do with Vatican II but was an organism that routinely attended to questions of discipline in the clergy, as well as the supervision of brotherhoods and associations, of Catholic Action, foundations, and ecclesiastical benefices and property.) The prefect of this congregation was Cardinal Ciriaci, future "protector" of Opus Dei. The next year Julián Herranz joined it, in the "section of pastoral and catechistic activity" (*Anuario Pontificio,* 1961; the Ciriaci-Portillo-Herranz trio took on the direction of the Committee on Discipline in the Clergy, as we saw earlier). Salvador Canals (juridical office of the secular institutes) and Javier de Silió (commissioner) continued to work in the Congregation of Religious, while Escrivá was consultant to the Congregation of Seminaries and Universities.

According to the data in the *Anuario* of 1961, in addition to Julián Herranz's entrance into the curial world, which he would never leave, Portillo joined the Supreme Sacred Congregation of the Holy Office. Salvador Canals left his work in the Congregation of Religious after many years and became consultant to the Pontifical Committee on Cinema, Radio, and Television; he was also "auditor" of the Tribunal of the Holy Rota.

The following year Opus Dei appears for the first time in an area that would later have fundamental importance for it: Msgr. Escrivá was named consultant to the Committee for the Authentic Interpretation of the Code of Canon Law. Simultaneously, Canals's name appears again in the Congregation of Religious, but now as a consultant, along with Portillo. Javier de Silió continued to work in that congregation, while another Opus member, Severino Monzó, also joined it; all this took place at the moment when Opus Dei was officially negotiating to leave the aegis of the Congregation of Religious. . . . To his post in the congregation, Salvador Canals added the job in the Holy Rota and another in the Committee on Audiovisual Communications Media, and he also joined the Congregation of the Council as a consultant, where, as we have seen, Portillo and Herranz were already present (*Anuario Pontificio,* 1962).

This situation lasted until the end of Vatican II. These years, then, were years of expansion for Opus Dei within the world of the Curia. Immediately after the council, Opus Dei would participate very little in the new organisms created in the Vatican as an outcome of the most reformist aspects of the council (secretariats for unity, for non-Christians, for nonbelievers, council of laypersons, etc.), which were primarily occupied by the representatives of the most "open" or "advanced" currents, who were enjoying a wave of euphoria at that moment. These same "progressive" sectors, however, undervalued the importance of another area, which they found unattractive and did not want to devote any time to: canon law. Opus Dei, on the other hand, fomented the cultivation of this discipline; the University of Navarre became the seedbed for a school of canonists, and from the very moment of the creation of the Committee on Revision of the Code of Canon Law, Opus Dei took an active role in it.

Initially the secretary of the committee was a Basque Jesuit, Bidagor, hardly a friend of Opus; but the associate secretary was a Belgian professor from Louvain, who was soon awarded an honorary doctorate from Navarre; and one of the first "officials" to work on the committee was Julián Herranz, an Opus priest, at the same time that Portillo was acting as consultant (*Anuario Pontificio,* 1966). Both remained on the committee until the approval of the new code (meanwhile, in Navarre, Fuenmayor was made dean of the faculty of canon law). Herranz then became secretary of the Committee for the Authentic Interpretation of the Code (*Anuario Pontificio,* 1964), later changed to the Papal Council for the Interpretation of Legislative Texts (ibid., 1989), a kind of job that usually leads to the concession of the episcopal miter, which Herranz obtained in 1990.

The other curial organisms in which the presence of Opus members is a constant are the Congregation of Religious—in spite of the Work's expressed desire to be independent of it—and the former Committee on Cinema, Radio, and Television, changed after the council to the Committee for Social Communications. In the Congregation of Religious, Álvaro del Portillo acted as consultant until 1967, and Salvador Canals until the 1970s, while Julio Atienza replaced Severino Monzó in 1965 in the Office of Secular Institutes. As for the Committee for Social Communications, in addition to Salvador Canals, Ángel Benito joined it as a lay consultant (*Anuario Pontificio,* 1966); at the end of the eighties, Portillo appeared as a consultant, while Joaquín Navarro Valls occupied the post of director of the Press Office of the Holy See.

According to Grootaers's thesis, after the close of the council the most conservative sector of the Curia, which earlier had employed "diverse strategies to make reform impossible, began to use indirect attacks: they denounced abuses that were being committed to the Pope, and begged him to react" (Grootaers, 95). This provoked growing unease on the part of Paul VI, which turned to alarm in 1967, reaching its most critical point in the following year after the hostile reaction to the encyclical *Humanae vitae*. As far as the Roman cupola of Opus can be associated with these most conservative sectors of the Curia, this was exactly the context in which one must view Msgr. Escrivá's irritation when he found "the passivity and vacillation unacceptable, and nobody dares to find a remedy for the situation" (Vázquez, 363). But this was also the moment when Msgr. Benelli, Pope Paul VI's principal collaborator, became the sponsor of the restructuring of the Curia (Grootaers, 128 n. 1).

The accession of Benelli—an all-out enemy of Opus Dei, as we will see in the next section—to an important position in the Department of State of the Vatican coincided with the moment of Opus Dei's maximum implantation in the organisms of the Curia: Portillo as consultant and judge of the Congregation for the Doctrine of the Faith (formerly the Holy Office), consultant of the Congregation of the Clergy, and of the Committee on Revision of the Code of Canon Law; Canals was commissioner of the Congregation of Discipline of the Sacraments (section of causes of matrimonial dispensation), auditor of the Rota, and consultant of the Congregations of Religious, of the Clergy, and of the Committee on Social Communications; Herranz worked in the Committee on Revision of the Code; Atienza worked in the section of Secular Institutes of the Congregation of Religious; Benito was consultant in the Committee on Social Communications. Monsignor Escrivá himself, however, was no longer consultant to the Congregation of Seminaries and Universities and, for the first time in many years, held no post connected to the curial organisms (*Anuario Pontificio,* 1969).

The restructuring of the Curia promoted by Benelli meant putting the brakes on this process of Opus Dei's constant expansion, and it gradually began to lose positions; in parallel fashion, the process of the resolution of its juridical status was stalled. It was precisely at the moment of the unblocking of this process, which coincided with the election of John Paul II, that a new modification of its presence in the curial organisms took place, this time in different places and with different objectives from those of the sixties.

In fact, after the deaths of Canals and Escrivá (1975), Benelli's departure to Florence (1977), the deaths of Paul VI and of John Paul I (1978), and the election of Cardinal Wojtyla, Opus Dei managed to definitively push the matter of its juridical status toward recognition as a "Personal Prelature." Once this objective was attained, Opus came under the aegis of the Congregation of Bishops.

In addition, the introduction of the process of beatification of the founder came under the jurisdiction of the Congregation for the Causes of the Saints. In the eighties and nineties, these two curial organisms, together with the committee that dealt with the Code of Canon Law, constituted the basis of Opus's institutional presence in the Curia.

In the *Anuario Pontificio* of 1982 Opus Dei was still registered as a secular institute (*twenty-five years after* Escrivá declared that "in the future that name cannot be applied to us"; Fuenmayor et al., 564). The next year it was listed in the *Anuario* as a Personal Prelature (approved on November 28, 1982). Javier Echevarría, vicar general of the Prelature, was a consultant of the Congregation for Causes of the Saints. The prelate, Álvaro del Portillo, was too. He and Herranz were also on the Committee for the Revision of the Code of Canon Law (*Anuario Pontificio,* 1983). Somewhat later Herranz was named consultant to the Congregation of Bishops, to which the Prelature was subordinate (*Anuario Pontificio,* 1984).

By the end of the eighties and the beginning of the nineties, the obstacle presented by the not-too-friendly Msgr. Benelli had been overcome, and Opus Dei's Roman cupola was once more solidly planted in the Curia. Monsignor del Portillo, prelate, bishop since 1991, and grand chancellor of the Universities of Navarre (Spain), Piura (Peru), and La Sabana (Colombia), is also grand chancellor of the Roman Atheneum of the Holy Cross and consultant to the Congregation of the Causes of the Saints, the Congregation of the Clergy, and the Council for Social Communications. Julián Herranz, also made bishop in 1991, is secretary of the Council for the Interpretation of Legislative Texts and consultant to the Congregation of Bishops. Javier Echevarría, vicar general of the Prelature, is consultant to the Congregation of the Causes of the Saints. So is Joaquín Alonso, while José Luis Gutiérrez is one of the seven members of the College of Relators in the same congregation. In the Congregation for the Doctrine of the Faith we find Fernando Ocáriz and Antonio Miralles, who is also dean of the Faculty of Theology of the Roman Atheneum of the Holy Cross. Amadeo de Fuenmayor is consultant to the Council for the Interpretation of Legislative Texts. Ignacio Carrasco de Paula is on the Committee for Catholic Education, as well as being consultant to the Pontifical Council for the Family and rector of the Roman Atheneum of the Holy Cross (whose vice rector is Valentín Gómez-Iglesias, the coauthor, with Fuenmayor and Illanes, of *El itinerario jurídico del Opus Dei*). Javier García de Cárdenas is rector of the International Roman College of the Prelature (*Anuario Pontificio,* 1990, 1991).

This account does not pretend to be exhaustive. It is limited exclusively to those priests of Opus Dei who are Spanish and who have some job in the world of the Vatican; my intent is to demonstrate first that Opus's implantation is substantial and second that the cupola of the Prelature is still basically Spanish.

Opus Dei and the Papacy of Paul VI

The "official" literature's efforts to present relations between the highest directors of Opus Dei and Paul VI as friendly and positive cannot hide the troubled, and frequently conflictive, nature of these relations. No matter how many times the Padre is quoted as saying that Msgr. Montini was "the first friendly hand I found in Rome" (Sastre, 483), or how many times it is claimed that Paul VI "was profoundly convinced of the extraordinary importance of Msgr. Escrivá de Balaguer in the history of the Church" (Berglar, 236), it is clear that Paul VI's attitude was in fact quite a bit less favorable to the interests of Opus Dei and that, correspondingly, Escrivá's attitude toward Paul VI was hardly characterized by understanding and sympathy.

According to Antonio Pérez, then an Opus Dei priest, the election of Paul VI provoked the indignation of Escrivá, who exclaimed that "everyone who cooperated in this election was going to be condemned to hell" (Moncada, 1987, 27). The same author claims that as archbishop of Milan Msgr. Montini had denied authorization for the opening of an Opus residence (ibid., 27). It is not easy to reconcile these assertions with the version according to which "we know that Paul VI used *The Way* for his personal meditation" (Berglar, 249). In any case, it is beyond question and perfectly well documented that Msgr. Montini felt little sympathy for Franco's regime, and that he viewed with concern the fact that certain sectors of the Spanish Church undertook to play the role of legitimizer of that system. This happened at a time when the presence of some members of Opus in the Spanish government was notorious and visible. In the words of the *Historia de la Iglesia en España,* "the naming of Montini was received, especially in the broad sectors of the regime, with cool respect, and without any hint of the patriotic-religious enthusiasm of the times of Pius XII" (García-Villoslada, 686).

Furthermore, relations between Paul VI and Escrivá were strongly conditioned by Opus Dei's desire to modify its juridical status, dispensing with the format of the secular institutes. During the fifteen years of Paul VI's papacy Opus's plans on this issue were paralyzed.

In fact, independent of the manner in which the enigma of the scope, diffusion, and date of Escrivá's letter *Non ignoratis* may someday be resolved, the truth is that during the papacy of John XXIII Msgr. Escrivá made his first formal attempt to change the juridical status of his secular institute. Early in 1962, he simultaneously sent a letter to the pope and to the cardinal–secretary of state in which—albeit in much less radical language than *Non ignoratis*—he asked that the possibility be considered of conferring on his secular institute a new status which would avoid any comparison of its members to the religious, since this gravely prejudiced "the Institute's apostolate of penetration" (Fuenmayor et al., 570; six years later Msgr. Escrivá would declare that this formula of the apostolate of *penetration* "does not apply at all to the apostolate of Opus Dei"; *Conversations,* no. 66). The concrete solution he proposed—a prelature in which all the priests of the Institute would be incardinated—was denied by the Holy See, which sent a written reminder to Msgr. Escrivá that at the time of its approval Opus Dei was granted a set of "privileges which would be difficult to concede in the present," and that independence from the Congregation of Religious would automatically entail "the loss of those privileges," which it possesses because it "is a state of perfection" (Fuenmayor et al., 572).

The reasons the negotiations came to a dead end seem rather diaphanous. On the one hand, Opus Dei did not want to continue to be a secular institute and a state of perfection, nor did it want to remain under the aegis of the Congregation of Religious; on the other hand, it absolutely did not want to renounce the privileges it enjoyed, which basically gave it autonomy with respect to the ecclesiastical diocesan structures. And the Holy See gave Escrivá to understand, without mincing words, that if Opus ceased to be a secular institute and a state of perfection, "it would be reduced to a simple association of priests and laypersons, and would lose its privileges" (Fuenmayor et al., 572). "The refusal was, for Msgr. Escrivá de Balaguer, a cause of deep pain" (ibid., 338).

The subject was not raised again until the beginning of the papacy of Paul VI. And judging by the way it is treated in the "official" literature of Opus Dei, one may clearly deduce that the issue remained on terms totally unfavorable to its interests. In spite of the allusions to the "friendly hand" and the "amiable smile" of Paul VI, the truth is that any possible juridical transformation of Opus was totally paralyzed for the rest of his papacy and that, as Artigues says, "it might be said that the first months of 1964 were an especially difficult period for Opus Dei" (Artigues, 134).

One of the axes around which these difficulties revolved can be traced to an audience granted to Escrivá by Paul VI on January 24, 1964 (six months after his election as pope). The first symptomatic item is this: commenting on the "excellent" relations between Escrivá and the Holy See, Berglar writes that different popes "granted important private audiences to Msgr. Escrivá de Balaguer"; but in his list of such audiences, he omits the one in January 1964 (Berglar, 412 n. 27). Other authors mention it but limit themselves to mentioning the "affectionate embrace" (Gondrand, 232; Sastre, 483), without referring at all to the content of the interview. A few days later, Escrivá received a letter from Paul VI: "There are words of praise in it" (Vázquez, 332). In addition to the praise and embraces, it also dealt with "the institutional problem of the Work" (Fuenmayor et al., 350).

From a letter sent by Escrivá to Paul VI (February 14, 1964, reproduced in Fuenmayor et al., 574f.), we know that the Holy Father had expressly asked him for the text of the Constitutions of Opus Dei. We also know that Escrivá added several other documents; among them was the letter *Non ignoratis,* according to the "official" version, although this is not confirmed but rather covered up (see the contradiction between the list of documents as it appears in Fuenmayor et al., 350, and as it is described by Msgr. Escrivá in the same book, 574f., a contradiction noted in the previous chapter). Furthermore, the only allusion made in the letter to the "institutional problem" is to say that "we are not in a hurry" (ibid., 575). He was in so little of a hurry that three months later Escrivá was "a little surprised" to find out indirectly that Paul VI had had the documents the Padre had sent him studied (ibid., 351).

The lack of hurry seemed to be due to the fact that at that time Opus Dei was going through a crisis. The highly calculated ambiguity of the language used in the "official" literature confirms this. Curiously, the *same day* he wrote the letter in question to Paul VI, the Padre said, according to Vázquez de Prada, that "the persecution was growing stronger," that "there was no other word in the dictionary to express what was happening," and that he felt he was being treated as if he were "not just a piece of rubbish, but everybody's spittoon" (Vázquez, 512f.).

In López Rodó's account, Msgr. Escrivá told him personally that Paul VI "said things to him that even his mother never said" (López Rodó, 1990, 160). Certainly, mothers sometimes give vent to praise of their children; but usually it is when the children are not listening. And just as certainly, no one can scold a child as his mother can.

A few weeks before Escrivá's audience with Paul VI, the Swiss theologian Hans Urs von Balthasar had published a concise and important article in which he reviewed the course of the main integrist currents in the Catholic Church, and he stated that "no doubt, the most powerful of all of them today is Opus Dei" (von Balthasar, 1963, 742). A few months later, while according to Escrivá "the persecution was growing stronger," and while Paul VI was having the documents Escrivá had sent him studied, von Balthasar published a second article in which, addressing himself directly to Opus Dei, he wrote: "That you have great power, a great deal of money, many political and cultural positions; that you use intelligent and discreet tactics toward the end of attaining these positions by the fastest and most direct route; there is nothing to say. In itself, power is not evil. The whole question, the decisive question, is this: why do you want power? What do you mean to do with it? What is the spirit you are attempting to propagate with these methods?" (von Balthasar, 1964, 117).

Several authors agree that the stormiest phase of the crisis came to an end in October 1964, when Escrivá was again received by Paul VI. Was there a truce? In a way, yes: the offensive ceased, but only in exchange for compliance with certain demands. After this Opus Dei yielded on the question of secrecy or discretion, began to speak profusely about freedom, and replaced its former silence with an increasing amount of publicity about its activities and social works (Artigues, 136; Hermet, 1980, 1:267; Piñol, 41ff.). Furthermore, Opus agreed to a period of waiting before raising the juridical question again (this is admitted even in the "official" literature; Fuenmayor et al., 353). Concretely, this meant that Opus Dei would continue to be a secular institute for eighteen more years.

After the audience of October 1964, Paul VI sent a letter to Msgr. Escrivá, apparently full of praise. Since it has never been published (it is not in Fuenmayor et al.'s appendix, nor do they even mention it), one wonders whether it was exclusively a letter of praise for Opus Dei. While we are asking, we can raise the possibility—as we have on other occasions—of whether it was in fact a letter of praise, but not in praise of Opus Dei; in fact, in the first paragraph (quoted in Vázquez, 333), Paul VI makes reference to *"the members of this Priestly Society of the Holy Cross."* But other authors quote a different passage (Sastre, 484; Seco, 58), in which Opus Dei is mentioned. Only a reading of the whole letter, however, would enable us to know if it contains other, perhaps not so positive statements, or whether the pope—the "friendly hand" that Escrivá found in Rome in 1946, when the entity was officially recognized as the Priestly Society, not Opus Dei—was still making some kind of distinction between the Priestly Society and Opus Dei. It has been said in this regard that even at the moment of Msgr. Escrivá's death in 1975, Paul VI sent some lines of condolence to Opus headquarters in which he alluded to the deceased as the founder of the Priestly Society, and nothing more. And the truth remains that although all the biographies of the Padre mention the existence of the letter, not one publishes it in its entirety.

Finally, these observations on the greater or lesser degree of cordiality in the relations between Opus Dei's Roman cupola and the *terzo piano* of Saint Peter's Square demand some comment on Paul VI's principal collaborators, against whom the Opus authors discharge aggressions that they would never discharge—or would not dare to discharge openly—against the figure of the pope. We refer, of course, to Msgr. Benelli, to whom we referred earlier as the "promoter of a restructuring of the curial services" (Grootaers, 128) at the end of the sixties, a restructuring that slowed the gradual expansion of Opus leaders into the organs of the Vatican bureaucracy.

The language used by some Opus authors in speaking of such a high-ranking ecclesiastical dignitary as Msgr. Benelli is notable for its rarity. In a chapter of *El franquismo y la Iglesia*—which is about the deterioration of relations between the Franco regime and the Church—Gómez Pérez accuses Benelli of "indirect, but effective support" of the conflicts led by an "active minority of priests and religious." He even thinks that "Msgr. Benelli encouraged—or at least did not hinder—the confrontation of the movements [of Catholic Action] with the hierarchy (Gómez Pérez, 1986, 160). Further on, he quotes several notices published in the international press on the occasion of Benelli's death in 1982, saying that he was "a man with conservative points of view," who had "a great deal of power in the Vatican," and who "read all the letters sent to the Pope and all the reports; he organized the audiences, proposed the appointments of nuncios and bishops" (ibid., 160; see also 162f.).

In the *Memorias* of the diplomat and politician López Rodó, Msgr. Benelli fares no better. One of Franco's ministers lashes out at him: "You give orders in the Church and I'll give them in Spain" (López Rodó, 1991, 234). When Paul VI was ill in 1967, the Spanish ambassador said that he did not feel that Benelli had "the necessary experience and maturity" to take charge of certain things (ibid., 237). Two years later, the new Spanish ambassador asserted that one had to tell "this Monsignor [Benelli] to stop meddling in political questions and giving his opinion about things he doesn't understand" (ibid., 481).

What is especially significant is not so much the judgments about (or against) Msgr. Benelli in themselves, but the fact that a group of authors who are usually so meticulously prudent in their opinions when it comes to ecclesiastical personalities would allow themselves in this case to express such hostile observations and commentaries. (Note that in the various biographies of the Padre the name of Benelli is never mentioned, with the exception of two instances which I will mention later.)

The truth is that the history of the confrontations between Benelli and

Opus is a long one, and that its origins go far back. Before he was named by Paul VI in 1967 to serve in the Vatican secretary of state, Benelli had spent a few years in Madrid as counsel to the nunciature after the nuncio Antoniutti (a friend of Opus, and a friend of Franco) was replaced by the nuncio Riberi (not so friendly with either). In a book published in 1985, Paul Hofmann, who was the *New York Times* correspondent in Madrid, reports that Msgr. Riberi told him that in the nunciature it was impossible to speak openly in the presence of the household help, that the nunciature was full of "nuns" from Opus, and that he had been obliged to replace several telephone operators because he was convinced that they were listening in on all the conversations (Hofmann, 229f.). In an interview I conducted during my research, a bishop told me exactly the same thing: he was a personal friend of Benelli, and they had both decided to give false names when calling the nunciature, because they were convinced that the conversations were being overheard by the "in-service" members from Opus Dei. Also, at table, conversations on "delicate" topics were automatically interrupted when the associate members of Opus Dei serving as waitresses came into the dining room.

Without going into a discussion of the objective truth of such information, what we wish to stress here is the *feeling* that Msgr. Benelli and the others had that they were being spied on, and hence the hostility of the relations between Benelli and the directors of Opus Dei. It was this Msgr. Benelli who became a close associate of Paul VI in 1967 and who—according to Gómez Pérez, an Opus member—"read the letters and reports addressed to the Pope, organized the audiences," and so on. When Msgr. Escrivá died (1975), the personal representative sent to the funeral by Paul VI was precisely . . . Msgr. Benelli! (Sastre, 632, 635; Gondrand, 293).

Even though the "official" literature tries to convince us of the contrary, relations between Paul VI and Opus Dei were always troubled and tense. And this explains why seventeen years after having solemnly written "we are not a secular institute, nor in the future can that name be applied to us" (in the letter *Non ignoratis;* Fuenmayor et al., 564), Msgr. Escrivá was still, when he died, president general of the Secular Institute of Opus Dei.

Opus Dei, Personal Prelature

After Escrivá's second audience with Paul VI in 1964, the topic of the juridical status of Opus Dei entered a long waiting period, but shortly after the death of the Founder things began to move forward. In 1977 Msgr. Benelli was named archbishop of Florence, and he left Rome. In the year that followed Paul VI died and John Paul I was elected pope and died within a month. Among many other things, the future of Opus Dei probably hung in the balance at the second conclave of 1978. Monsignor Benelli was one of the cardinals most likely to be elected pope, and he nearly had the necessary two-thirds majority (Grootaers, 124–33). But the man who was elected finally was the Polish cardinal Wojtyla, who almost immediately declared that he deemed it necessary to resolve "the question of the juridical configuration of Opus Dei."

From this moment on, the process leading to the establishment of the

Prelature of Opus Dei is well documented, and we shall not spend time on it, although it should be said that it took four more years and was not exempt from certain difficulties. There is also an abundant bibliography on the Personal Prelatures and their characteristics, produced above all by members of Opus Dei, the only Personal Prelature in existence to date. (See Rocca, 1985, 103–28, with many bibliographical references and information about some juridical aspects not entirely resolved, as well as a documentary appendix, pp. 207–27; Fuenmayor et al., 421–503, and, especially, the documents on pp. 594–627; see also Walsh, 83–113, and Hertel, 132–41, as "nonofficial" synopses, in contrast to "official" synopses in Berglar, 361–84, Le Tourneau, 62–68, and the studies included in Rodríguez et al., 403–65.)

With the publication of the papal document establishing Opus Dei as a Personal Prelature, and the publication of its Statutes ("Code of Law Particular to Opus Dei"), the long and complex "juridical itinerary" of the movement founded by Msgr. Escrivá de Balaguer comes to an end. The "critical" authors emphasize the fact that the directors of Opus Dei asserted, on this occasion, exactly the same thing they had asserted in 1950, when it was approved as a secular institute: that is, that Opus Dei had found its *definitive* juridical solution.

♦ ♦ ♦

But the juridical itinerary of Opus is not the only thing that thereby came to an end. This was also the end of the extraordinary exercise of *alternation* which we have witnessed, whose progress we have tried to follow throughout these pages.

In the apostolic constitution *Ut sit* (which establishes the Prelature), after recognizing that Opus Dei was founded in Madrid on October 2, 1928, by José María Escrivá de Balaguer, guided by Divine inspiration, it states that "from its beginning, this Institution has striven, not only to illuminate with new lights the mission of laypersons in the Church and in human society, but also to put it to work; it has striven likewise to put into practice the doctrine of the universal vocation of holiness, and to promote among all social classes the sanctification of professional work and sanctification by means of professional work" (in Rodríguez et al, 395f.).

For an ecclesiastical organization, it is hard to conceive of an exercise in alternation that culminates in a more thorough success than one that could place its own reinterpretation of the past and its own adaptation to the circumstances of the present on the lips of a pontiff. It is not the Padre's disciples, but John Paul II, who attributes the founding to "divine inspiration," on a specific date in "the year 1928," realized by "Escrivá de Balaguer." Let no one dare to cast doubt on the name, the date, or the inspiration!

We began this historical review of Opus Dei taking as our point of departure the thesis—which I consider to be elementary and commonsensical—that Opus Dei evolved gradually, adapting itself to new circumstances as the context changed in Spain (from the republic and the civil war to Francoism and the post-Franco era), internationally (from fascism and the world war to the cold war and détente), and in the Church (from the precouncil era to the postcouncil era). From this perspective we can understand why Opus Dei speaks today of "the mission of the layperson in the Church," of "universal vocation for holiness," of "sanctification of work," and of "all social classes," and why during the first years it spoke instead of "caudillos," of "academics and intellectuals," and of "holy intransigence, holy coercion, and holy shamelessness." But we have been able to verify that the literature we have here called "official" systematically refuses to adopt this perspective, asserting that Opus Dei has never changed, and will never have to change. After contemplating the numerous vicissitudes of every sort traversed by Opus Dei during its brief but complex history, we can say that the assertion in a solemn papal document that "from its beginnings" Opus Dei has said and done what today it says it does presumes not only the success of an exercise in alternation, but also the triumph of the "holy scheming" in which Msgr. Escrivá de Balaguer so distinguished himself.

Part

II

The Opus Dei Ethic and the Spirit of Capitalism

12

Franco's Spain, Between Fátima and Brussels

Max Weber's Thesis and Opus Dei

In my introduction I explained that the original idea for this research project—a study of the Opus Dei ethic and the spirit of capitalism—could be found in a sentence from my introduction to the Catalan edition of Weber's famous thesis (Estruch, 1984, 10).

The actual pursuit of this possibility made it necessary to examine and present the web of paradoxes that make up the history of Opus Dei. At the end of this historical journey, to which the entire first part of the book is devoted, we can now surmise that the answer to our initial question about the possible existence of a relationship between "the Opus Dei ethic" and "the spirit of capitalism" will have to be inconclusive and full of shades of meaning: *yes and no;* or, as people sometimes say, "neither yes nor no but just the opposite."

First of all we should mention that the question has been posed before and studied from various perspectives and different points of view. In fact, many authors have pointed out the parallels between Opus Dei and the Puritans Weber studied, and have consequently suggested the possibility of applying Weber's thesis to Opus.

These authors have nearly all been Spanish, because, as we shall see in a moment, the change in the orientation of Spanish economic policy in the sixties (and the role played in that change by some members of Opus Dei) was universally recognized as the determining factor that produced these parallels. Let us look at a first set of examples, representative of the general tenor of these writings and also illustrative of the kind of difficulties we will have to overcome in our attempt to provide more shades of meaning to our answer.

Alberto Moncada affirms that "the bourgeois religion of success in business which presages predestination came to Spain from the hand of Opus Dei nearly three centuries after it flourished in northern Europe," and that Opus Dei presents itself as "a symbiosis of fundamentalist Catholicism and modern Protestantism" (Moncada, 1982, 67, 99). In addition to the fact that this "bourgeois religion" came *without* the doctrine of predestination, we must not forget that in the territory officially considered as Spanish, for better or for worse, regions which did not experience the industrial revolution until the twentieth century coexist with nations like the Basque Country and Catalonia, which have a bourgeoisie that antedates the birth of Opus Dei, and which have been concerned with "success in business" for a long time. The problem of the homogenization of the plural reality camouflaged under the label "Spain" is recurrent in practically all the texts, and must be kept in mind. On another level, we must not lose sight of the fact that "fundamentalism" is typically *Protestant* in origin, not Catholic.

In his biography of Franco, Brian Crozier—after reminding us that what attracted the most attention in the governmental change of 1957 was the inclusion of two Opus members as ministers (Crozier, 2:243)—writes that "Opus Dei teaches a philosophy of success which is not very Spanish, offering the alternative of hard work in place of the passive acceptance of the will of God." This is exactly the contrast between "modern mentality" and "traditional mentality" in Weber's work. In this manner, Crozier continues, Opus Dei creates professional elites by stimulating "those who join the association to dedicate themselves firmly to triumph in the service of God, in the way that Opus Dei understands such service" (ibid., 244f.).

Vilariño, in a more recent book, writes that "Opus Dei brings two fundamental things to ecclesiastical organization and to Spanish society: first, the sense of the preeminence of business and of secular professionalism . . . and second, a revised form of Puritan morality in the sphere of private life. The confinement of religion to private life, and the preeminence of economic success through a pure cultivation of professionalism—these are the two fundamental contributions of Opus Dei to Spanish society" (Pérez Vilariño and Schoenherr, 453). The distinction between the business sphere and the sphere of private life is surely, in the case of Opus Dei, very important, although I believe, particularly in the case of Opus Dei, that one cannot speak of a "confinement of religion to private life." Vilariño poses the interesting hypothesis that the present Spanish government of the Socialist party "has continued the process of modernization promoted by Opus with its dual components of economic rationalization and a secular ethic," and that it is therefore not surprising "that the Spanish Socialist party has replaced Opus as the new path for upward mobility in the eighties." The text concludes—à la Weber—that "Opus's modern contribution to the Church and to Spanish society can be defined as the delayed but effective introduction of the spirit of capitalism, in the form of an ethic of individual vocation as professional achievement" (ibid., 454).

Carandell uses a similar argument in speaking of "Escrivá's stimulating economic doctrine." But according to him, "the idea of the sanctification of work and through work, the idea, definitively, of the comparability of professional and economic success with spiritual perfection, did not arise by spontaneous generation in Escrivá's mind. It is an idea with Protestant roots (and here we must unavoidably recall Weber's book) which came to him through who knows what complicated route" (Carandell, 182).

The main problem with all these texts is that they offer a view of Opus Dei which, even though it is a view "from outside," *starts from an uncritical acceptance of certain stereotypes* diffused by the "official" Opus Dei literature. Keeping in mind the historical evolution of the movement as we have analyzed it here, we have to specify that the hypothetical "contribution" of Opus cannot take place prior to the fifties, and that probably it is wrong to speak of Escrivá's "economic doctrine," stimulating or otherwise.

We have a caricature of this type of approach in the frivolous and superficial treatment of Opus Dei often seen in the press, especially when the articles carry the byline of some professor of sociology. Some consider Opus Dei to be "the first Protestant sect of Catholicism," and they ask if "the thirteen laymen whom Escrivá chose to found Opus Dei in 1928 were familiar with Weber's writings on the sociology of religion" (*El País,* November 25, 1982, pp. 11f.). The serious issue here is not the writer's ignorance of the fact that in 1928 Escrivá did not "choose" any laymen, and that it was many years before he had thirteen at his side. The serious issue is his lack of awareness that the "miracle" of Opus is that its intentions were one thing and the consequences were something else, because this means that he understands neither Opus nor Weber.

As for Opus being "the first Protestant sect of Catholicism," this sounds something like the fulfillent of Montesquieu's prophecy: "The Catholic religion will destroy the Protestant religion, and then the Catholics will become Protestants" (quoted in Aranguren, 1952, 217).

Even so, the caricature does not lack some degree of interest, since we find it also in some texts of the "official" Opus Dei literature which directly or indirectly allude to Weber's thesis. On Msgr. Escrivá's message, Berglar says, "What is it that he wants, many ask? What is this 'sanctification of work'? That sounds like something for Protestants . . ." (Berglar, 223). In the "official" literature, in fact, the prevailing impression is that the authors do not understand Weber, or rather they know him only through hearsay. Here is just one example: "It is common in certain kinds of literature about Opus Dei to compare it to Calvinist spirituality, in matters relating to the great value placed on work. . . . In reality, in Calvinism, dedication to work—and, above all, success in it—is a sign of predestination, of salvation. For Opus Dei, within Catholic spirituality, the value of work derives from its being the primordial vocation of man and its having been practiced by Christ" (Gómez Pérez, 1986, 252).

However, the usual strategy of the Opus literature is different. It consists of happily dispensing with any reference at all to the Reformation, to Calvinism, to the Puritans, or to Weber. The same author who defensively places Opus "within Catholic spirituality," in another, less committed text says only that "Msgr. Escrivá de Balaguer, since October 2, 1928, offers a new way in the history of Christian spirituality" (Gómez Pérez, 1978, 19). And the other author quoted above, when he is not afraid someone will tell him that "that sounds like something for Protestants," explains that "one of the characteristics of human and Christian solidarity consists of not hoarding wealth, and making it productive" (Berglar, 314).

Another author says, "Probably, the idea of work as a universal means of sanctification is one of the richest contributions Msgr. Escrivá de Balaguer made not only to Christian doctrine, but to universal culture" (García Hoz, 1988, 201). History is eliminated. Opus Dei branches off directly from the first Christian communities (Portillo, 1978, 40), or to go even further back, from the beginning of all history: "in many aspects, Opus Dei represents a return to the original plans of God: the first command he gave to Adam and Eve, in Paradise . . ." (Helming, 21).

But of course, beyond these simplifications, we can also find—at both extremes—much more rigorous approaches to the phenomenon that interests us here. For example, there is Carlos Moya's study *Las élites del poder económico en España* (The Economic Power Elites in Spain), which argues that the spirituality of Opus Dei "has fulfilled, for the development of a bureaucratic-entrepreneurial ethic in Catholic Spanish society, the same driving function that Max Weber assigns to the Calvinist ethic in relation to the development of the spirit of capitalism" (Moya, 1984, 137). Above all, there is José V. Casanova's thesis, debatable on certain questions of interpretation but substantially parallel to the thesis of this chapter, which presents "the ethic of Opus Dei as a Catholic version of worldly asceticism, that way of 'being in the world' which leads to a rationalized style of life and, according to Weber, presents elective affinities with modern rational capitalism. Opus Dei introduced for the first time in the history of Catholic Spain the typically Protestant notion of the sanctification of work in the world through the professional 'vocation'" (Casanova, 1982, 5).

Spanish Economic Development in the Sixties

Before continuing we must ask, what are the factors that make a group of authors concur in viewing Opus Dei as "an organization that plays a role very similar to that which Weber attributes to the Protestant sects in his thesis" (Casanova, 1982, 3)? To be able to answer this question, the parallel must be placed *in its proper historical context,* which is Spain in the 1960s. Opus Dei is never referred to as the vehicle or "carrier" (*Träger*) of the economic culture or of the "spirit" of capitalism in the Philippines or in Peru, much less in Australia, Japan, or even Ireland. Similarly, Opus is never mentioned in these terms in a discussion of Spain in the 1940s.

This phenomenon has very precise coordinates in space and time. It is produced specifically at the moment when *a change in the orientation of the economic policy* of Franco's Spain coincided with the entrance of *two Opus Dei members as ministers* into the government, in positions that were specifically in the economic sphere.

Briefly, the sequence went as follows. At the end of 1956 Laureano López Rodó was named technical secretary general of the presidency of the government. The basic aim of López Rodó was to be, according to some writers, the modernization and rationalization of public administration; others tend to consider him as the ultimate artificer of the new economic policy; he himself, over time, tended progressively to present his own trajectory as fundamentally oriented toward guaranteeing the peaceful transition from the Franco regime to the present monarchy (see especially his book *La larga marcha hacia la monarquía,* 1977).

At the beginning of 1957 there was an overhaul of the cabinet (twelve of the eighteen ministers were replaced), the greatest changes being the appointments of Alberto Ullastres as minister of commerce and Mariano Navarro Rubio as minister of the treasury; both were members of Opus Dei. In addition, according to Ramón Tamames, Opus Dei had a director general in the Ministry of Information and undersecretaries and directors general in Public Works and Education (Tamames, 1973, 511; from the "official" point of view, of course, this is not correct: Opus Dei does not *have* ministers or directors general; what may happen is that some ministers *are* members of Opus Dei).

The government of 1957 was the government of the economic "stabilization plan," which ended the autarchic period of Francoism and made the first opening to the outside world with the establishment of links with the International Monetary Fund, the World Bank, and economic advisers from the White House (Gallo, 342f.; on the "stabilization plan" and the new economic policy of 1957, see Tamames, 1973, 464ff., a book fiercely criticized by certain organs of Opus Dei, for example, the magazine *Nuestro Tiempo,* no. 236, February 1974).

After the government of the stabilization plan came the government of the "first development plan" (1962–1965): López Rodó was commissioner of the plan; Navarro Rubio was mainly occupied with the rationalization of financial institutions and contacts with international capital; Ullastres was in charge of business development and the rationalization of the market. In 1962 a new member of Opus joined the government, Gregorio López Bravo, as minister of industry (in 1959 he had been named director of the Spanish Institute of Foreign Money). In February 1962 Spain tendered its first request for membership in the European Economic Community.

In 1965 Navarro Rubio left the government to become governor of the Bank of Spain, and he was replaced in Treasury by Espinosa Sanmartín, also an Opus Dei member. The scandal of the Matesa affair (July 1969) exploded in the face of the latter, and several members of the government took advantage of this occurrence to try to unseat Opus (Tamames, 1973, 528). Ullastres was named ambassador to the European Community and was replaced by another Opus member, García Moncó.

In reference to both the 1965 and 1969 governments, the authors do not agree on the number of ministers who belonged to Opus Dei. This is particularly true of 1969, which Tamames calls "the monochromatic government" (Tamames, 1973, 529), because he thinks that eleven of the eighteen ministers belonged to Opus. Hermet, on the other hand, thinks that only seven ministers were from Opus, but he admits that most of the rest were very "close" (Hermet, 1981, 1:464f.). The argument of "closeness" is extremely ambiguous, however, and Hermet contradicts himself when, in trying to show that the government of 1973 was marked by a clear retreat on the part of Opus, he identifies as "falangist" certain ministers who in 1969 he had categorized as "allies of the technocratic clique" (ibid., 473, 466); Carrero Blanco himself, who according to Hermet was a great protector of the Opus people, in 1973 heads Hermet's list of "Francoists and extremists" (ibid., 473). Be that as it may, the government that was formed in January 1974 after Carrero Blanco was assassinated by Basque separatists did not have a single Opus member left in it (ibid., 474f.), thus bringing to an end the period of its most *visible* presence in the world of Spanish politics.

Knowing what we now know, we cannot expect that any explanation will be forthcoming from official Opus sources for the significant presence of Opus members in Spanish political life of the sixties. In the words of Msgr. Escrivá, "the presence in public life of some members of Opus Dei is a natural consequence of the growth and expansion of the Work. Today it does not shock anyone, and everyone accepts the public activity of Opus Dei members as a normal thing, as citizens with the same rights as everyone else" (statements made in 1960, assembled in López Rodó, 1990, 226). The only element Escrivá seems to forget is that, unlike the majority of the "rest of the citizens," the members of Opus Dei are ruled by a set of constitutions that make the "exercise of public jobs a particular means of apostolate" (*Constitutions,* 1950, art. 202).

On another occasion, Escrivá himself declared that "those who attack the Work attack because we are Catholics. They would like us to shut ourselves up at home, they would like the Catholics to disappear from the public scene. It is logical for us to be present in all human activities. We Catholics have the right, just like other citizens, to practice our profession or trade. Those who attempt to deny us this right—what do they want? That we not eat? We work to earn our bread" (quoted in López Rodó, 1990, 295f.). It is curious that on this occasion Escrivá does not remember that his "children" work not exclusively "to earn their bread" but above all to "sanctify themselves in work and to sanctify others with work." Here, on the other hand, this essential apostolic dimension does not appear: "the only thing that interests me is your personal holiness," Escrivá said to López Rodó (ibid., 296). When a person performs an important public job, is he supposed to worry only about his "personal holiness"?

If for Escrivá "those who attack the Work attack because we are Catholics," for Franco, on the other hand, "if they are attacked, it is because they serve the regime"; the attack on Opus was part of what Franco liked to call "the international conspiracy" (Franco Salgado, 1976, 475). In Franco's opinion, this showed that Opus Dei "did not interfere in politics and dedicated itself exclusively to the service of God through the exemplary conduct of its members" (ibid., 412). The general saw in the people of Opus loyalty, honor, and competence, and if he appointed members of Opus to be ministers, it was justly due to their integrity. All the Opus members he knew were—he said—"honorable and worthy gentlemen" (ibid., 475).

We find this same official argument in statements by Carrero Blanco when he asserts that the "supposed penetration by Opus Dei, as a group or organization, is totally false. I assure you categorically, having heard it from the lips of its own founder, of whose veracity I have not the slightest hint of doubt" (quoted in Ynfante, 180). William Ebenstein makes a similar point in his study *Church and State in Franco's Spain,* when he says that one of the great triumphs of Opus Dei "is the general recognition that its members who hold high government or professional positions have never been suspected of having participated in the corruption that was so rife in the Franco regime" (Ebenstein, 1960, 40). Of course, this statement was made before the outbreak of the first financial scandals involving Opus members.

Personal integrity and the absence of any wish for organized penetration are, it seems, the identifying characteristics of this presence of members of Opus in Spanish public life. Nevertheless, López Rodó himself indirectly recognizes that some of Franco's ministers did not share their points of view on this matter with Franco himself and with Carrero Blanco. Thus, when in 1962 López Rodó was named commissioner of the "development plan," Castiella, the minister of foreign affairs, commented "this demonstrates that Opus Dei would like to take over everything" (López Rodó, 1990, 311); the falangist minister Solís Ruiz "tried to upset Franco by telling him that it meant the control of the economy by Opus Dei" (ibid., 311); and Fraga, minister of information, "believed that the diversity of opinions and conduct he observed in Opus Dei members obeyed a coordinated plan and that, at any given moment, various persons were moving to opportune spaces on the political and economic playing board" (ibid., 378).

This leads us finally to a paradoxical conclusion which is both curious and amusing. Assume that the application of Weber's thesis to Opus Dei was justified. Suppose, therefore, that the growing rationalization of the economy and of Spanish public administration and the access to political power of certain members of Opus Dei were something more than pure coincidence: then it follows that their systematic denial of any responsibility or collective participation as an organization means that *Opus Dei never could take credit for the role that in fact it played in the modernization of Spain.* To put it differently: paradoxically, only someone writing "from outside" and not connected to the official positions of Opus Dei would be able to recognize—going contrary to the "official" version—the positive historical role that Opus Dei collectively played in the Francoist Spain of the sixties.

Opus Dei and "Technocracy"

This group of individuals who promoted economic change in Spain in the sixties were dubbed "the technocrats," and even today they are known by this name. But except for unanimous agreement on this label, everything, or practically everything, is subject to dispute when it comes to analyzing and evaluating the policy this group put into action.

As we do not have time to examine all sides of the debate, which would render our task interminable, we will discuss only a few points that have more or less direct bearing on the question of the role of Opus Dei (or the role of certain members of Opus Dei, or the role of certain personalities who "just happen" to be members of Opus Dei)—points which, curiously, largely reproduce numerous aspects of the discussion at the beginning of the century immediately after the publication of Max Weber's book.

In the first place, obviously not all the Spanish technocrats of the sixties were Opus members. On this everyone agrees. What happened, some authors argue, is that public opinion came to associate the two labels so closely (Opus Dei = technocrats) that the importance of the role of Opus members was automatically exaggerated, and the contribution of the technical team of advisers from different economic ministries was minimized; the latter were people who had nothing to do with the spirituality of Opus Dei, and who often came originally from the ranks of the Falange. "If one keeps this group in mind," asserts Manuel J. González in one of the weightiest studies I know on the topic, "attributing the swing in Spanish economic policy to the exclusive influence of Opus Dei seems a grave error" (González, 1979, 29). The economist Joan Martínez Alier makes a similar point: "It seems to me that in Spain Opus Dei did not command so much as the economists did, or, better said, Opus Dei commanded because Opus Dei recruited economists or accepted them as fellow travelers: orthodox economic theory serves as a doctrine of social harmony as good or better than the social doctrine of the Church" (Martínez Alier, 1978, 44f.).

José V. Casanova counters this interpretation with an argument that seems worthy of consideration: with examples taken from the Hispanic sociological literature, he demonstrates a relative inability to distinguish between "the technician" and "the technocrat." "The technician is the one who is capable of carrying out the decisions and putting into practice the political projects of the politician. The technocrat, on the other hand, is the one who imposes on the politician certain political decisions based on technical criteria" (Casanova, 1982, 113).

Consequently, Casanova asserts, although not all the technocrats of the sixties were from Opus, the ones who weren't were advisers to the Opus ministers, who had the power and the initiative. It was they who little by little transformed the old paternalistic bureaucracy into a rational modern administration (*relatively* rational and *relatively* modern, I would hedge). And they did it in opposition to Franco's style, generally without the others being aware of the consequences of certain actions (and perhaps even *without themselves* being fully aware, I would venture to say). In this regard, while authors like Fanfani (1934), Robertson (1933), and other critics of Weber, are right in stressing that modern capitalism was born before the Reformation, and that it occurred in Spain as well as in Italy, it is no less true that Spanish capitalism was not transformed into the "rational modern capitalism" of which Weber speaks until the middle of the twentieth century. In Spain there was capitalism, but there was no industrial revolution or rationalization of the economy until the arrival of the technocrats. (We must not forget—I feel justified in pointing this out—that the case of Catalonia is different: in Catalonia the industrial revolution occurred well before 1960, but it happened during a regime of political dependency which hindered, at least in part, the rationalization of the economy.)

For Casanova, it was not necessary to be from Opus to be a technocrat, but the prototypical technocrat was nevertheless from Opus, and this type of person was most likely to occupy the highest positions in Spain in the sixties. Was this the result of a deliberate policy? Not necessarily, says the author (although we are not tired of insisting that something comparable to a "deliberate policy" was going on in the case of an organization that saw in the accession to these jobs "a peculiar means of its apostolate"). Did the one thing have absolutely nothing to do with the other, as the "official" Opus Dei version claims (Herranz, 1957)? Rather, replies Casanova, it was a logical consequence of Opus Dei's ethic. The technocrats have a mystique of professionalism, a vision of the future that conforms to their ideology of development, and a conception of social change expressed in their consciousness of being its "managerial elite" (Casanova, 1982, 63).

What is this conception of social change that the Opus Dei technocrats defended? Its basic objective is the rationalization of the state apparatus, with the goal of placing it at the service of the economy (contrary to the strategy of the falangists, who throughout the history of the Franco regime tried to subordinate the economy to politics). If we pair this affirmation with Casanova's definition of a technocrat—an individual who imposes specific political decisions based on technical criteria (Casanova, 1982, 113)—we necessarily must arrive at the conclusion that the quintessential technocrat was Laureano López Rodó, so much so that in the end we do not know if it is because he is the one who best fits the definition or because the definition was made to order to fit him.

In an article published a year later, the author states that "no one better than he symbolizes and represents the model technocrat" (Casanova, 1983, 31), and he places the figure of the commissioner of the economic development plan at the center of his discussion. But giving this central—if not exclusive—position to the figure of López Rodó leads to a dead end when the time comes to explain why the economic change achieved by the technocrats was not accompanied by the corresponding political change (the democratization of Spanish society). In spite of having held that the political liberalization "was a dependent variable of economic modernization" (ibid., 41), in this field López Rodó and his ilk failed spectacularly.

On this point the analysis of Manuel González (1979) seems more accurate. Not all the so-called technocrats conformed to the same pattern, *not even when they were members of Opus Dei,* and López Rodó is not representative of the whole, contrary to what Casanova would say a few years later. Specifically, with respect to the correlation between economic change and political change, it is not that López Rodó failed in his attempt to achieve democratization of the regime. What happened is that he did not attempt to do that, and he did not propose it. His objective, at least at the beginning, was to implant "modern methods of work and of business organization," without altering the political nature of the Spanish state (López Rodó, 1970, 137).

The "two Lópezes," López Rodó and López Bravo (named minister of industry in 1962), were much less liberal and much more interventionist than Ullastres and Navarro Rubio. It was the economic policy of the second pair that led to "a process of rationalization of the productive apparatus, an increase in competitiveness, an improvement in business practices . . . and technology, and a transformation of the structures of agricultural production" (González, 1979, 353). This reform implied free trade, realism in currency exchange, monetary policy without direct intervention, discipline in public spending, and financing of business without overprotection: in short, *a capitalist market economy*.

This policy has to lead directly to political change. But Ullastres and Navarro Rubio were supported only by Catalan businessmen and a few banks. The liberating impulses of these men would have jeopardized López Rodó's objectives (González, 1979, 299). After these two left the government he was preparing, with a second generation of technocrats, for the perpetuation of the Franco regime in the short run, and for the "continuation" solution of the monarchy of Juan Carlos I in the long run.

In any case, even without provoking political change, the Spanish economic growth of the sixties was not only indisputable but was really spectacular (although Manuel González, consistent with his thesis, states that it would have been much greater if it had not been held back: "if it could have grown more and didn't it is due to the economic policy practiced by the successors of Navarro and Ullastres"; González, 1979, 299). People started talking about the economic "miracle." From 1961 to 1964 Spanish industrial growth outstripped that of Japan and the United States (Gallo, 343). The percentage of the population engaged in agriculture, 50 percent of the total in 1950, declined to 28 percent in 1968.

Between 1960 and 1970, more than four million inhabitants of the Spanish state moved; sixty-six hundred of the eighty-six hundred townships lost population. A little more than 20 percent of these individuals entered the construction industry, about another 20 percent entered industry, and 60 percent entered the tertiary sector (see *Información Comercial Española,* no. 496, 1974). A significant contingent emigrated to foreign countries; the foreign currency they earned, and that brought in by tourism (the number of tourists increased from seven and a half million in 1961 to seventeen million in 1966) was decisive in reestablishing the balance of trade, permitting imports, and putting a definitive end—economically and culturally—to the autarchy of the early Franco years.

In the face of these changes, which the public linked directly to the influence of Opus Dei, although the spokesmen of Opus Dei kept insisting that "Opus Dei does not intervene in politics," a question began to take shape among the intellectuals of the anti-Franco opposition which has not disappeared to this day: how was it possible for a "traditionalist," "reactionary," and "integrist" group to play a role that was "progressive" (comparatively speaking), more or less "liberal" (even if it was the peculiar "liberalism" of technocracy), and in favor of "modernization" (although it had an undeniable streak of "enlightened despotism")? There seems to be no doubt that this is exactly the context in which the myth of Opus as "a secret society of a political character," a "holy mafia," arose.

Several authors—not too familiar at the time with the thesis of Max Weber, who in the final analysis was nothing more for them than the contemptible "father of bourgeois sociology"—began to talk about Escrivá's *The Way* as a "Kempis for technocrats" (comparing him to the author of *The Imitation of Christ*). They did not realize that they were thus turning Escrivá into an important prophet, because if *The Way* was really a Kempis for technocrats, they would have to admit that it was written at a time when there was not a single technocrat in Spain.

On other occasions they minimize the role of Opus members, pointing out that in fact the policy of the new economic ministers appointed in 1957 followed "an economic philosophy quite similar" to that of the rest of the decade of the fifties (Clavera et al., 1978, 315f.). Some go even further and deny the importance of Opus's role, defending the thesis that the economic transformation was produced not because of, but *in spite of* the economic policy of the technocrats (Tamames, 1976; Ros, 1975, especially the article by José Luis Sampedro).

In a good part of the "leftist" economic literature of those years we find a kind of psychological block against admitting the possibility that economic expansion and progress can take place in a nondemocratic regime. Tamames, for example, begins with the assumption that a capitalist model of development *can never* lead to "authentic" development. This would lead finally to a rather surrealistic situation which can be summarized as follows: the technocrats ("rightists" all their lives, and anti-Marxists all their lives) propose economic growth as their objective and consider that well-being, and perhaps even democracy (but they are not sure of this, and in any case they are not in any hurry) will be the natural, if indirect, consequences. On the other hand, the opposition to the Franco regime (the "leftists," accustomed to Marxist analysis) holds that social justice and democracy have to be the primary objectives, and that economic growth will in every case occur as a consequence. Thus these writers considered the policy of the technocrats as doomed to failure, and later tried to minimize its successes.

If we were allowed to extrapolate, we would say that this is not going to be the last time the authors closest to a Marxist orientation are, paradoxically, the ones who come closer to supporting "glasnost" than "perestroika," while the right-wing technocrats seem to have a clearer perception of the dynamics of the relations between the "infrastructure" and the "superstructure." In any case, the question remains: in this context, who are the "moderns" in the Weberian sense? In our view, the answer would have to be: the creators of business schools like IESE (Opus Dei) or ESADE (Society of Jesus), which we will discuss in the next chapter.

13

The Education of Businessmen and the Management of Business

The Starting Point

(*a*) During the sixties, the sociologists Juan J. Linz and Amando de Miguel did several studies on businessmen and special interest groups in Spanish business and on the official elites confronted with the process of administrative reform (Linz and Miguel, 1966, 1968; see also Beltran, 1977, and Moya, 1975, 1984). Their research led to the conclusion that the strategy of economic liberalization and administrative reform adopted in Spain in 1957 implied the progressive replacement of the old financial aristocracy by the new executives and technocrats. An ideology favorable to growth was constructed around "entrepreneurship."

At the same time, an article connected to these research projects presented the following results on the education levels of Spanish businessmen (Miguel and Linz, 1964, 33): 22 percent had only a primary education, 43 percent had been to secondary school, and 32 percent had some higher education; among the last group, the majority had degrees in engineering (16 percent) and law (7 percent); only 1.5 percent had degrees in economics. In Barcelona the average level was noticeably higher: 36 percent of businessmen had higher education, 48 percent had attained the secondary level, and only 10 percent had just a primary education. In a study on Catalan businessmen in the same period (Pinilla, 1967, using data based on a survey done in 1964), Esteban Pinilla specifically names 1958 as the year when awareness of business problems arose and the year of the start of a movement of rationalization and modernization.

As Pinilla also points out, 1958 is precisely the year of the creation, in Barcelona, of the first two college-level business schools in Spain.

(*b*) The year 1957 saw the reprinting in Madrid of a book written by a Spanish essayist who was shot in 1936, and who had been one of the *maîtres à penser* of the traditionalist right-wing intellectuals who adhered to Francoism. The author was Ramiro de Maeztu, and the book has an odd title: *El sentido reverencial del dinero* (The Reverential Feeling for Money). Maeztu wrote the book after a trip to the United States (1925) that made an impression on him comparable, to an extent, to Weber's impressions twenty years earlier. The difference was that Maeztu was familiar with Weber's thesis, so he returned from America convinced that the German sociologist was right in establishing a relationship among Protestantism, the creation of wealth, and capitalism (Artigues, 191f.).

The economic prosperity of the United States was explained, according to Maeztu, by the authentically religious "reverence" people there had for efficiency and for wealth, whereas in the Latin countries indifference toward money had unfortunate consequences from the economic point of view. It was essential, therefore, to replace this indifference with a "*reverential feeling for money,*" which did not necessarily have to be incompatible with Catholicism and was in any case the price that has to be paid if one wanted to have access to wealth (Velarde, 1967, 131ff.).

Maeztu wrote:

> My ideal would be to increase the number, in the Spanish-speaking countries, of captains of industry, model agriculturalists, great bankers, men of business. . . . It is much more difficult to carry on a business that creates wealth than it is to distribute our fortune among the poor and go into a monastery. The latter requires nothing but abnegation, courage, and charity. The former requires the same courage, because one risks one's fortune in business; it requires more abnegation, because business is not content to ask for the sacrifice of a moment, but requires the sacrifice of a whole life of work; and although it does not seem to require as much charity, in reality it gives much more, because the poor to whom one gives charity continue to be poor after they receive it, while the people occupied in business are ennobled by work, apart from finding a way to improve their position. . . . In the roots of economic life one always finds the moral. The economy is spirit. Money is spirit. [Maeztu, 1957, 139]

"Time is money," said Benjamin Franklin, quoted by Weber. "Money is spirit," said Maeztu. And Escrivá said, "Those who are engaged in business say that time is money. That seems little to me: for us who are engaged in affairs of souls time is . . . glory!" (*The Way,* no. 355).

(*c*) Years ago Alberto Moncada argued that the entrance of Opus Dei into the world of business was not intentional and deliberate, but rather the unanticipated consequence of a situation which Opus Dei itself did not control. Moncada's thesis cannot be disregarded, if we keep in mind that he was not only a member of Opus but as secretary of the Association of Friends of the University of Navarre he traveled to Peru, Colombia, and Venezuela in 1966—in the name of and on account of IESE, according to Ynfante—to interview representatives of the planning departments and schools of public administration in those countries (Ynfante, 346).

In his first book written about Opus after having left it, Moncada says that during the first years the question of business was of little concern, and that "it was more or less dismissed from higher, more intellectual perspectives" (Moncada, 1974, 78). Eight years later, he explains the appearance of this concern as *the result of three fears:*

> It seems certain that Opus's entry into the areas of political and economic power was initially due to fear. Fear of being isolated by the most powerful groups in Spanish Catholicism, such as the Jesuits and the propagandists, who received the followers of Monsignor with unsheathed claws. Fear of the animosity of the Falange, who were just as hostile. And fear, above all, of the debts that Escrivá had recently incurred, obliging his children to find a way to pay for the buildings in Rome and Pamplona. This put the group in contact with financiers, especially Catalans and Basques, and then with the Franco administration through the serpentine passageways of government. [Moncada, 1982, 106f.]

Therefore one first possible interpretation about the relation of Opus Dei with the business world would be that this relationship was fundamentally conditioned by the situation and the internal needs of the organization itself. However, this type of proposition easily leads to the sort of discussion of "the economic empire of the work of God" (the paradigmatic example of which continues to the present day to be Ynfante's book, 229–95), which, while it can offer more or less interesting information, suffers from the drawback of never distinguishing the Opus enterprises from the businesses managed by members of Opus, or even from those in which members of the Work had jobs (Casanova, 1982, 330). Inevitably one is led into a real dialogue of the deaf with those who insist that all this has nothing to do with the fact of belonging to Opus Dei, and that "any presentation of Opus Dei as a center of temporal or economic commitments or orientations lacks basis" simply because "Opus Dei is a supernatural and spiritual organization" (Escrivá, *Conversations*, nos. 52, 53).

But we can offer a second possible interpretation, without necessarily disregarding all the elements of the first, which has the advantage of being less complicated and more pragmatic. Even assuming that Opus Dei was "a supernatural and spiritual organization," this does not mean that its directors could not see the advantages of having a solid economic base, not in order to construct an empire or to pay the debts of the architectural enthusiasms of the founder, but as a rational and indispensable means for their "apostolate" at the moment of the expansion and internationalization of its activities.

Specifically with respect to the creation of the Opus's Institute of Business Studies, there is yet another possible interpretation, which is that the school was not a response to any internal need of the Work, and it did not represent any kind of "center of economic commitments and orientations." It was simply that the Spanish situation in 1958 required an initiative of this type (not contemplated at that time within the framework of official university education): the business situation demanded it, and the social and economic context also lent itself to it. Thus it is no coincidence that the Jesuits and Opus Dei once again coincided—and competed—by founding similar educational institutions at the same time in the same city (the choice of Barcelona was no coincidence either).

The First "Business Schools": The Proposal of the Founders

In this section we will make an exception—the only one—to the rule stated at the beginning not to use material from the interviews made during my research as a source to justify the information being offered unless I could also refer to a written document. Consequently—and this is also an exception—I will suspend the criterion of anonymity in the sole instance of the lengthy interviews with Professor Antonio Valero, the founder of IESE, and with Father Lluís Antoni Sobreroca, a Jesuit, founder of ESADE.

Interview with Professor Antonio Valero

The founder of IESE is originally from Zaragoza, lives in Barcelona, is from a monarchist family, is an engineer and professor of economic policy at the Polytechnic University, and is a member of Opus Dei. During the first half of the fifties he had professional relations with two groups. The first was basically a group of Spanish friends with whom he started a business to perform productivity evaluations for businesses. The other group, which was international, centered around what was then called the European Agency of Productivity of the Organization for Economic Cooperation and Development (OCDE).

From these two experiences arose the initial plan for the future IESE. Valero thought an initiative was urgently needed that would contribute to the training of Spanish businessmen; but he thought that the studies did not have to be in economics or strictly in business administration, but rather the emphasis should be on business management. He knew that in the French city of Lille there was such a school (the École d'Administration des Affaires, under the aegis of the Catholic faculties of Lille), and through its director he made contact with the group of experts at OCDE. This group shared his idea, up to a point, although they thought Valero's project excessively ambitious.

Valero's design was to start with a program for "top management," to be followed by others for "general managers" and "operative managers," as well as perhaps a program offering a master's degree. In no case would the institute offer studies at a level lower than postgraduate. In fact, IESE began its activities with a "program of top management" for businessmen (1958). Then it offered a "program of management development" (1959) for middle management and a "program of general management" (1961) for young businessmen just starting out. In 1964 it started its master's program, which was established, among other things, to guarantee full-time, as opposed to part-time, positions to the institute's faculty.

In spite of criticism, and with no guarantee that it was going to work,

Valero wanted to try it. Then he found out (1956–1957) about the existence of the plan to create ESADE by the Jesuits, but he did not share Sobreroca's idea of having an undergraduate program. The group of friends Valero worked with in productivity studies agreed with his views. They thought the institution should be on the graduate level and be recognized as such, but that the project would hardly be viable within the framework of the state universities, and this led them to lean toward one of the very few existing private universities. Without Valero's knowledge, a member of the group approached the authorities at the University of Navarre, who consulted Valero. When they found out he was fully involved in the project, they put him in charge of carrying it out. But the University of Navarre did not clearly comprehend the project as Valero presented it—it was ambitious, and not well-adapted to the traditional mold of university education—and at first they rejected it.

From Navarre, the dossier was sent to Rome: Msgr. Escrivá, who was also grand chancellor of the university, supported it right away and asked them to reconsider their initial decision. Contrary to what has sometimes been said—Valero stresses—the initial idea for IESE was not Msgr. Escrivá's. But the founder of Opus favored it, and he played a decisive role in the final approval of the project.

As for the choice of Barcelona as the location of the institute, it was not an a priori decision but the result of a series of explorations Valero made in several large Spanish cities. The atmosphere was a bit more open in Barcelona, and the Catalan reality was the only one that would work as a laboratory for IESE. At the beginning of 1958 the creation of the institute was approved, and at the end of that year the first classes were held. Success came immediately and repercussions throughout Spain soon followed.

In sum, Professor Valero concludes:

- The establishment of IESE was a response of a strictly empirical proposal and had nothing to do with ethics.
- IESE could have been created perfectly well in a framework completely separate from Opus Dei from an institutional point of view; its connection with the University of Navarre was little more than a coincidence.
- The early criticisms of the project had to do with its ambitious character; ethical considerations had nothing to do with it.
- The ethical connotation appeared for the first time when Msgr. Escrivá lent his support to the project and made it his own; and it only extended to the obligation that IESE's teaching conform to the social teaching of the Church.
- The moment of the creation of IESE coincided with the period of economic liberalization under Navarro Rubio and Ullastres, López Rodó's development plan, and the growth perspectives of the 1960s in Spain and all of Europe. IESE was not part of any strategy, but it was born in this context, and the context obviously facilitated its success.

Interview with Father Lluís Antoni Sobreroca

In 1926 one of the Jesuit schools in Barcelona created a section of business studies alongside the classical baccalaureate. Father Sobreroca comes from this institution, the Immaculate Trade Institute; having completed his training, he became a teacher there.

In 1952 the superiors of the Society closed several facilities, among them the Trade Institute, in order to be able to send Jesuits to work in Bolivia. Sobreroca, however, was sent to Deusto to study social sciences. The Commercial University of Deusto, created in 1916 alongside the classical faculties of the University of the Jesuits in this Bilbao complex, was the first institution in Spain specifically directed to the training of businessmen.

In 1956 the Jesuits founded a Higher School of Business Techniques in San Sebastian, whose mission was to train businessmen with a humanistic foundation (the following year the Jesuits organized, in Madrid, a Master of Business Management degree program). Also in 1956 a group of Barcelona businessmen approached the Jesuits with a petition to create "a Deusto in Barcelona," because the state university trained economists, but not entrepreneurs. The provincial of the Society called Sobreroca and suggested the possibility that he might take charge of the project. After contacting the San Sebastian school, which seemed to him to offer the optimal combination of economic training and humanistic education, Sobreroca took a course (1957–1958) abroad to study English and to learn about other international projects in business education.

In Barcelona, meanwhile, the promoters formed a corporation (in which there was not a single Jesuit) and created the name of the future institution: Higher School of Business Administration and Management (ESADE). Sobreroca had wanted them to use the expression "business sciences," which, in fact, is what the state universities later adopted. The model agreed on by the sponsoring corporation distinguished the academic part, which was the responsibility of the Jesuits, from the economic part, which was separate from the Society. A joint academic committee (Jesuits and laypersons) was created, presided over by the director (a Jesuit), and an economic committee, also joint but presided over by a layperson.

On his return to Spain, Sobreroca assumed the directorship of the school, which began its operations with the school year 1958–1959 (the same as IESE) with an initial class of thirty-four students. ESADE's sponsors knew about IESE, and in the planning stages, in order to avoid duplication, they learned something about it. Sobreroca found out that IESE wanted to work exclusively with graduate students, while the initial plan of ESADE was to give an undergraduate degree.

In any case, by the beginning of the second year ESADE's activities had already diversified: seminars, studies, research, and training courses for businessmen. And after six years had passed there would be specialized courses and a master's program (first offered the same year as at IESE). One justification for this diversification was that it allowed the institution to employ a full-time faculty. This is the same reason given by Professor Valero with respect to IESE. In general terms, the parallels between the two institutions are remarkable, and it is clear that even though they began from different starting points (one with top management, the other with undergraduates), they progressively converged and developed very similar activities.

The initial plan of studies at ESADE was structured around an economic core and a humanistic core (psychology, social doctrine of the Church, ethics, theology). The school had a religious adviser and organized religious activities and retreats, which it would gradually abandon as the years went by. But there was no religious indoctrination of students, and the school was characterized from the beginning by pluralism and by lack of connivance with Francoism. Although not marked by a confessional spirit, the school nevertheless had a Christian philosophy. Through the years all these aspects have evolved, and it is possible that at one time the institution even ran the risk of a certain loss of identity, although this danger has disappeared. Now the school defends both pluralism and the values of the pursuit of justice.

IESE and ESADE

Up to this point I have tried to convey the content of the two interviews with maximum fidelity, withholding any observations or comments. Overall, based on interviews with other professors and former students of both institutions, I believe that we can distinguish both a series of parallels and a series of differences.

(*a*) There is a clear similarity of objectives, and a nearly total coincidence of chronology. In spite of the fact that IESE and ESADE at first were aimed at different publics, the gradual expansion of both produced a wide band of convergence even in this aspect.

(*b*) The success of both is indisputable. They were pioneers in the field of education in business management, and even though the state universities have for years now offered courses in business sciences in the economics departments, ESADE and IESE have done extremely well against the competition.

(*c*) The cost of the education is, of course, much higher than in the state universities. But both the prestige and the network of contacts in the two centers lead to a common perception of these costs as a profitable investment, since a job or promotion is considered guaranteed. Some former students in the master's program at IESE, however, feel that the connection of the institute to Opus Dei, given the suspicions and hostility it provokes in broad sectors of Catalan society, carries with it more disadvantages than advantages in the corporate world and big business. Not once did I hear a parallel argument expressed by former students of the Jesuit ESADE.

(*d*) Certainly the ideological load (in the double sense of baggage and, as it were, mortgage) associated with ESADE is much greater than at IESE. One must not lose sight of the fact that the Society of Jesus has a moral and casuistic tradition that does not exist (or at least did not exist in the fifties) in Opus Dei. In this respect, ESADE represents without doubt an innovation when compared to the thought of Jesuit authors like Joaquín Azpiazu (*La moral del hombre de negocios,* 1944) or Martín Brugarola (*La cristianización de las empresas,* 1947). But the innovation does not presume a total rupture. The initial idea of ESADE was clearly to train businessmen in accordance with the social doctrine of the Church (a few years later they would add: "and in accordance with the teachings of the Second Vatican Council"; Nicolás Cabo, 101). Thus the initial plans of study at ESADE and other similar centers directed by the Jesuits contain courses that "develop complementary studies in modern theology to plant the roots of Christian understanding in individual and social life" (ibid.). The general directives refer to "five courses in theology and another five in Catholic Social Thought" as the usual norm (ibid., 347), althought in the case of ESADE the original plan of study provided for only four courses in Christian thought and one in "professional deontology" (ibid., 147ff.).

(*e*) Increasingly this dimension of the educational offerings at ESADE has been centered in courses like social philosophy, sociology, and the history of economic thought, to the detriment of the social doctrine of the Church conceived of as a study of the papal encyclicals. The doctrinal aspects remain on a secondary plane, and currently both professors and students readily admit the possibility of expressing attitudes that are critical of certain positions taken by the Catholic authorities.

In this sense, there is more room for ideological pluralism in ESADE than in IESE, a reflection in good part of an ideological evolution which, in the Society of Jesus, has been undoubtedly more public and more explicit than in Opus Dei. Thus during the sixties ESADE presented itself as a center oriented toward "the training of men capable of creating and running businesses, *with economic, social, and Christian standards in the service of the common good*" (Nicolás Cabo, 147; emphasis mine). The information/publicity brochures currently in circulation speak of "the scientific and humane training of men capable of creating and running businesses, *with critical sense of man's realization and the transformation of society*" (emphasis mine). In an equivalent brochure from IESE, by contrast, the reference to the religious dimension is still explicit: after speaking of education that enables the promotion of an authentic social development, it says that "IESE understands that this development is only possible when it includes the perfection of all dimensions of the person, *in accordance with the Christian view of man,*" and goes on to specify that "*the Prelature of Opus Dei is in charge* of all the doctrinal and spiritual aspects of the formative activity" (emphasis mine).

(*f*) Teachers at both centers agree—frequently even using identical words—that there is no effort to indoctrinate the students, and that the important thing is "the spirit which imbues the various activities," "the upholding of certain principles, without any wish to impose them and respecting the freedom of all," "exemplary testimony," and so on. They are also alike in showing themselves to be extremely prudent in evaluating the real influence of this "spirit" on the later professional behavior of their former students.

As for these former students—and admitting the very limited scope of these statements, which could easily be overgeneralized—we would dare say that those who studied at IESE *and are members of Opus Dei* tend to recognize both the nature and the positive character of this influence; on the other hand, former students of ESADE and former students of IESE *who are not Opus members* tend rather to perceive the existence of a dual level and a dual discourse—the technical and the ideological—without too many reciprocal connections.

Students admit practically unanimously that the level of technical competence of the faculty and the education as a whole is more than satisfactory. But frequently students from both schools have difficulty explaining what specific contribution was made by what is usually called the "moral discourse." At IESE there is a clear difference between the master's students and the managers who took highly advanced courses, who are more inclined to ignore all the ingredients of an ideological character offered by the institution. Altogether, however, among the alumni of IESE there seems to be a clearer perception that the institution in fact proposes and defends certain specific elements of an ideological type, which the individual accepts or rejects. In the case of ESADE, on the other hand, this perception is frequently much more blurred, possibly as a consequence of the greater pluralism mentioned earlier.

On those occasions when these perceptions are verbalized in a critical tone, and if we may be permitted to reduce the accusations to an inevitably simplistic formula, I would say that the reproach directed at IESE is that they are "people with fixed ideas," while the criticism directed at ESADE is that "ideologically they're not clear themselves." "There is always a priest from Opus nearby, ready to give orientation chats and suggesting spiritual direction," the former would say. "A Jesuit can go to a revolution in Central America, if he wants; and if he wants, he can dedicate himself to training businessmen in Barcelona; but if what he wants is to do both things at the same time and doesn't know how to marry them, the problem is his and not ours," the latter would say.

14

The Worldly Asceticism of Opus Dei

Preliminary Observations

On his great "catechistic" journey through the Iberian Peninsula in 1972, Msgr. Escrivá de Balaguer visited IESE, where he met with a large group of businessmen. This is what the biographers have deemed worth saving about this visit and his conversation with the businessmen:

> Do not forget the Christian meaning of life. Do not take pleasure in your successes. Do not feel hopeless if something should fail. Anyway, if you have a hundred things going, one has to go wrong, because the other ninety-nine are going well. Remember those who have less than you. [Sastre, 551, who adds that Escrivá "encouraged them to use money with the generosity required by the Gospel"]

"The fact of managing money, or of having it, does not mean that a person is attached to wealth," is the sentence another biographer (Bernal, 328) puts in the mouth of the Padre, who went on to tell the story of a poor man who, in spite of being poverty-stricken, felt rich, and that of a woman who, although very rich, lived poorly and generously gave away her possessions.

In their simplicity, these examples are the best way for us to pick up the thread of the parallels between the Puritan economic ethic and the ethic of the members of Opus Dei. What interests us fundamentally here is not so much the elaboration and learned formulation of the ethical discourse, but its practical effects—the psychological legitimation and impetus they contributed to the adoption of specific kinds of behavior. Let us look at a few examples.

Michael Barret is a New York priest. He joined Opus in 1972 while he was a student at Columbia University. He explains it this way: "The idea of sanctification in the midst of the world excited me. It seemed something fantastic. So Opus Dei was fantastic: to be a doctor or something and at the same time to try to make oneself holy and help others to be holy" (West, 20).

Héctor Reynal is another Opus Dei priest, a professor of Christian ethics for businessmen at a business school for managers—similar to IESE—in the Philippines: "That was one of the things that Msgr. Escrivá wanted to inculcate in the minds of many: that it is not good to give up the fight, that one has to *try to stand out, excel, be successful.* Not for selfish reasons, but because in this way one can spread the evangelical word further" (West, 151; emphasis mine).

Eduardo, an engineer who pursued graduate studies at IESE, and who now works in Latin America for a company specializing in public works and construction technology, singles out from his experience at IESE "the kind of people at IESE, the seriousness of work, and the sincerity of the people who belong to the Work" (Moncada, 1982, 136). He states that in Opus Dei he found "certain doctrine, true friendship, and morality without ambiguity (ibid., 136). Of Msgr. Escrivá he says, "The Padre is the point of contact between Catholic orthodoxy and Protestant practice. He reconciles the fecundity of well-kept faith, which belongs to Catholicism, with the boldness and social efficacy of the Reformation" (ibid., 146).

Manuel is a Catalan industrialist, father of an engineer who joined Opus Dei while still a student. After visiting one of the Work's centers and discovering its spirituality, Manuel said: "Today I have discovered the best thing of my life; this suits me; *I have worked like a dog, frantically, doing nothing else, and now I discover that by working I can also sanctify myself. This is fantastic!*" (Seco, 81; emphasis mine).

Finally, José Barco Ortega, who has a master's degree from IESE, asked in an article published in the magazine of the University of Navarre, *Nuestro Tiempo,* if a "leader" has to be a technical expert, a politician, or an ideologist, and he thinks that all three things are equally necessary, while singly they are insufficient. "One achievement of our age is direction by competence, leaving behind direction by those whose only qualification for command is that of bloodlines or money" (Barco, 29). This is clearly the Weberian definition of the bureaucratic leader, who bases his authority on rational-legal legitimation. But the rest of this article is particularly interesting, since the author describes the five requisites for any good leader as follows:

- He has to have a "strong, whole" personality, and be sure of himself, "so that through faith in himself he can achieve knowledge of reality."
- He has to be someone "who in governing achieves the fullness of his being," in such a way that the only persons to attain this "are those who through exercising it are going to be content, fulfill their vocation, achieve the fullness of their development."
- He has to be a man with "a will to serve," oriented "toward others instead of toward himself."
- He has to be a person "in control," who does not allow "his ambition

and desire for pleasure make him act in an excessive or unnatural manner," because only in this way will his actions be "rational."

- He has to be prudent, in the sense that "wanting and working" have to depend on his "knowledge of the truth," and not the other way around; he is not someone who makes "the truth depend on the desires and acts of the will" (Barco, 32ff.).

All these examples except Moncada, who is not out of tune with the others, were taken from the "official" literature of Opus Dei, and all of them, while recognizing the differences which undoubtedly exist between the Protestant Puritans and the followers of Msgr. Escrivá (let us not forget that Weber himself emphasized the differences existing even among Calvinists, Pietists, Methodists, Baptists, etc.), illustrate that the fundamental parallel between "Protestant ethic" and "Opus ethic" is more than justified.

The Rational Asceticism of Opus Dei

It would not make much sense to follow Weber's argument step by step in order to find out exactly where the points of convergence and divergence are. In this chapter we are going to adopt a broader perspective, centering basically on an analysis of Opus Dei; at the same time we will utilize some of the principal Weberian notions, especially those which revolve around the phenomena of worldly asceticism (*innerweltliche Askese*), the asceticism of work (*Berufsaskese*), and professional duty (*Berufspflicht*). In other words, we will take the least controversial point from Weber's theory: the religious legitimation of constant and methodical professional activity or, in the terminology typical of Opus Dei, the "sanctification of work." Even one of the most radical critiques of Weber's thesis, Herbert Lüthy's, admits that the transposition of asceticism onto temporal life, life "in the world," legitimates economic activity, since work is converted into an exercise of piety equivalent to prayer and is seen not as punishment but as "for the greater glory of God" (Lüthy, 1965, 62). From this a morality of success gradually evolves, and work becomes the bourgeois virtue par excellence (ibid., 67).

Moreover, it is evident that it would not make sense to pose the question of the relation among the spirituality of Opus Dei, its economic ethic, the professional behavior of its members, and the historical role they might have played in specific contexts in terms of strict causality. Obviously we are looking at a history of *elective affinities* (*Wahlverwandtschaften*), just as Weber was, and just as Goethe had been. (On the origins of the notion of elective affinities from the eighteenth-century Swedish chemist Torbern Bergman through Goethe's novel and up to Max Weber, see Howe, 1978.) The story of the elective affinities of Goethe is not, as has often been said, the story of adultery. It is the story of adultery desired but not consummated: the story of fantasies of possession. Above all, it is the story of the unforeseen consequences of an action, unleashed unintentionally by the protagonists, in which the "elective affinities" constitute the factor that is at once uncontrolled and decisive.

Likewise, *the whole history* of Opus Dei is *a history of the unforeseen consequences of an action:* an action in whose origins we find a desire, a fantasy of possession. "When will we see the world ours?" (*The Way,* no. 911). "You can feel the pride of a conqueror of a hundred worlds. The world is ours because it belongs to Christ" (Urteaga, 1948, 113). It is the desire and the fantasy of the "re-Christianization of Spain" and the "kingdom of Christ": "We want Christ to reign, it is right for Christ to reign," Escrivá repeated to his "children" (Vázquez, 156). A large part of humankind and earthly realities *still* are not "presided over by the sign of Christ's cross"; and "the Word Incarnate goes on wanting men and reality to be His" (Casciaro, 1964, 756f.).

"For Christ to reign in the world there must be some people who, with their eyes fixed on heaven, seek to acquire prestige in all human activities, so that they can carry out quietly—and effectively—an apostolate within their professions" (*The Way,* no. 347). Perhaps nowhere but in this passage from *The Way* do we find—in a single sentence—three words that so aptly define Opus: "prestige," "quietly," and "effectively." The history of Opus Dei is the history of the consequences of the will to acquire efficacy and prestige, and the decision to act with discretion, "quietly."

These three words seem a condensation of the peculiar rationality of Escrivá's initial proposal which later on will develop—very logically when looked back at a posteriori, but not foreseen a priori—into a morality of success linked to the conscience of selected members of Opus, and a whole theory of the sanctification of work that the disciples will pretend to discover, at least in its essence, in the original thought of the Founder.

In the case of Protestantism, likewise, the appearance of the phenomenon of the communities of the chosen, of "saints," of "true believers," is a late but inevitable consequence of its historical development. In Opus Dei, we have already seen that the immediate origins of its "elitism" must be sought in the Jesuits' "vocation of the training of elites" (Charentenay, 1990, 70), in what P. Ayala called the "training of the select" (Ayala, 1940). Unlike the Society of Jesus, however, Opus Dei wishes to reproduce the structure of the Church (priests and laypersons, men and women; Fuenmayor et al., 119), and this enhances the danger of its becoming an exclusive group (exclusivist and exclusionary).

This is where the parallel between Opus Dei and certain Protestant sects is most obvious. Membership in a church, says Weber, does not in itself constitute any test of the quality of the individual, whereas a sect is "an exclusive voluntary association of persons who are qualified from the religious and ethical point of view": only those ethically qualified are admitted into the sect, and consequently the mere fact of belonging to it automatically amounts to a certification of ethical qualification (Weber, 1920, 211). Of course this does not suffice to define a sect sociologically, and many other characteristics of Opus Dei distance it—radically, in my view—from this model of religious organization. As the former archbishop of Valencia, Marcelino Olaechea, protector and friend of Escrivá, said at the Congress of Perfection and Apostolate to which we referred in an earlier chapter (Chapter 10), one cannot speak of sects in Catholicism: "Ours are facets of the same discipline, the same submission, the same apostolate: facets" (*Actas,* 1957, 317). But, the archbishop continued, in a sentence that could be interpreted as a kind of *mise en garde* to the Opus people present at the congress, "we must take care not to let these facets become so broad that they seem like sects. . . . Let us not make our poor non-Catholic brothers right in any respect. We do not have sects" (ibid., 317).

The elitism of Opus Dei, Escrivá's references to "caudillos," the conviction of being "the small remnant of Israel," the insistence on being more perfect than the rest—these might give rise to the formation of what Weber himself called "conventicles," and what Joachim Wach, in a particularly happy turn of phrase which is unfortunately little used, called *ecclesiola in ecclesia* (Wach, 1944). It is the same danger that Cardinal Baggio, prefect of the Sacred Congregation of Bishops, in his request for information from Álvaro del Portillo before the approval of Opus Dei as a Prelature, called the "parallel church" (Fuenmayor et al., 613).

Parallel church or *ecclesiola in ecclesia:* over and above the question of what label one attaches to Opus Dei, what must be emphasized here above all is its tendency toward religious aristocracy. Especially in the early years of its development, the years of the gestation of its ideology, Opus Dei—then focused exclusively on the apostolate among intellectuals and academics—elaborated a *Virtuosenethik* (Weber, 1920, 260ff.), an ethic of heroes and of caudillos. Jesús Urteaga (in *El valor divino de lo humano,* 1948, translated into several languages and reprinted thirty times in Spanish) is probably the best exponent of this tendency; for him, those who live their religion in a more coherent and methodical manner become a "select minority" in the bosom of the Church, and are far above the ordinary religious life of the great majority.

These "select few," or "virtuosi," are at the same time active individuals who have accepted a mission, who feel they have been called and that they must respond to this call ("calling" in *The Way,* nos. 902–28; "vocation" in later Opus literature). They are *instruments* at the disposition of the divine, rather than *receptacles* of divine grace; that is to say, their religiosity has a much more ascetic than mystical orientation (Weber, 1920, 538f.).

As instruments chosen to contribute to the triumph of a cause, the members of Opus Dei are to ensure the success of the same, "with prestige, effectively, quietly." For this reason Msgr. Escrivá insisted that "one must strive to stand out, to excel, to be successful" (West, 151), while his "caudillo ethic"—the Hispanic version of Weber's *Virtuosenethik*—creates the necessary conditions for this success. And when one has effectively managed to "excel and stand out," one sees the success achieved as confirmation or proof—the *Bewährung* of Weber's Puritans—or the fact of having been chosen.

The mechanism, then, is the same as in all rational asceticism: the secret of Opus Dei's success is the same as that of the medieval monasteries (Weber, 1920, 544f.), and the same as the one intuited by Wesley, the founder of Methodism (Weber, 1904, 256f.). In 1939 Escrivá said: "If you are a man of God, you will seek to despise riches as intensely as men of the world seek to possess them" (*The Way,* no. 633). In 1972 he would say to the businessmen gathered at IESE: "the fact of managing money, or of having it, does not mean that *one is attached to wealth*" (Bernal, 328; emphasis mine).

Between the statement of 1939 and the one of 1972 intervenes the whole history of the *unforeseen consequences of action,* to which we referred above. "Asceticism itself has generated the wealth that was previously despised" (Weber, 1920, 544). Also intervening is the discovery of the possibilities opened by a certain interpretation of the "parable of the talents" (Matthew 25:14–30), the same discovery, in fact, that the Puritans made. "The rich man is not condemned for what he has, but for the bad use he makes of the resources God has entrusted to him," we read in a recent text by an Opus author (Fuentes, 1988, 15); when Jesus said, in the Sermon on the Mount, "Blessed are the poor," it was a spiritual, not a socioeconomic affirmation (ibid., 22f.).

"It is not riches or honors that make us of the world or not. These things are not prohibited to Christians, as long as they do not love them immoderately. One can be rich without being attached to riches." The sentence could be Escrivá's, were it not for the fact that it was written nearly four hundred years ago by the Jesuit Bellarmino. And it is the same idea held by the Puritan moralist Baxter, who did not want a preoccupation with material things to weigh any more heavily on the shoulders of his "saints" than "a light overcoat which could be taken off at any moment" (Weber, 1904, 264).

If Escrivá adds any original element to this idea, which we can find phrased in a thousand ways throughout the history of Christian spirituality (see Troeltsch, 1912), it is this: to renounce the use of wealth—in the service of a good work—would be equivalent to allowing only the enemies of the good cause to make use of it, those who do not want "Christ to reign."

But this kingdom of Christ must be a kingdom *in the world.* "I will reign in Spain," was the slogan of those who urged devotion to "the Sacred Heart of Jesus," who filled the country with monuments and religious posters after the Spanish civil war. The active asceticism of Escrivá's followers is an asceticism whose will is to influence the world and, in the end, to dominate it. The individual establishes a relation of domination with the world, says Víctor García Hoz, commenting on "the anthropology of Msgr. Escrivá de Balaguer," since man "is made to dominate the world through knowledge and through action" (García Hoz, 1976, 9). In other words, the asceticism of Opus Dei is a *worldly asceticism.*

This is the point that places Opus Dei at odds with the monastic tradition and at the same time reinforces its similarity to Protestant Puritanism. Remember that for many years one of Escrivá's major worries was preventing his "children" from being absorbed into the religious. While respecting and venerating the religious state, "we are not religious, nor do we resemble religious, nor is there any authority in the world that can oblige us to be religious" (Escrivá, *Conversations,* no. 43). And the members of Opus are radically distinguished from religious and from monks because they do not withdraw from or leave the world, but they stay embedded in it and in it they are sanctified. This is exactly the same as the case of the representatives of ascetic Protestantism: "This special religious life," Weber wrote, "did not develop in monastic communities away from the world, but in the world and its institutions. A like rationalization of conduct in this world was, in the end, the consequence of the *Berufskonzeption* of ascetic Protestantism, that is, the conception of *the profession as vocation*" (Weber, 1904, 218).

> The idea of vocation, of calling, occupies (in the thought of Msgr. Escrivá de Balaguer) a central place but—and this is the important thing—that vocation assumes calling, yes, but not exit or flight from the personal human condition; rather it looks with a new light on the reality in which one is already living, and, therefore, it perceives an invitation to perform the acts one is already performing with a new spirit. "Know well (said Escrivá) that on the outside nothing has changed: the Lord wants us to serve him precisely where we carry on our human vocation: in our professional work." [Illanes, 1984, 80f.]

We have worldly asceticism, then, and the rationalization of conduct as a consequence of the conception of professional work as vocation. Thus far, the parallel is nearly perfect. But there are differences too, significant differences, between the type of worldly asceticism analyzed by Weber and that of Opus Dei.

If the otherworldly asceticism of the monk rationalizes life so as to make it appear to be *a sign* of the future Kingdom, and if Protestant worldly asceticism sees the believer's work life as his contribution *to the construction* of the Kingdom of God, Opus Dei's asceticism tends rather to conceive of work as a collaboration in the establishment or *the implantation* of the Kingdom as a reality in this world—present, not future. "Persevere in your place, my son: there . . . what work you can do to establish our Lord's true kingdom!" (*The Way*, no. 832). At bottom, the worldly character of Opus Dei's asceticism is reinforced even *more*, as it endeavors "to construct the earthly city in peace and before God" (Portillo, in Fuenmayor et al., 445). As Berglar would say, "the birth of Opus Dei was a giant step across the threshold of the new stage the Church was initiating: the Christianization of the world from within"; a little further on, totally euphoric, he states that Escrivá's message presumed "the beginning of a new era in Christian life" (Berglar, 79, 97).

Without going to such extremes, we might perhaps consider, more modestly, that Opus Dei represents *a new modality* of worldly asceticism. But first we must present a brief analysis of the fundamental ideas on which it is based.

The Sanctification of Work in Opus Dei

The worldly asceticism of Opus Dei can be summarized in the words of its founder: "To be holy means to sanctify work itself, to sanctify oneself in work, and to sanctify others with work" (*Conversations*, no. 55). Although those in Opus Dei often speak simply of the *doctrine of the sanctification of work,* this "unitary expression" always implies this triple dimension (Rodríguez, 1971, 10). Consequently, it signifies that in Escrivá's conception personal sanctification through work is inextricably linked to the apostolic vocation of the Christian, to "the apostolic mission through work" (Rodríguez, 1965, 237).

The Notion of Sanctification

This presence of the "apostolic" dimension as an indispensable ingredient linked to the search for "personal sanctification" constitutes the first characteristic fundamental element of Opus Dei's asceticism.

In ascetic Protestantism, the discussion that initially arose around the conception of professional activity as a vocation was an attempt to respond to the individual's concern for *his own* salvation, to his often anxious search for signs that would allow him to reach something resembling *certitudo salutis,* the certainty of salvation. Only later, with the appearance of the sects, the communities of "true believers" and of "saints, chosen and saved," was the preoccupation with one's own salvation, by now practically superfluous, replaced by a preoccupation with *the salvation of others.*

(*a*) In Opus Dei's case, on the other hand, this "apostolic" concern was present from the very beginning. Obviously, this was largely because the Opus Dei member was not psychologically burdened by the assumption of the *doctrine of predestination.* Although Escrivá at times used expressions that seemed to approximate it, as when he said that "all men can and must aspire to holiness, following the particular will that God has for each one of them" (quoted in Fuenmayor et al., 274), and especially when he wrote that "*God has you all numbered from eternity!*" (*The Way,* no. 927), there is no doubt that the spiritual solitude and uncertainty of the faithful Calvinist is at the opposite extreme from the situation of the Opus member. The Protestant must travel the whole slow and laborious path from the anguished doubt of the individual who believes in the "double decree" and does not know what his destiny will be, to the secure and reasonable hope of the Puritan who verifies that "God blesses him in his tradings"; the followers of Msgr. Escrivá are spared all this from the beginning, because the Padre guarantees their salvation from the moment they join the Work.

"A messiah," writes Hutch, "is one who at some moment feels capable of answering the question: What do I have to do to have eternal life?" (Hutch, 69). Using the nice formulation Hertel chose as the title of his book, this is what the Founder told those who joined Opus Dei: "I promise you heaven" (Hertel, 1990).

As a group chosen by God to realize "his work," therefore, the members of Opus Dei consider personal sanctification to be a question of perfection (*The Way,* no. 291) and perseverance (*The Way,* no. 983: "To begin is easy; to persevere is sanctity"). But the call of God is simultaneously a call to sanctity and to the apostolate (Alonso, 1982, 231f., 247, 253). The sanctification of work is simultaneously the sanctification of others through work: "sanctified" work is "sanctifying" work (Rodríguez, 1986, 202).

In fact, the Opus Dei texts of the early years insisted more on the apostolic dimension of work than on work as an ascetic means of personal sanctification. The earlier chapter devoted to a more detailed analysis of some aspects of *The Way* (Chapter 6) allowed us to see that the doctrine of the *sanctification of work in the world* was a later development that did not appear at all in the subject index of the first editions of the book, which had instead whole chapters devoted to "apostolate," "apostle," and even "proselytization."

The expression "sanctification of work" was used by Escrivá on very few occasions during the early years, and then in very imprecise if not to say elementary contexts: "Add a supernatural motive to your ordinary work, and you will have sanctified it" (*The Way,* no. 359).

When in later years he began to use it in a more habitual manner, practically always in connection with questions relating to the apostolate, one has the impression that he was referring to what was intitially a very secondary dimension of Opus Dei's spirituality but which gradually became more central. Nevertheless, the "official" literature presents the following programmatic text as dating from 1934: "To unite professional work with the ascetic struggle and with contemplation—something which might seem impossible, but which is necessary, in order to help reconcile the world with God—and to convert this ordinary work into an instrument of personal sanctification and apostolate" (in Fuenmayor et al., 43).

But it was in the sixties that the theme of the sanctification of work actually begins to be presented as central in the spirituality and asceticism of Opus Dei. (In Mateo Seco's bibliographical study, in Rodríguez et al., 541–51, for example, the earliest reference to it is 1965.) In 1948 Urteaga, in a book that is, as we have said before, an excellent reflection of the points of view held by Opus Dei in those times, dedicated very few pages to the topic (168–76), which does not seem basic at all. "Love your profession like crazy, the instrument God put within your reach to gain heaven on earth and save souls" (Urteaga, 1948, 217). After a study published originally in 1966, however (Illanes, 1980), book after book began to appear on the theme, as well as all kinds of articles. Even the former minister of the treasury from the years of the Spanish economic development plan, Navarro Rubio, published a little book called *Sobre el trabajo,* whose last section was dedicated to "Work in the spirituality of Opus Dei" (Navarro, 1987, 133–40). And one of the functions of all this literature was (consonant with the pattern repeatedly observed and analyzed in our chapters on Opus's historical evolution) to attribute exclusive paternity of the doctrine of the sanctification of work to Msgr. Escrivá *from the very moment of Opus's foundation.* The literature thus affirms that

> Opus Dei arose from the first moment with a clearly defined theological and apostolic content, founded on the universal call to holiness and the apostolate inherent in Baptism, and on the personal commitment of its members. Through the faithful fulfillment of their family and social duties and, in particular, in their own professional work, the members of Opus Dei seek their own sanctification and that of others, each one remaining in the canonical state to which he belongs, with a specific spirituality, clearly secular. [Fuenmayor et al., 424]

Regardless of historical vicissitudes, this is the official and *actualized* version of the doctrine, as expressed in numerous articles of the current Statutes of the Prelature (see especially articles 2, 3, 22, 82, 86, 92, 94, 113, 116, and 117). The thing that in *every* case remains constant and unchanged is the connection between "call to sanctity" and "call to the apostolate," and between one's "own" sanctification and that "of others." Over and above the obvious parallels between this and the ascetic Protestant conception of work—to work always, everywhere, hard, and well—in this link between sanctity and apotoslate we have, then, the first peculiarity of the kind of worldly asceticism represented by Opus Dei.

(*b*) Opus Dei was not only spared the initial burden laid upon Protestantism by the doctrine of predestination in its "harshest" Calvinist version. In addition, the link between work as vocation and sanctification is much easier to establish in the case of Opus, thanks to the value placed upon *the works* in traditional Catholic theology, as opposed to the principle of "justification through faith" (*sola fide*) in the churches of the Reformation.

Here we need quote only one passage by an Opus author:

> In its ethical aspects, a theology of sanctity connotes or presupposes the theological study of the value of the works of the Christian. The works are, in the first place, a sign of the justification, its external manifestation: the Christian cannot keep following a pagan or mundane existence, but must live according to Christ, manifest his belonging to God and his faith in his commitments. But that is not all: the justification is not only in God, but in the justified; the works have not only the value of a sign, but they form part of the very process of justification and more concretely of its increase. [Illanes, 1984, 31]

In ascetic Protestantism, the rejection of salvation through works, or of works as a means of attaining salvation, obliges the moralists to take a long detour. In every case works constitute a means, not of "buying" salvation, but rather of freeing oneself from anxiety about salvation. That is, work does not lead directly to salvation, but indirectly to the hope of having been "counted among the saints." This "detour," this roundabout route, Opus Dei can serenely dispense with.

What the asceticism of Opus Dei has in common with Protestant asceticism is thus not this element, but exactly the element Weber considered *specific* to Calvinism: the necessarily *constant and methodical* character of work. Instead of the old traditional Catholic practice of isolated good works, Weber says, Calvanism is "a whole life of good works built on a system." This constant and methodical control is likewise what Opus Dei introduced with its doctrine of the sanctification of work.

In Calvinism, the logical consequence of predestination ought to have been fatalism; but the psychological result was exactly the opposite, thanks to the introduction of the notion of "confirmation," or *Bewährung* (Weber, 1904, 161). In the case of Opus Dei, on the other hand, *logic and psychology coincide* and mutually reinforce each other at the moment of legitimating the active intervention of man in the world and, thereby, the development of the capitalist spirit. Like the Reformed Protestants, however, Msgr. Escrivá too availed himself of a *latent* logic and not—as some critics pretend—of a logic inscribed in his manifest intentions.

Remember Weber's celebrated assertion:

> Let us make it clear once and for all: . . . the only thing that really concerned [the Reformers], the only axis of their life and their work, was the salvation of the soul. All the ethical objectives, and all the practical effects of their doctrine revolved around this axis and were, therefore, nothing more than the consequences of purely religious motivations. Thus we must be willing to admit that by and large the cultural effects of the Reformation have been unforeseen or even *undesired, unwanted,* consequences of the work of the Reformers: consequences frequently far from what they intended, and sometimes even contrary to what they were attempting to achieve. [Weber, 1904, 125]

In the case of Escrivá de Balaguer we would also like, as far as possible, "to make it clear, once and for all."

(*c*) In addition to the themes of predestination and works, there is yet a third important factor differentiating Christians of the Reformation from Catholic Opus. This factor contributes to the internal cohesion of Opus Dei as a group, and at the same time shows that its worldly asceticism is based—in certain significant areas which we shall study later—on certain assumptions that are very different from the Calvinist case. We refer to the fact—considered by Weber to be the most decisive point of rupture with Catholicism (Weber, 1904, 142f.)—of the radical abolition of any kind of *mediation* among the predestinationists. Neither the Church, nor the sacraments, nor the clergy could help the individual "in his path to the encounter with a destiny predetermined since eternity" (ibid., 142); Reformation man was obligated to travel this path completely alone.

In Opus Dei, on the contrary, these *mediations* are not only present, as in all of Catholic tradition, but even seem particularly accentuated. "Cum Petro ad Jesum per Mariam": devotion to the Virgin, devotion to the Holy Father (whom Escrivá called the "vice-Christ on earth"), devotion to the eucharist, devotion to the saints, to the guardian angels; frequent observance of the sacraments, with special emphasis on confession; spiritual direction. Nothing could be further from Weber's "extraordinary internal solitude of the individual" (Weber, 1904, 142) than the situation of an Opus Dei member. His world is full of mediations and mediators, to the point where these condition his conception of his relation with God, the Church, and other men.

The Opus member is part of a gigantic alliance, to which he can appeal at any time for supernatural forces to aid and support him in achieving his objectives. "Magic," any good Calvinist and faithful Puritan would have to say of this. But it is a "magic" of tremendous psychological effectiveness for the individual determined to sanctify others while he sanctifies himself. The Founder himself, in the meteoric race toward recognition of his "heroic virtues" and his "beatification" by the Church, is a palpable demonstration that the members of Opus are in fact able to achieve their objectives while at the same time they obtain, in passing, *a new mediator,* who will of course look after all their affairs with very special attention.

Under such conditions, even though the "official" Opus Dei literature, when referring to the "laical" character of its spirituality, resorts to the formula of "the universal priesthood of all the faithful," it is clear that this expression possesses very different connotations from the Protestant case (see, for example, Portillo, 1981). When all is said and done, "When a layman sets himself up as an expert on morals he often goes astray: laymen can only be disciples" (*The Way,* no. 61). Similarly, even though Berglar affirms that "I do not know any association of Christians that is more *egalitarian* than the Work (Berglar, 96), the truth is that the organizational structure of Opus is strictly hierarchical, to the point where anyone who disagrees with a superior is advised, "Never contradict him in the presence of those who are subject to him, even if he is in the wrong" (*The Way,* no. 954). Assuming that one could speak of the existence of an ecclesiology of Opus Dei, even though its authors make use of the council formula of the Church as "the People of God" (*Conversations,* no. 21), it is clear that Opus's self-definition as a "big family" (Bernal, 313) refers basically to the model of the extended patriarchal family, and not to the community understood as a voluntary association of free individuals who constitute, according to Weber, the very basis of democracy.

In all its aspects, furthermore, Opus Dei belongs, in the last analysis, within the framework of the most traditional Catholic ideas. We are not pointing out anything new here, but simply showing that there are significant differences between the worldly asceticism of the Puritans and the worldly asceticism of Opus Dei. There remains the fact, however—and for us this is what is really most important here—of the equally worldly character of both types of asceticism.

The Notion of Work

Whenever Opus Dei members talk about the sanctification of work, they usually add immediately: *no matter what the occupation is*. Monsignor Escrivá said that it was the same to him whether one of "his children" was a government minister or a street sweeper, a university professor or a construction worker, an entrepreneur or a laborer, and that the only thing that really concerned him was the sanctity of all of them.

This proposition comes from an organization which, especially during its early years, was oriented toward "the intellectuals and the leadership part of society" (remember how the objectives of Opus Dei were defined in the Constitutions of 1950, and how they specified these objectives in the *Anuario Pontificio* while Opus Dei was a secular institute; today the Statutes of the Prelature still recall this past in specifying that it is oriented toward "persons of every condition, and principally the so-called intellectuals"; art. 2.2). But the insistence on "sanctification, whatever the professional activity," is never accompanied by any reflection on the different—and unequal—working conditions experienced by different people; instead there is a silence which amounts to an implicit acceptance of the status quo.

Weber stressed that "the sign which permits recognition of the sanctity of the worker is awareness of a job well done." If for the bourgeois entrepreneur a job well done consisted precisely in looking out for his own economic interests, "the power of religious asceticism placed workers at his disposal who were austere, conscientious, who had perseverance and were unusually industrious, and who identified with work they considered an end desired by God." In a footnote Weber added the following comment: "You will notice that there is a somewhat suspicious concordance on this point between the interests of God and those of the entrepreneurs" (Weber, 1904, 258).

(*a*) This concordance is strengthened even if, as happens in Opus Dei's conception, the sanctification of the individual through his work is linked to the sanctification of others through the work. Obviously, a person who occupies a preeminent job, with subordinates under him, is more likely to have a beneficial influence on others; hence the "elective affinities" of the doctrine with specific social sectors, from the homemaker who, in addition to children, has a servant, all the way to those who teach or occupy distinguished positions in the world of politics, and to those who work in management positions in the world of work and the corporation. In a work environment governed by social harmony and the convergence of interests—which is apparently the ultimate goal of much of the discussion of "business ethics" (see Llano et al., 1990)—the well-integrated worker will be enriched by his work, at the same time developing "intelligence and will" (ibid., 31), while the entrepreneur sacrifices by saving, so that he can amass the necessary resources to "place himself at the service of others through investing" (Gilder, 63). But after all is said and done, this sacrifice does not presume more than a temporary postponement of satisfaction, since the enrichment of the other opens up prospects of "greater profit spirals" for the entrepreneur. In other words, "the entrepreneur wants the poor to prosper, since they are the largest untapped market in any capitalist society" (ibid., 66).

The entrepreneur is altruistic and the capitalist is someone who started like this: "the first capitalist acts were gifts of men . . . who decided, simply, to give a gift, make an offering, a loan, without any preestablished reciprocity" (Gilder, 62). The profit made by the capitalist in the long run "is an indicator of the altruism of the gift" (ibid., 64; the word "indicator" is quite similar, in this context, to the Puritan notion of *Bewährung*). Similarly, the apostle who sacrifices himself while procuring the sanctification of others at the same time adds to his own santification: "Just compare: a hundredfold and life everlasting! Would you call that a poor bargain?" (*The Way,* no. 791).

From this perspective—which is not that of the initial ascetic Protestantism, which was unable to "bargain" for eternal life or to enter into "negotiations" with heaven—the prolific Rafael Gómez Pérez, an Opus member, asserts in a book on business ethics that there is no need to resort to explanations like those of Max Weber on the relationship between the Calvinist ethic and the spirit of capitalism. For him the ethic is simply "an element of the enterprise's good functioning. It is not a sufficient condition for good business, but it fosters it" (Gómez Pérez, 1990, 20, 47ff.). Thus throughout the book he defends the thesis that "in sum: ethics pays; vice is not profitable"; moreover, "although one does not act primarily because ethics is profitable, neither can one exclude the impact of this motivation" (ibid., 18, 19). Finally, "it must be said that good ethical behavior brings profits to the enterprise" (ibid., 94).

(*b*) When the insistence on the principle of the sanctification of work—whatever the occupation—is not accompanied by a reflection on the inequalities inherent in different kinds of work, everything leads us to anticipate a resort to Saint Paul's famous dictum: "let each remain in the state [or in the vocation, or in the profession] to which he was called" (1 Corinthians 7:20). In fact, both Msgr. Escrivá and the "official" Opus literature refer to it, and in so doing they run into the same kinds of problems that Luther had with a text that was, according to Weber (1904, 109ff.), one of the precipitating factors of the whole semantic evolution leading to the conception of professional activity as a vocation.

Curiously, on this point disagreements—important ones—can be found among the Opus authors themselves. Thus Pedro Rodríguez criticizes Luther's interpretation, saying that by reducing "the vocation" to "the profession"

Luther converts it into "the germ of a radical secularization of the divine vocation," which led to "the modern theology of secularization" (Rodríguez, 1986, 39).

"The divine vocation, according to Saint Paul, is not—is not identified with—the secular profession: this is completely foreign to the thought of the Apostle" (Rodríguez, 1986, 40). Monsignor Escrivá, on the other hand, uses this same text of Saint Paul to affirm that "the Lord wants us to serve him precisely where our human vocation leads us: in our professional work" (quoted in Illanes, 1984, 81).

Luther's interpretation of the text also evolved, so that—according to Weber—"he became more and more entangled in the occupations and preoccupations of this world" (Weber, 1904, 118). In the case of Opus, I believe that the problems arise from the fact that they want to use the text for two different ends. On the one hand, the desire to affirm that "each should remain in *the same professional activity* to which he was called" responds perfectly to Escrivá's idea of sanctification in all occupations. Opus's big family includes people from all walks of life, and the only thing that matters is for each to do a good job at what he does, thus sanctifying the work and sanctifying himself in the work. All activities can be equally holy.

In passing, it is clear that the argument favors a harmonious and nonconflictive concept of social life, in which the entrepreneur, the politician, and the intellectual have to be content with sanctifying themselves in their professions, in the same manner that the laborer, the cabdriver, and the bricklayer do. And if it turns out that in Opus Dei, although members come from all walks of life, the first group is more predominant than the second, glory be to God! One only need never mention the context in which Paul's letter to the Corinthians is written, which is an eschatological context: "little time is left." The so-called Pauline indifference to the temporal world—which the Puritans and Opus Dei replaced with an attitude precisely opposite to indifference—is exactly what leads Paul to state that, given that the present life is so short and ephemeral, *it's all the same* whether one continues to do the same thing, because *it isn't worth the trouble to change it*. In its eschatological context, Paul's statement denotes indifference; but used in the context of secularity, of affirmation of the worth of the temporal world, it oozes conservative ideology.

But this text can also be used in another sense, which is curiously close to the sense in which Luther used it at first. For Luther Paul's statement brought out the fact that in order to follow a vocation it was not necessary to be a monk, but that any *state* "has exactly the same value before God" (Weber, 1904, 113; this dimension is also recognized by Rodríguez, 1986, 38). This coincides with Msgr. Escrivá's interest in stressing that the vocation of Opus Dei is not the vocation of monks. And here the translation: "let each remain *in the same state* to which he was called" best fits the assertion that "the members of Opus Dei seek their own sanctification and that of others, each remaining in the canonical state that belongs to him, with a specific spirituality, clearly secular" (Fuenmayor et al., 424f.). Thus on another occasion when Escrivá quotes the same verse from Paul's letter to the Corinthians, he does not say as before that "the Lord wants us *to serve him in our professional work*" (quoted in Illanes,

1984, 81), but rather that "the spirit of Opus Dei has as an essential characteristic the fact of not taking anyone from his place, but of letting each person fulfill the tasks and duties *of his own state*" (*Conversations,* no. 16).

Last, when the reference to the biblical verse is not explicit, the texts often limit themselves to simply uniting both dimensions, just as they earlier joined the dimensions of the calling to holiness and the calling to the apostolate. Thus the decree of introduction of the cause of beatification of the Founder says that the members of Opus are to "seek holiness and practice the apostolate among their associates and friends, each in his own surroundings, profession, and work in the world, without changing their state" (Berglar, 76). In a book by Msgr. Portillo we read: "The Lord has called us all, each remaining in our own state of life and in the exercise of our own profession or trade, to sanctify ourselves in work, to sanctify work and to sanctify with work" (Portillo, 1981, 197). Many other passages in this style could be cited (see, for example, Fuenmayor et al., 191f., and the references in n. 128 on p. 192; see also *Conversations,* nos. 20, 66; etc.).

This almost obsessive preoccupation with clearly establishing that belonging to Opus Dei does not in any way require a change of state seems a little strange. Some critics have claimed to see in it a certain defensive attitude on the part of the "official" literature, a defense against the accusation that persons committed to celibacy who live together in residences (in the case of the regular members) could hardly be considered simple "ordinary Christians," nor could those who in their vocation to Opus Dei are called to the priesthood and thus leave their former occupations. Our hypothesis, however, is that their insistence is due to the fact that, even though members of Opus have never been part of what is juridically called "the canonical state of perfection" (or "the religious state"), for many years it was felt that membership in the organization really did involve a certain change of state, which implied a susceptibility to a change in one's professional activity.

In other words, the question has less to do with a commitment to chastity than with the interpretation of the *commitment to obedience,* above all during the long period during which Opus Dei was, de jure, a secular institute (1947–1982). In *The Way,* the Padre wrote, "Your obedience is not worthy of the name unless you are ready to abandon your most flourishing personal work, whenever someone with authority so commands" (*The Way,* no. 625). Artigues cites an article published in 1965 in the *Revue Franco-Espagnole,* whose author, a member of Opus, stated that in the association "a doctor who has shown particular talent as an administrator might be appointed to a job which separates him from medicine; in general, any member may be assigned to a better apostolate where he is judged more useful than he was in a profession which he will now have to give up" (Artigues, 101). And in his book *Institutos seculares y estado de perfección,* published in 1954 and reprinted in 1961, Salvador Canals, Opus priest and expert on juridical subjects, states that whereas other secular associations "do not change the fundamental character of the lives of their members in such a way that one could say they have changed their state, the secular institutes, on the contrary, require total consecration of their members' lives to the acquisition of perfection . . . and total and complete dedication to the apostolate," so that these institutes—for whom Opus Dei was the *model*—constitute a true state of perfection, which should not be confused with the religious state but which "clearly is very distant from the mere secular state" (Canals, 1954, 85).

In accordance with our hypothesis, therefore, the present insistence by Opus Dei that its members do not change their state is not so much a defense against certain outside critics as it is—in an exercise of alternation similar to the many others we observed in the first part of this book—a defense *against its own past.*

At any rate, it does seem indisputable that these words of Saint Paul (1 Corinthians 7:20) played a primary role in the foundation of Opus Dei's asceticism, just as they did in the asceticism of the Christians of the Reformation. And it seems equally undeniable that the worldly character of this asceticism is strengthened by the fact that, in interpreting the text, the Opus Dei literature does not frame it in an eschatological context of "holy indifference" toward the things of this world; much to the contrary, the Opus member, rather than being a *pilgrim* passing through this world, is a *citizen* of the world, with all a citizen's rights and duties.

(*c*) Earlier we observed that, unlike ascetic Protestantism, Opus Dei sees work as not so much man's contribution to the construction of a future Kingdom of God, as his collaboration in the establishment of "Our Lord's true Kingdom," here and now (*The Way,* no. 832).

"Work, all work, is a testimony to the dignity of man, or his dominion over creation. It is the occasion for the development of one's own personality. It is the link joining a man with other human beings, the font of resources to sustain his own family; the means of contributing to the bettering of the society in which he lives, and to the progress of all humanity" (Escrivá, homily of 1963, *Christ is passing by,* no. 47).

The two biblical texts which seem to provide the foundation for all the "theology of work" of Msgr. Escrivá and his followers and which establish the origin of the divine commandment (vocation) to work are, first, "Be fruitful and multiply; fill the earth and dominate it" (or subdue it, master it; Genesis 1:28); and second, the statement that God created man "so he would work" (Genesis 2:15).

In the case of the second reference, the text of Genesis actually says that "God took man and placed him in the garden of Eden that he might cultivate and watch over it." All the versions agree on these two verbs; but Escrivá used to use only one of the two Latin verbs in the Vulgate—*ut operaretur*—and to translate it as "to work," not "to cultivate." "The Lord placed the first man in Paradise, *'ut operaretur'*—that he might work" (*Surco,* no. 482); or simply, "God created man that he might work" (*Conversations,* no. 24).

We may assume that the use of this verb, *operaretur,* obviously related to the noun which gives its name to *Opus,* must have been very pleasing to Padre Escrivá. The problem is that the text of the Vulgate, according to which God placed man "*in Paradiso voluptatis, ut operaretur et custodiret illum,*" does not allow the kind of translation Escrivá gave it. The verb in question requires an object and, consequently, the only possible way to translate it is to say "that he might cultivate *it*" or "that he might work *it*" (Paradise). If one wants to see in this text the origins of man's vocation, it would have to be, not working in general, but specifically cultivating or working the *Paradisum voluptatis*.

The other aspect that inevitably attracts attention is that both this text and the preceding one ("fill the earth and dominate it") come *before* the expulsion from Paradise, that is, *before* original sin. Therefore, it is logical that Escrivá saw work not as a curse or a punishment, because it is not the consequence of sin ("with work you shall eat all the fruits of the earth all the days of your life"; "with the sweat of your brow shall you eat your bread"; Genesis 3:17–19). Once again, Opus Dei leaves out the long "detour" the Puritans had to make in order to give a positive connotation to work, at the end of which they saw professional activity as the vocation to which God called them. Escrivá reached this point by the direct route, skipping, as it were, the embarrassing episode of man's sin. Helming, then, rightly states that "in many ways Opus Dei represents a return to God's original plans, to the first commandment he gave to Adam and Eve in Paradise" (Helming, 21).

"Let us recognize God not only in the spectacle of nature, but also in the experience of our own labor, our own efforts" (Escrivá, *Christ is passing by*, no. 48). It is the principle of asceticism in action, in contrast to the contemplative attitude of the mystic. (For Opus Dei "healthy" contemplation is that of Saint Joseph; it is not "that of the monks, but that of the worker who, in the midst of his daily chores, does not cease to be aware of the ultimate meaning of things at every moment"; Suárez, 61.)

Thus work is at once a vocation and a blessing. Furthermore, work is not only an ascetic means, but it even holds "a positive value of cooperation with the creative work" (Aubert, 218). "The mission of human work is precisely to serve as a means for man to attain participation in divine life through participation in creative action, prolonging it and putting it in relief in the glorification of God. The greatness of this intuition [Escrivá's, of course!] must be well understood: man being the image of God, nature, transformed by work, can be made in this way more similar to man; and, humanized, it becomes at the same time more similar to God" (Aubert, 216).

This is the antithesis of any empirical analysis of the conditions in which most human work is actually carried out, not to mention any conception of man as alienated from nature and his work. The only thing the Opus Dei authors constantly stress is the vocation of the "dominion" of man. Sin is provoked in any case by human weakness, by base passions, by the weakness of the flesh; if Adam's sin had been a sin of pride, of defiance of God and a wish to be like God, man would tread more carefully before self-appointing himself as a "co-creator" with God and a "co-redeemer" with Christ. According to Escrivá, "work presents itself to us as a redeemed and redeeming reality" (Escrivá, *Christ is passing by*, no. 47). Any representative of ascetic Protestantism would say, no doubt, that this is "idolatry of the creature," pure and simple.

It seems strange for an author like García Hoz, writing about "the pedagogy of the ascetic life in *The Way*," to recall that for Escrivá "true poverty consists in being detached, in voluntarily renouncing one's dominion over things" (*The Way*, no. 632, in García Hoz, 1988, 187), while he asserts elsewhere that "human beings establish a relation of dominion over things since man is made to dominate the world through knowledge [would this be the knowledge of "the tree of good and evil" in Genesis?] and of action." And this "relation between man and things has a name much loved by Monsignor Escrivá de Balaguer: it is called work" (García Hoz, 1976, 9f.). On this view, work would appear to be exactly the *opposite* of "true poverty." Through the message of Opus, God calls man to work, to affirm this relation of dominion; on this point, the elective affinity between the "Opus ethic" and the "spirit of captialism" could not be more evident.

At the same time, the rest of this article by García Hoz offers us the key to two fundamental differences between the worldly asceticism of Opus and that of Protestantism. "Work is not simply a duty of man with things, but a participation by man in *the creative work of God* [that is, in the "Work of God"—"Opus Dei"—not in the "construction of the Kingdom"], a participation also in divine power and sovereignty, given that human beings were made to dominate the world. And the most patent dominion of the world is that realized precisely through work, because through work things are placed and modified *to the service of man*" (and not "to the greater glory of God").

A "Catholic" Worldly Asceticism

In the eternal—and beautiful—Christian dialectic of "already yes, but not yet" (we have already been redeemed, but we are not yet saved; the sin has already been pardoned, but we have not freed ourselves from it yet), Opus Dei places all the stress on the first part of the formula and tends thus to deny tension and conflict. The world in which the Opus member lives is not a "vale of tears" but a "garden of delights," a *Paradisum voluptatis*. It is a garden in which God placed him "to cultivate and watch over it," according to Genesis, but "to work and dominate the earth," according to Escrivá's version.

In this world, God is the "entrepreneur"; the Opus member is the "manager." When Berglar writes that Escrivá's message assumes "the beginning of a new era in Christian life," he explains it this way: without being "worldly," we are "children of God in this world"; we are to love the world as a *work of God* (the world = Opus Dei, Work of God), "and to work in it as his collaborators, but without wanting to possess the fruits of this work," which we will offer "to the Lord of the earth, to the divine *entrepreneur*" (Berglar, 97).

If starting from the Calvinist ethic the capitalist entrepreneur saw in accumulation the sign of divine choice, in a later phase of capitalist development the Opus member is rather the professional expert who sees the sign of this choice in the success of the organization he directs (Casanova, 1982, 450), whose management has been entrusted to him directly by God himself, who chose him as an "instrument" for this job (Portillo, 1976).

Compared to the extremely distant and inaccessible God of Calvinism, who from the point of view of the simple believer presents himself with the terrifying features of what we would almost dare to call an "autistic" God, the image of God in Escrivá's writing is no longer that of the Heavenly Father of the New Testament, so humanly understanding that he is pleased by the repentance of the sinner" (Weber, 1904, 141), but that of friend (*Friends of God* is the title of one of Escrivá's collections of homilies), and almost that of "colleague." This is a God who could even be manipulated and bribed (remember that supplication of Escrivá's: "Lord, look after my mother, since I am busy with your priests," quoted in Sastre, 280), if it were not that the interests of the "divine entrepreneur" and those of his "manager" are, logically, the same. This is a God with whom the members of Opus are "collaborators," with whom they are "co-creators" and "co-redeemers," a God to whom Escrivá could pray: "Lord, do one of your things: let us see that you are You" (quoted in Fuenmayor et al., 347). If from the point of view of the simple Calvinist believer we dared to say that God, in his inaccessibility, presented himself as an "autistic" God, from the point of view of the simple faithful of Opus Dei the "Father who is in the heavens" (God) and "our Father in heaven" (Escrivá) (Portillo, quoted in Vázquez, 486), are literally *compadres*—"co-fathers."

◆ ◆ ◆

If we had intended to be exhaustive, we would no doubt have taken into consideration many other aspects of the peculiar style of worldly asceticism represented by Opus Dei. There are many questions raised by Weber in his thesis, especially in his last chapter, which would have permitted us to go deeper in our analysis of the parallels and differences between the Puritan ethic and the Opus ethic. To cite only one example, it is nearly certain that a comparative analysis of Richard Baxter's *Christian Directory* and the works of Escrivá de Balaguer would be enormously interesting and very useful.

But while recognizing its interest and utility, I do not believe such an undertaking is indispensable for our purposes. I did not intend to produce a systematic application of Weber's thesis; neither was I guided by any pretensions whatever to exhaustiveness. I wanted to show that the same type of "elective affinities" Weber detected between ascetic Protestantism and the spirit of capitalism played a similar role in the case of Opus Dei, and that even though they started from very different premises, the end results were more of less the same.

Hidden in the small print of one of Weber's endless footnotes (arguing with Brentano) is a particularly interesting explication of a "basic assumption of my whole thesis: that is, that the Reformation brought Christian asceticism and the methodical life out of the monasteries and implanted them in professional life in the midst of the world" (Weber, 1904, 168 n. 84). Four centuries later, the creation of the juridical figure of the secular institutes sanctioned this same process in the bosom of Catholicism; and Opus Dei was considered its perfect model.

No doubt there is some basis for the ironic commentary of some critics who observe that the supposed novelty of Opus Dei, with its proclamation of secularity, lay spirituality, and the like, was nothing more than another step in the "Protestantization" of the Catholic Church. But such comments are often made with malicious intent—we must keep in mind that for the Padre Escrivá of the forties the Protestants were not children of God, and they had "dry hearts," the same as atheists and Masons (*The Way,* no. 115; remember that this statement is found only in early editions of the book).

At the same time one must recognize that the allegation in question is, in large part, false, because although the effects are the same, the propositions are not. It is false because the characteristics of Opus Dei as a movement and as an organization, its spirituality and its trajectory, cannot be explained outside the framework of Catholicism, and the most traditional Catholicism at that (in the best *and* the worst senses of the word "traditional)."

One of the most frequent misunderstandings to which Weber's thesis has led, both among those who know it only from hearsay (the majority) and those who have read it and have not realized what it is about (and the authors of Opus belong, in the best cases, to the second group, either because they don't understand it or because they don't *want* to understand it), finds its expression in the argument that the Protestant Reformation was the immediate cause of a relaxation of morals and customs. By itself alone, the very term "Puritanism" should have been enough to dispel that notion. But just in case it wasn't, right at the beginning of his book Weber says the following:

> We ought to keep in mind a fact that today tends to be forgotten: the Reformation did not represent the elimination of ecclesiastical control over life, but rather a replacement of the then prevailing form of this control by a new and different one . . . penetrating in a much more direct way in all spheres of public and private life, and subjecting individual conduct to meticulous and burdensome regulation. . . . For us, today, the worst imaginable form of ecclesiastical control over individual lives would be the rule of Calvinism just as it was enforced between the sixteenth and the eighteenth centuries. [Weber, 1904, 59]

It is because this fact "tends to be forgotten," as Weber writes, that Robertson could attack the Jesuits, accusing them of having relaxed the discipline of Christians in economic matters even more than the Calvinists had (Robertson, 109); the Jesuit Brodrick replied that Robertson wanted only "to defend the Puritans against Weber's accusations" (Brodrick, 2). As if Weber had ever accused the Protestants of a relaxation of morals!

Without meaning to go to the other extreme, and also with no desire to accuse anyone of anything, we might nevertheless ask: to what extent can we today apply to Opus Dei the Weberian characterization of the control that penetrates in all the spheres of public and private life of its members, and which subjects "individual conduct to meticulous regulation"? Can one speak, in the last years of the twentieth century, of "ecclesiastical control over individual lives" exercised by Opus Dei?

The theme of "freedom" versus "individual control" in Opus Dei has already appeared many times in these pages. This is one topic where the points of view expressed by the "official" literature and the views of the "critical" literature are most irreconcilable. Almost all the texts written by persons who have left Opus Dei (Moncada, Moreno, Steigleder, Tapia, etc.) refer to it with singular insistence, describing the internal functioning of Opus Dei in terms which sociologically are very close to those that characterize the functioning of any "total institution" (Goffman, 1961). These texts are speaking, in fact, of strict multiple control, basically in the case of the regular members: obligatory changes of residence, dispossession of property, limitations on relations with family and the outside world in general, control over reading; psychological controls based on the creation of strong links of dependency, through spiritual direction and the perpetuation of old customs of the world of the convent, for example, the "confidence" (outside the framework of confession), the "chapter of faults" (where the individual accuses himself) and "fraternal correction" (where the individual has to accuse another); in addition to the diverse prescriptions relating to corporal mortification (hair shirts, flagellations, cold showers, etc.). One can find a good summary of this set of practices, described from a critical perspective, in Michael Walsh's book (Walsh, 117–24).

The "official" literature, on the other hand, avoids mention of such details. The Regulations, and other texts which govern all these matters, have never been published; and there has never been a satisfactory official explanation of the most controversial points of the Constitutions of 1950 (art. 58, and arts. 147–54, all relating to obedience). There are only generic and imprecise declarations, limited to mention of Opus's scrupulous respect for the freedom of its members (see Herranz, 1962, a text which sets the pattern for all the subsequent literature), affirming, in brief, that Opus "does not organize the work life of its members; it does not direct or control their activities, their relationships, their influences" (Gómez Pérez, 1978, 24).

At certain moments the "official" explanations even seem rather ridiculous, as when they say that although there is an obligation to ask for advice, there is no obligation to pay any attention to the advice received (Fuenmayor et al., 244 n. 32; this statement is, furthermore, in flagrant contradiction with what Fuenmayor himself said in 1956 at the congress described in Chapter 10; see *Actas,* 1957, 1203–5). Another example is Escrivá's own statement, in an attempt to deny the existence of any kind of control, comparing the Work to "a sports club or an association for good works" (*Conversations,* no. 49), a comparison that Franco's former minister López Rodó (1991, 18) apparently found completely justified.

Clearly, then, a comparison of the "official" literature to the "nonofficial" literature will not help us clarify anything. The question, however, is an important one, not only in and of itself (from the point of view of the former members who have left Opus, traumatized by the experience of the control to which they felt subjected, the problem is evidently very serious), but also because from a sociological perspective it seems to cast doubt on one of Berger's central theses in his study on the capitalist revolution.

In fact, precisely at the root of Berger's analysis of Weber's *The Protestant Ethic* (Berger, 1986, 99–102) is the thesis that "bourgeois culture has engendered in the West, and especially in Protestant societies, a type of person clearly distinguished both by the value and the psychic reality of individual autonomy" (ibid., 103). Is it possible that here lies another of the meaningful differences between ascetic Protestantism and Opus Dei? Is it possible that the indisputable "modernity" of Opus Dei in certain areas, including anything relating to economic ethics, coexists with a notable "traditionalism" in other fields, particularly in the sphere of individual autonomy? This is basically the question we will try to address in the next chapter.

15

Conclusion: The Traditionalism and the Modernity of Opus Dei

The Paradoxes of Opus Dei

To me the entire history of Opus Dei seems to be a succession of paradoxes, from the complex personality of its founder and the origins of the movement to the consolidation of the organization in postwar Francoist Spain, its official recognition and internationalization, and the long process of its gradual retreat from the juridical category of the secular institutes, which ended with its acceptance as the first Personal Prelature in the Catholic Church.

In good part this paradoxical history can be explained by the lack of correspondence between the intentions and wishes of its protagonists and the real objective consequences, unforeseen and often undesired, of their actions. We have discussed this at length throughout these pages, and there is no need for repetition here.

But there is another kind of paradox, which has cropped up on various occasions, into which we should delve a little deeper. This is the apparently contradictory perception, which has always been held by many outside observers, of Opus Dei as an institution that is *at the same time* both "reactionary" (or "integrist," in classical Catholic terminology) and innovative. To many it appears to be an institution that presents a set of "sectarian" (or "fundamentalist," to use the currently fashionable term) features, while *at the same time* no one has been able to prove that it has ever had "schismatic" whims or temptations to separate itself from communion with the Church of Rome.

We are faced with the apparent contradictions of an "elitist" organization that nevertheless says it is made up of "ordinary Christians," the contradiction of a movement that claims to profess a "clearly laical spirituality" while inducing many of its most valuable members to enter the priesthood, a movement that wants to be fully integrated into the ecclesiastical structures but which nevertheless has always been "a special case," with its own institutions, its own seminaries, and its own priests, a movement which pretends to participate 100 percent in all the affairs ("noble" ones) of society, in the name of its radical "secularity," but which at the same time, in the name of "discretion," wants to pass unnoticed in society, and frequently in fact utilizes all the means at its command to preserve this "invisibility." This is a movement, finally, which says that the spirit of "true poverty" consists of *renouncing* control over things while it simultaneously asserts that its vocation is one of *establishing* a relation of control over things through work. (Which is why it is frequently accused of preaching "the spirit of poverty" and practicing exactly the opposite.)

We could go on like this almost indefinitely. We are speaking, in the last analysis, of the apparent contradiction between the ideals of the movement ("you are free, my children," according to Escrivá) and the structure of the organization, turned in on itself like an authentic "total institution" in Goffman's sense, that is, an institution enforcing on its members a whole new process of socialization.

Probably this contradiction is not usually experienced as such by the Opus member who has been well socialized and has internalized the ideals but is not integrated into Opus's organizational structure, being simply an ordinary member trying to "work and sanctify himself in the midst of the world." Nor, certainly, is this contradiction perceived by the member who *is* fully integrated into the organization—usually a priest—who lives in it and for it, and is less immersed in the external world. The contradiction tends to be apparent to the outside observer, however, and to be experienced by the Opus member who occupies a position midway between that of the "simple member" and that of the "member of the organizational structure." At any rate, it seems as if most of those who abandon Opus Dei in troublesome and conflictive circumstances are persons who occupy this kind of intermediate position.

What is the explanation of this paradox of Opus Dei as an institution that is simultaneously modern and traditional, innovative and close to integrism? To answer this question we must consider more closely which of its activities cluster around the first pole, and which are clearly oriented toward the second.

A first distinction—perhaps a little simplistic and crude, but with the advantage of being clear and forceful—is that one's first impression of Opus members is usually that they are very technically competent people but that they have a frankly elementary religiosity. José Casanova goes even further and refines the point a little more—not far enough, in my opinion—when he says, "what constitutes the essence of Opus Dei is a combination of the most characteristic elements of modern religiosity ["early modern," in the typology of Robert Bellah, 1964], particularly its worldly asceticism, with the elements most typical of traditional Catholicism, especially its uncritical acceptance of Catholic authority" (Casanova, 1982, 189).

I would say, rather, that if Opus Dei presents itself anywhere as strongly acculturated to modernity, it is with respect to the adoption and utilization of *modern techniques* in the world of economics, politics, and the communications media. In this area—which includes the formation of lobbies, pressure groups, and the like—it shows that it is as technically competent and as innovative as any other group. IESE is one of the best examples of this, but it is not the only one: Opus has proved extremely skillful in the use of the communications media, both to disseminate specific kinds of information and to contradict others. It has been most effective in managing the process of the beatification of the "Founder," in the midst of an atmosphere which has certainly not been unanimously favorable.

But at the same time that Opus so successfully makes use of such techniques, I would say that in general the members tend to reject the *dominant values underlying* their common use in modern societies. Perhaps this is the point we should look to for the key to an understanding of this paradoxical combination of traditionalism and modernity: flexibility at the level of norms, and consequently of practical conduct, where solutions are valid when they are technically correct; but inflexibility and traditionalism at the level of values, and consequently also of the belief system, the legitimations on which those values rest.

Might we interpret the Padre's famous maxim that "the standard of holiness that God asks of us is determined by these three points: Holy intransigence, holy coercion, and holy shamelessness" (*The Way,* no. 387) in light of this distinction?

On the level of values we have "intransigence." "To compromise is a sure sign of not possessing the truth" (*The Way,* no. 394). And, "a man, a 'gentleman,' ready to compromise would condemn Jesus to death again" (*The Way,* no. 393). At this level, "tolerance" is synonymous with surrender, and "compromise with error" is equivalent to "fornication of the truth" (Urteaga, 1948, 96).

On the level of legitimation, the level of the belief system, we have "coercion," When it comes to "saving the Life (with a capital) of many who are stupidly bent on killing their souls," coercion becomes "holy coercion" (*The Way,* no. 399).

And on the level of norms, we have "shamelessness." "If you have holy shamelessness, you won't be worried by the thought of 'what will people say?' or 'what can they have said?'" (*The Way,* no. 391). When it comes to norms, which are the means to attaining fixed objectives, the slogan is "God and daring!" (*The Way,* no. 401).

This kind of interpretation, moreover, is consonant with the programmatic formula put forward by one of the most brilliant Opus Dei ideologues of the early Franco period, Florentino Pérez Embid, who in 1949 proposed this slogan: "Spanification in the ends and Europeanization in the means" (Pérez Embid, 1949). In a country that was living through a total "regime of autarchy" Pérez Embid's proposal seems a prophetic omen of what, eight or ten years later, would become the historical role of the "technocrat" ministers, Ullastres and Navarro, and especially Laureano López Rodó. In the Spanish context of the time, to speak of "Spanification in the ends" was equivalent to saying intransigence in values, while "Europeanization in the means" implied flexibility, and even a broad acceptance of norms, without having to have a bad conscience or worry about "what people would say," thanks to the sanctification of "shamelessness" converted into a virtue.

If we go one step further, and replace the word "Spanification," a synonym for intransigence, with the Weberian *Wertrationalität,* or rationality oriented toward values, and at the same time replace the word "Europeanization," the synonym for compromise for the sake of effectiveness, with the notion of *Zweckrationalität,* or instrumental rationality (Weber, 1922, 12–26), we will have advanced considerably in our comprehension of the paradoxical combination of traditionalism and modernity in Opus Dei.

Let us remember that for Weber a *wertrational* action is one which implies a conscious belief in the *absolute* value of ethical, religious, and so on, behavior, while in *zweckrational* action the expected behaviors of others are taken as *conditions* or *means* in order to reach the ends, or rational objectives, of the actor (Estruch, 1984, 146 n. 30). The traditionalism of Opus Dei consists in the fact that in its beliefs and in the defense of values, its members behave like intransigent persons, in accordance with the *wertrational* pattern of action, while its innovative character derives from the flexibility and the standard of instrumental rationality they adopt in the use of the characteristic technical means of modernity.

The Ethics of Conviction and the Ethics of Responsibility

In the last years of Max Weber's life he developed a second distinction, which in large part runs parallel to that between the two types of rational action, but which for us offers the additional advantage of being situated in the field that interests us specifically here: the field of ethics. We refer to his distinction between "the ethics of conviction" (*Gesinnungsethik*) and "the ethics of responsibility" (*Verantwortungsethik*), which is developed especially in his lecture "Politics as Vocation," delivered in Munich in 1919 (Weber, 1921).

The ethics of conviction is an absolute ethic, based on absolute fidelity to certain principles and on defending them at all costs, without taking into account the possible consequences. The ethics of responsibility, on the contrary, obliges one to be very aware of the possible consequences of action, to such an extent that the individual, faced with a choice, might on occasion consider it preferable to temporarily sacrifice his principles in order to avoid the greater evil of the foreseeable undesirable consequences of an action exclusively guided by convictions.

Weber's distinction is so clear that anyone can instantly recognize, in his own conduct, actions belonging to one or the other of these poles. Even if some authors would disagree with this (for example, Spaemann, 74ff.), I consider that the distinction is enlightening and useful, and on the question of understanding the paradox of Opus Dei specifically it is basic and definitive.

Of course we must keep in mind that Weber is operating (as always, but this is exactly what Spaemann seems to want to ignore) with "ideal types": analytical constructions that do not *describe* a specific reality but are formulated to help in *understanding and explaining* reality. The ethics of conviction, Weber says, is the ethics of the "saint"; in the case of a real saint, it is perfectly meaningful, and an expression of great dignity (and he cites as examples Jesus, the apostles, and Saint Francis), but when it is only half a saint, it is an ethics of unworthiness. The ethics of the politician, on the other side, must be basically an ethics of responsibility. A politician whose acts are guided exclusively by an ethics of conviction would be a very dangerous person: a visionary and a fanatic.

The ethics of conviction and the ethics of responsibility are not mutually exclusive terms for Weber. Weber states expressly that the first does not imply a lack of responsibility, and the second does not imply a lack of principles or convictions. On the contrary: an ethics of conviction that was not accompanied and tempered by an ethics of responsibility would approach fanaticism, while an ethics of responsibility which sacrificed all conviction would lead to pure cynicism totally lacking in principles and scruples. Precisely for this reason, Weber concluded his lecture "Politics as Vocation" by asserting the complementarity of the two kinds of ethics, and declaring that the "authentic individual," the mature person, is one who is capable of integrating them in his attitudes and his behavior.

I wish to conclude this inquiry into Opus Dei with the thesis that its paradoxical combination of modernity and traditionalism consists in the fact that in specific spheres it proclaims and applies this complementarity of ethical positions (and so is perceived as "modern"), while in other spheres it is guided and obliges its members to be guided by an ethics of conviction that is not complemented or tempered by the ethics of responsibility (hence its "traditionalism" or "integrism").

The Ethics of Responsibility and the Ethics of Conviction, Integrated

(*a*) The adoption of the first perspective is most obvious in *economic and business activity*. Here the members of Opus clearly opt for the standards of instrumental rationality, calculate the relation between means and ends, and try to foresee the consequences of their actions at all times. The model is clearly that of the ethics of responsibility. This is why an institute like IESE enjoys international prestige and can be legitimately presented as an emblem of Opus Dei's modernity.

In the literature on "business ethics" one often detects a preoccupation with integrating the ethics of conviction and the ethics of responsibility. "Ethical conflicts in business tend to appear when persons who have to make business decisions discover the apparent impossibility of choosing actions which will simultaneously satisfy both their criteria for economic rationality and their ethical criteria" (Pérez López, 1990, 33). The language is not Weber's, but what the author wants to express with his language is the same. He tends to reserve the word "ethics" for principles or convictions and proposes the development of what he calls an "entrepreneurial asceticism" (ibid., 35), whose function would be approximately the same as what Weber called *Verantwortungsethik,* or ethics of responsibility.

We also observe a certain "traditionalist" inflection in Pérez López's language, in that he tries to base his propositions on Aristotle and Aquinas, a line which inevitably leads to a type of argument that generally makes sociologists nervous: the argument of "natural" ethics, based on the given, indisputable character of a preexisting reality and on the "ascetic" exercise of adaptation to that reality, which is reified and not perceived as socially constructed. But in spite of this tendency, the complementary nature of the two types of ethics seems to be reclaimed when the author denies that business ethics can be reduced to "normative ethical theories" (Pérez López, 39f.). Such normative theories can correctly point out unacceptable practices (in our terminology, those principles which *cannot* be sacrificed, not even from an ethics of responsibility) but cannot provide a positive orientation for action (so that it becomes *necessary* to resort to an ethics of responsibility). (Juan A. Pérez López, a professor at IESE, kindly allowed me to use material from a book of his that is still in progress; for that reason it is not quoted here, but I found it extremely useful.)

(*b*) The same effort at integrating both kinds of ethics can be detected in the actions of certain Opus members in the *world of politics*. The best known example, and the one most often mentioned in this book, offers a perfect illustration. From our perspective, Opus Dei's "technocrat" ministers in Francoist Spain of the late fifties introduced a kind of behavior governed by the ethics of responsibility under a dictatorial and authoritarian regime—a regime which, since the end of the "civil" war, had based its actions on the ethics of convictions. (We say this without making any value judgments about the class of "convictions" involved, or about possible semantic abuses that might arise at times from such a generic and "neutral" use of the word "ethics.")

The Ethics of Responsibility without the Ethics of Conviction?

This concern with the relation between the means and the ends of action is equally clear in another text on business ethics by a member of Opus (but not of IESE). Gómez Pérez—again in the language of traditional casuistry—examines the circumstances that may affect and modify a moral act: "who, what, by what means, why, how, and when" (Gómez Pérez, 1990, 33). In this case it is not always clear which "convictions" cannot be renounced by one who acts according to the ethics of responsibility. The author's thesis is that "good ethical behavior brings profits to the enterprise" (ibid., 94), that is, "ethics pays," "ethics is profitable" (ibid., 18f.). However, he justifies "bribery" when one is trying to protect a business from "serious losses that threaten its survival" (ibid., 120), and he forgives a lack of fiscal conscience (ibid., 83) or tax evasion when the state "does not satisfy the common good" and when the tax evader dedicates "those funds to the defense of the common good" (ibid., 125).

The same author presents, as a case study of his own elaboration, the example of an entrepreneur who has become rich and has evaded taxes. "A book on business ethics" makes him aware that his behavior, in addition to being illegal, has been immoral. But to declare the truth to the Treasury now might threaten the survival of his business and cost two thousand jobs. "Furthermore, he is not willing to give so much money to an administration which is characterized by excessive spending and which spends a lot of money on activities which to him seemed completely illicit." Finally, he discusses his case with the author of the book he has read on business ethics (without specifying if it is an Opus Dei author), and in the end he decides to restore the money that he did not pay in taxes. How? "With the bulk of these funds he will create a foundation with undisputably beneficent ends" (Gómez Pérez, 1990, 136f.).

This goes far beyond Franklin's "Necessary Hints to Those That Would Be Rich" and "Advice to a Young Tradesman," which Weber quotes as examples of what "the spirit of capitalism" means (Weber, 1904, 70ff.). The ethics of responsibility leads in this case to a cynicism lacking in the scruples to which we alluded earlier, or to Padre Escrivá's "holy shamelessness."

When an author who is a member of Opus presents such arguments in a book entitled *Business Ethics* published by a press closely connected to Opus, can we accuse a person of evil thoughts if he sees it as a legitimation of the possible diversion of funds toward the "corporate works" of Opus Dei?

The Ethics of Conviction and the Ethics of Responsibility in Conflict

In 1980 the same author published a volume titled *Problemas morales de la existencia humana* (4th ed., 1987), which is used as a textbook in some of Opus's high schools. We will use it as a basic reference in this section and the next, comparing it to two other books that are not directed at the school population: *Introducción a la ética social,* by the same prolific author (Gómez Pérez, 1987), and *Catecismo de doctrina social,* edited by Juan Luis Cipriani, an Opus priest who is a bishop in Peru (Cipriani, 1989).

Besides the economic and political spheres we have just mentioned, there are other areas where these authors take into account what we have called "the ethics of responsibility." They do so in various ways and degrees, and frequently it comes into conflict with what would appear as an "ethics of conviction." These are areas in which the two ethics seem poorly integrated, or even contradictory. Our range could be broader, but we will limit ourselves to three examples: war, conscientious objection, and the death penalty.

War. War, with which humanity has had long historical experience, "always entails an immediate threat of actual death" (Gómez Pérez, 1980, 107). Opposite the desire for war to "be proscribed as a means of resolving conflicts" (the level of principles), he places the concrete reality of war as a defense against unjust aggression (the ethics of responsibility). The author analyzes the conditions of the "just war" and concludes that "while defensive war is almost always just, other types of war are morally difficult to justify" (ibid., 108).

Since in spite of all the efforts made till now the phenomenon has not been eradicated, ethics "must, meanwhile, be concerned with all the systems that make the consequences of war less grave" (ibid., 109).

In sum, principles are very nice, but the facts are stubborn. Although Opus members hate to hear Freud quoted to them, I would say that in reference to the phenomenon of war they would consider that the "reality principle" must be imposed on the "pleasure principle," or that the ethics of responsibility must prevail over the ethics of conviction. (While this is not the place to discuss the subject, I believe that the parallel between the Freudian dichotomy and the Weberian one is not entirely absurd.)

In connection with this topic of war, the other Opus texts adopt the same approach (Cipriani, 259; Gómez Pérez, 1987, 149f., with a more positive evaluation of disarmament policies). In some other cases, support of "just war" even leads to such statements as "sometimes war is not only legitimate, but it is a duty to participate in it"; and in spite of the fact that "one does not go into war to kill the enemy but to defend oneself," it is "allowed" and "legitimate" to kill in a just war (Herrera, 254, 258).

Conscientious Objection. In 1974—a year before Franco's death—the magazine of Opus's University in Navarre, *Nuestro Tiempo,* published a curious article in which the author, José Zafra, argued that every Spaniard had the obligation to serve his country under arms, and that it was politically inconceivable not to accept the idea of military service. Zafra then posed the possible consequences of a hypothetical objection to this basic principle:

> It would only be consonant with *good logic* to follow one of these lines of conduct: to take the lives of the objectors; to strip them of their nationality and leave them in the condition of resident aliens until they found another country to accept them; to declare them mentally deficient and shut them up in appropriate establishments; or to reduce them to the condition of half-citizens, that is, subject to some form of "*capitis diminutio.*" [Zafra, 64]

This is the "pure logic" of José Zafra to this point: the ethics of conviction, pure and simple. It is even more interesting—once the goose bumps have gone away and the panic subsides—to observe how, in the second part of the article, the author presents the possibility of using not "pure logic" but rather "prudence"—the possibility, in other words, of introducing a tiny dose of the ethics of responsibility into a discussion governed by the ethics of conviction. Thus he explicitly asserts: one can govern while "forgetting the *principle*" (Zafra, 65). Then, after observing that the mere imprisonment of the objector seems not severe enough to him, Zafra finds that the solution of civil degradation makes sense; moreover, it should not be limited to the suppression of political rights or a prohibition on working in public administration or teaching but should include "even a prohibition on owning real estate" (ibid., 70), because a person who refuses to bear arms for his country has no right to own even one little piece of it. Finally, a last possible solution would be alternative service. This would have to be an activity that was "no less arduous or risky, more harsh, and equally beneficial to the nation as military service" (ibid., 71).

A few years later, after the question of conscientious objection had been minimally addressed in Spanish legislation, Gómez Pérez offers suggestions giving more emphasis to the ethics of responsibility, although there is still a conflict with his own convictions (curiously, while these convictions are opposed to war, they favor military service). From a "subjectivist ethical standpoint," which affirms the autonomy of the conscience, objection presents no problem. "The objective conception of morality affirms, on the other hand, that natural moral law, engraved on human nature, obliges naturally" (Gómez Pérez, 1980, 194). In the last analysis, everything depends on the law: if the law is unjust, objection is a right and a duty; but if the law is just, "there is no place for conscientious objection." The law is just, obviously, when it is in agreement with "natural moral law." It would be hard to find a formula that better sums up the meaning of the sociological concept of *reification* (as analyzed in Chapter 1), and which in passing better illustrates the uses of this reification in the service of the interests of legitimizers or ideologues who are able to impose the "official definitions" of reality. *Who decides* when a law is just or unjust?

"Compulsory military service is not in itself unjust," according to "an objective conception of morality," pronounces our author (Gómez Pérez, 1980, 195), the same person who would justify tax fraud on the part of a businessman because the state was spending money on activities "which to him seemed completely illicit" (Gómez Pérez, 1990, 136).

In sum, the recognition of the right to conscientious objection appears in these texts as a concession, made in the name of the ethics of responsibility, in full awareness that it implies a certain sacrifice of convictions. The two ethics are clearly in a state of conflict, not integration. The outlook will be totally different when, with an elegant pirouette, Gómez Pérez says that the problem of conscientious objection today goes far beyond the "specific case of military service" and, relating it to the growing "social pluralism" and to the "democratic view of politics" (Gómez Pérez, 1987, 135f.), he finds a new agreement between objective and subjective conceptions of morality. Thus he writes: "To what extent is a citizen obliged to pay taxes that will be used in order to fund a hospital practicing abortions?" (fiscal conscientious objection). And: "How can a doctor, a nurse, practice abortions that are each time contrary to her or his conscience, which condemns such an act as a crime?" (professional conscientious objection) (Gómez Pérez, 1980, 196).

The Death Penalty. On this subject as well the argument is the same. Our author says he does not want to criticize present-day "abolitionist tendencies" toward the death penalty, but it is proper not to exaggerate when condemning it. "In fact, if the death penalty were declared absolutely illicit (therefore, an immorality), one would have to conclude that throughout history when the penalty has been applied there has been an immoral action" (Gómez Pérez, 1980, 115). Let us flee, then, from the *absolute* character of the ethics of conviction and let us be *realistic* and responsible: always and everywhere there have been crimes and punishments (ibid., 111); "the evidence of punishments is closely linked to their utility, or, in other words, with their ends" (ibid., 112; in this whole chapter there is no mention of "natural moral law"); the death penalty has for many centuries been "the punishment *par excellence,*" and for many centuries "even the most serious and even-handed thinkers had no doubts about its utility and justification" (ibid., 113).

The author goes on to offer five arguments in favor and six against the death penalty, concluding that on some occasions it could be considered licit (Gómez Pérez, 1980, 114) but that, in the end, "the question remains open" (ibid., 115).

A dozen pages before, a chapter of the same book is devoted to a discussion of euthanasia. Did Gómez Pérez in that case also state that it was a matter of an "open question"? Was his reasoning the same as on the subject of capital punishment, to wit: "if euthanasia is declared absolutely illicit one must conclude that always in history one has acted immorally when it has been applied," leading to the conclusion that in some circumstances it might be licit? No. Although he recognizes that historically euthanasia has been a frequent recourse, he gives no arguments in favor of it, only those against it (Gómez Pérez, 1980, 103). "A minimal sense of humanity allows us to see that this is not progress, but regression, a step backward" (ibid., 105); when he discusses the death penalty he does not introduce this type of consideration, although he recognizes, as many do, that "it seems clear that the abolitionist tendency corresponds better to the humanization of the law and to the possibility of a pluralistic society" (ibid., 115).

Thus, while in the case of the death penalty there is no definitive and absolute pronouncement, in the case of euthanasia we enter fully into the field of a pure ethics of conviction. In any of its forms "euthanasia properly speaking is an immoral action, because the object of this act is intrinsically bad: the taking of a life" (Gómez Pérez, 1980, 103). War also involves "the immediate threat of actual death," but it is not therefore "intrinsically evil."

Gómez Pérez continues: "The immorality of euthanasia is deduced directly from *natural moral law,* when men confront its basis: the existence of God as the only master over life and death. Therefore, euthanasia—even with the consent of the victim—is an attack against moral law" (Gómez Pérez, 1980, 103). Here again is the reference to "natural moral law," which is absent in the discussion of the death penalty. Its ultimate and *absolute* basis, the existence of God, is not even mentioned in the chapters on war, conscientious objection, or the death penalty. In these three cases, in fact, the noun dignified with a capital letter is the word State, not the word God!

♦ ♦ ♦

War can be "just"; the law requiring military service is "not unjust"; the death penalty can be "justified" in some circumstances. In these three cases *the autonomy of the individual* is regulated, and limited, by his social position. In the three cases reference is made, therefore, to the variability of social and historical contexts. Historical and cultural evolution may oblige the reformulation of moral judgments. Thus, for example, "the specific facts—nuclear armament—have modified today the traditional position on the ethics of war" (Gómez Pérez, 1980, 110). The possible consequences of action, therefore, are taken into account; this is exactly the definition of the ethics of responsibility.

Similarly, the pluralistic character of modern society or the democratic conception of politics makes one tend to accept conscientious objection and favor the currents which argue for the abolition of the death penalty. In a "plural and democratic" society, at least as the Western world interprets the meaning of these words, it is scarcely conceivable that someone could propose as a solution the "pure logic" of punishing conscientious objectors with the death penalty, as Zafra was still doing in the Spain of 1974 (Zafra, 64).

In the three areas we have chosen to examine, then, we can see how the characteristic standards of the ethics of responsibility continue to play a role, but they are certainly in conflict with the criteria of the ethics of conviction; there is not a harmonious integration of the two, much less a situation where the predominance of the first is likely to threaten the survival of any kind of "principles." But neither is this a context where the defense of convictions acquires an absolute character, totally independent of the variability of historical contexts and the consequences of action, until we make the leap from the death penalty to euthanasia.

When we approach this last topic the discussion changes in a very radical manner. On all the topics relating to what the "official" literature of Opus Dei would call "the defense of life," topics which revolve around the family—which we would sociologically define as the social institution in charge of the regulation of sexuality, procreation, and a part of the process of the socialization of the individual—the ethics of conviction comes to play a preponderant role, so preponderant, in fact, that reference to the standards of the ethics of responsibility practically disappears. Thus we gradually move from the areas in which the "modernity" of Opus Dei is most in evidence to those that most clearly illustrate its "traditionalism."

The Ethics of Conviction without the Ethics of Responsibility?

It is actually in the fields of *sexuality* ("licit" behavior and judgments on sexual relations outside of marriage, homosexuality, etc.), *procreation* (birth control, contraception, abortion), and *family life* (divorce, education of children, etc.) where the Opus Dei literature resorts most systematically to an ethic based exclusively on principles or convictions. At this level Opus Dei tends not to make "concessions" of any kind in the name of the criteria of the ethics of responsibility. Here no account is taken of either the variability of social situations and historical contexts or the possible consequences of action. Principles acquire an *absolute* character: the slightest relativism becomes synonymous with moral laxity and, ultimately, moral degradation. This explains why Opus Dei is perceived, in these areas, as a highly traditionalist and nonmodern movement, a movement which from the perspective of "modern values" would be labeled conservative and reactionary.

Masculinity and Femininity. When we first looked at the origins of Opus Dei, the coexistence of men and women within it, and even certain features of Padre Escrivá's personality, we observed that underlying Opus's manner of posing the problem of sexuality there were specific biological, anthropological, and psychological readings of masculinity and femininity.

Problemas morales de la existencia humana devotes a chapter to each (Gómez Pérez, 1980, 129–44), but here we will refer especially to a more concise work by another Opus author. In his book *Ética del quehacer educativo,* Carlos Cardona includes an appendix, "About Women" (Cardona, 1990, 135–48), which offers a good illustration of what we might call the reification or "naturalization" of the cultural factor of gender, without distinguishing it from the biological factor of sex.

Cardona considers this appendix indispensable, although currently—he says—there are frequent assertions of an "equalization" which tends to deny the "difference" between men and women (Cardona, 1990, 135). There is an "erroneous feminism" which paradoxically ends by becoming a "masculinism of the worst sort." He adds: "There is nothing more disagreeable than a mannish woman; or better still, even *more detestable* is . . . a feminized man" (ibid., 141f.).

The distinct qualities, diverse and complementary, of men and women are not mere cultural artifacts. They are qualities that have been granted to them by God, and as such they have to be recognized "as vestiges, even more, as the image and semblance of God" (Cardona, 1990, 140f.). For example? Specific characteristics of femininity, "like that *instinct* which moves women to be kind, attractive (I do not refer here principally to the physical, but to the psychic and the spiritual: congeniality, tenderness, patience, pity, for example). And by the same token, one understands equally well the special *repulsion* inspired by a woman who is unfriendly, stubborn, aggressive, and, in the extreme, shrewish" (ibid., 144; emphasis mine). We might compare this characterization with the one Jesús Urteaga offers of virility (Urteaga, 1948, 63ff.), on which we commented in an earlier chapter. Or we might compare it to Víctor García Hoz's review of Escrivá's *Santo Rosario,* in which he asserts that the rosary is not "an idle pastime for old women," but a weapon, "something for men to use when they are occupied with matters of warfare" (García Hoz, 1945, 594). Or what about the negative connotations of certain feminine traits, as implied by Escrivá himself: "You are curious and inquisitive, prying and nosy. Are you not ashamed that even in your defects you are not much of a man?" (*The Way,* no. 50)? But for Carlos Cardona the values of femininity cannot be explained in psychological or phenomenological terms: "it is a matter of an ontological character. It is not that woman 'presents herself' thus. It is that woman *is* thus" (Cardona, 1990, 145).

The conclusion Cardona draws is that one must oppose "coeducation." During puberty and early adolescence, boys and girls have to be educated "separately and differently . . . *not for fear of assaults of the animal instinct,"* but in order to encourage the formation of the personality of each (Cardona, 1990, 147; emphasis mine).

Consistent with this approach, there is no ambivalence of anything open to discussion about what has come to be called "the philosophy of sexual liberation": it is simply to be condemned, and in order to condemn it, if necessary one must get out the "heavy artillery." In an article published in 1978, Rafael

Gómez Pérez considered that the expression "philosophy of sexual liberation" was nothing other than "a title of ennoblement for current phenomena, but not therefore less *infrahuman*" (Gómez Pérez, 1978a, 5). "We must not forget, lest we mean to nourish nationalist or racist prejudices, that Freud and Marcuse belonged to the *Germanic mental sphere* and the *Jewish genetic sphere*" (ibid., 6; emphasis mine). A little later, referring to Wilhelm Reich, Gómez Pérez recalls that Freud himself "saw to it that the *Mafia,* which by then already constituted the international psychoanalytical society, expelled him" (ibid., 11).

One might expect a much more prudent attitude on the part of a member of Opus Dei when it comes to using a word like "Mafia." Because if it is licit to employ it with such generic lack of rigor to describe the International Psychoanalytical Association, that makes it much harder to complain when others (Le Vaillant, Ynfante, etc.) apply it to Opus Dei. On the other hand, and more important, it is clear that the tone of the discussion has changed radically, and that this use of language is far from the field of the ethics of responsibility in which we had been moving.

Marriage and Sexuality. The fact that men and women have different and complementary characteristics through the work and will of God (Cardona, 1990, 140f.) implies as a direct consequence that "the only licit use of sexuality is that which is realized in true matrimony, that is, in the indissoluble union of a man with a woman" (Gómez Pérez, 1987, 73).

Therefore, relations outside of marriage are immoral (Gómez Pérez, 1980, 159), as is any form of homosexuality (ibid., 161) and also masturbation (ibid., 162). According to a Declaration of the Vatican Congregation for the Doctrine of the Faith (1975), any use of sexuality outside conjugal relations is immoral. When sociological surveys indicate the existence and frequency of such phenomena, "they are verifying facts. And facts do not constitute a criterion which permits one to judge the moral value of human actions" (quoted in ibid., 162). Remember that in the case of war we were told that it was facts, "concrete facts," which had modified "the traditional position on the ethics of war" (ibid., 110). Facts, on the other hand, do not oblige a change in their position on the ethics of certain sexual behaviors. This is crystal clear: in one case there is an appeal to the criteria of the ethics of responsibility, and in the other there is strict adherence to the ethics of conviction.

The only concession the author is at all willing to make—in a text, it must be said, on *social* ethics—is a statement to the effect that "peaceful coexistence implies, in this field as in others, *the toleration* of alien actions—even if they are objectively immoral—as long as they do not pose a threat to the common good" (Gómez Pérez, 1987, 75). The specific example he uses next is none other than prostitution, which "in general has not been defended, simply tolerated." He ends by saying that one must see to what extent coexistence is threatened by prostitution, and that "one cannot make general rules in this area" (ibid., 75).

Could the formula of "responsible paternity" have something to do, perhaps, with the ethics of responsibility? Yes, in the sense that a couple can reach a decision not to have more children, temporarily or permanently; it is obvious that such a decision is made taking into account the consequences of action. However, the only licit course of action in this situation consists of "following the natural methods of birth control prescribed by God with the biological rhythms of the woman" (Cipriani, 76). Explicitly condemned, therefore, are all contraceptive methods: sterilization, mechanical barriers, pharmaceutical products, and so on. The ethical judgment "has no other alternative than to be categorical" (Gómez Pérez, 1980, 88). No "specific facts" are provided in this case either that would oblige any modification in "the traditional position." For centuries epidemics have acted as a (natural?) means of birth control. Medical progress and better health conditions have produced a drastic reduction in infant mortality. Vaccination is certainly not a "natural method" prescribed by God. But some nonnatural methods operate in favor of the "defense of life," while others are the causes of death. Here lies the fundamental difference. The "just war" continues to be, according to all appearances, the only situation in which the "immediate threat of actual death" (ibid., 107) is not a sufficient reason for a "categorical" ethical judgment in the name of "natural law."

It is hardly necessary to say that abortion is condemned without palliative, precisely because it "constitutes a grave sin against the fifth commandment (thou shalt not kill)" (Cipriani, 81). In this case it is argued that even when abortion is not the objective, but rather is a means to an end (the mother's health, for example), it is still a morally illicit act, because "it is necessary to say that the good end (saving the mother's life) does not justify the bad act (the death of the fetus)" (Gómez Pérez, 1980, 82).

Again one must ask why "the good end" of defending the country can justify "the bad act" of killing a person (even if it is "the enemy"), or why "the good end" of creating a benevolent foundation can justify "the bad act" of tax fraud, or why "the good end" of saving a business justifies "the bad act" of bribery (Gómez Pérez, 1990, 136f., 120).

We are not concerned here—in this particular case or in this entire discussion—with presenting arguments for one position or another. The only thing we are trying to do—while recognizing the enormous complexity of the themes that enter into the ethical debate—is to highlight the apparent contradiction between the reliance on the criteria of an ethics of conviction (without taking any account of an ethics of responsibility) in certain situations, and in others the reliance on different criteria.

Finally, the subject of divorce is treated along these same lines. "There are no exceptions"; matrimony "is indissoluble by its very nature"; and "the mere legal possibility of divorce is already an incitement to it" (Cipriani, 71, 72, 73). Couldn't one say that the mere legal possibility of the death penalty is also an incitement to use it? Or that the mere existence of arguments that war can be just is likewise an incitement to war?

Furthermore, the question of divorce (not directly addressed in the books of Rafael Gómez Pérez) obliges the authors of the volume edited by Cipriani to perform considerable contortions in order to justify the practice of the Catholic Church on the matter. Thus while they condemn divorce "without exception," they admit that "if it is verified that there never was a marriage, because from the outset it was invalid, it can be declared annulled—certifying that it did not exist—by the competent ecclesiastical authority" (Cipriani, 71f.).

The civil authority has no power to regulate what the ecclesiastical authority regulates. Albeit only partially, the conflict between the ethics of responsibility and the ethics of conviction is in our case a problem of power. It is the problem of knowing *who* has the ability to issue judgments—legitimate and legitimizing—on ethical questions. And these authors' exclusive recourse to the ethics of conviction, without the intervention of the ethics of responsibility, presumes that the ecclesiastical authorities wish to "monopolize" the definition of legitimations (perhaps we should say, to perpetuate a former situation of monopoly), which is precisely what makes them look, in an ideologically pluralistic society, like nonmodern traditionalists.

Final Observations

This partially clarifies the paradox of the juxtaposition of traditionalism and modernity usually perceived in Opus Dei. If in spite of its worldly asceticism Opus Dei does not produce some of the consequences implied by Max Weber's thesis on *The Protestant Ethic and the Spirit of Capitalism,* if it does not fulfill Berger's thesis concerning the appearance of a "type of person strongly marked both by the value as well as the psychic reality of individual autonomy" (Berger, 1986, 103), this is because while in the sphere of economic life the members of Opus Dei clearly adopt the criteria of the ethics of responsibility, in other areas—fundamentally those concerning the value of individual autonomy, such as family life and sexuality—they continue to be firmly attached to an ethic strictly governed by convictions and principles ("natural moral law").

We saw that an author like Juan A. Pérez López recognized that business ethics could not be reduced to "normative ethical theories," which are capable of identifying unacceptable practices but not of providing a positive orientation for action (Pérez López, 1990, 39f.). This is, in fact, the very dead end to which the positions held by Opus Dei authors on questions like sexuality, contraception, abortion, and divorce lead. On such issues they clearly specify what is to be rejected on moral grounds, but the question of principles imposes itself in such a way on reasonable foresight of the consequences of action that when the time comes for "positively orienting action," families and married couples are required to exercise a degree of heroism never demanded of businessmen or politicians, who are conceded the legitimate possibility of using the criteria of an ethics of responsibility to guide their actions.

This is what creates the image—perhaps simplistic, but not altogether wrong—of the Opus Dei member as a kind of person who is "technically very competent but with a frankly elementary religiosity," as we said at the beginning of this chapter. All these observations would be a little unfair, however, if we did not add that where it adheres to "the ethics of conviction without the ethics of responsibility," Opus Dei is not defending its own distinctive points of view, but rather what is usually classified as the "official position" of the Catholic Church.

It is true that within the Catholic Church there is more than one view on these topics; it is true that if one were to contrast the pure principles of the "official doctrine" with the points of view and the actual practice of many Catholics, the divergences would undoubtedly be spectacular (and the ecclesiastical authorities must know it); it is true likewise that the application of the criteria of the ethics of responsibility is much more widespread among Catholics of the "rank and file," and that probably only in the "General Staff of Christ's Army" (*The Way,* no. 28) does a pure "ethics of conviction" persist; and finally, it is no less true that the ideologues of Opus Dei are singularly distinguished by their insistence on these themes, that they frequently figure among those who make a point of being "more Catholic than the pope." And in all of this we see, finally, that Opus Dei is not some sort of "sectarian" group, at a distance from the "official" points of view of the Catholic Church; on the contrary, Opus defends and identifies with the official Church better than anyone.

In a novel that ought to be required reading for every contemporary Catholic, *The British Museum Is Falling Down* (1965; Penguin Books, 1981, p. 12), David Lodge has his protagonist write this imaginary article on Catholicism for an imaginary Martian encyclopedia, after life on earth has been annihilated in an atomic war:

> Roman Catholicism was, according to archaeological evidence, distributed fairly widely over the planet Earth in the twentieth century. As far as the Western Hemisphere is concerned, it appears to have been characterized by a complex system of sexual taboos and rituals. Intercourse between married partners was restricted to certain limited periods determined by the calendar and the body-temperature of the female. Martian archaeologists have learned to identify the domiciles of Roman Catholics by the presence of large numbers of complicated graphs, calendars, small booklets full of figures, and quantities of broken thermometers, evidence of the great importance attached to this code. Some scholars have argued that it was merely a method of limiting the number of offspring; but as it has been conclusively proved that the Roman Catholics produced more children on average than any other section of the community, this seems untenable. Other doctrines of the Roman Catholics included a belief in a Divine Redeemer and in a life after death.

♦ ♦ ♦

In sum, if two of the main conclusions of the first part of this study were that Msgr. Escrivá de Balaguer was the child of a specific time and a specific country—Spain in the first half of the twentieth century—and that Opus Dei was the child of its founder, the two conclusions of this second part are that Opus Dei is a movement at once modern and traditionalist. Its modernity, largely the fruit of the worldly ascetic style it adopted, brings it noticeably close to some of the most characteristic features of ascetic Protestantism, from which it is separated, however, by certain doctrinal positions. Opus Dei's traditionalism is particularly noteworthy in those areas where it is guided by a strict ethics of conviction and not by the criteria of an ethics of responsibility. This traditionalism is basically shared by some of the official positions of the Roman Catholic Church and constitutes a good illustration of the difficulties of the insertion of Catholicism in contemporary Western societies, in spite of the efforts at adaptation of the Second Vatican Council, much opposed by significant sectors of the Roman Catholic Church—Opus Dei among them—without apparent success at the time, but in reality with notorious efficacy.

Afterword

Two years have elapsed since the original Catalan version of this book was written. And in this period (from Easter 1992 to Easter 1994) there have been a number of developments, either directly concerning Opus Dei history or indirectly connected with the argument of the last chapter, which call for some further comment and some updating.

1. In May 1992 the beatification process of the Opus Dei founder came to an end, and Msgr. Escrivá became officially the Blessed Josemaría. Given the scant attention accorded in the United States by the media to the whole process, and even to the solemn ceremony held in Rome, it is difficult for an American to grasp the intensity of the controversy aroused in other parts of the world, and particularly in Spain. All the old stereotypes, positive and negative, were rehearsed once again; some new titles were added to both the "official" and the "critical" literature on Opus Dei. At the same time, however, masses of members and "friends" congregated at Saint Peter's Square in Rome to attend the papal performance, thus showing to the Catholic world, and to John Paul II himself, both their enthusiasm and their force. Just one tiny shadow marred this magnificent ceremony: contrary to the wishes of Opus Dei officials, the Holy See decided that Escrivá was not to be the only "star" shining that day in the Roman sky. His beatification was proclaimed together with that of an obscure African nun—whose name everybody has by now forgotten, of course.

It has been argued in this book that the beatification process of the Padre was not, for the Opus Dei leaders, an end in itself, but just a necessary intermediate step on the road to his canonization, that is, toward his official acknowledgment as a saint by the Roman Catholic Church. In 1992 the Opus representatives proudly maintained—only off the record, of course—that this would be achieved in two years' time. Their estimate was in this case exceedingly optimistic. According to some Vatican officials—equally speaking off the record—John Paul II could not remain insensitive to the controversy aroused by the beatification, and in spite of his own personal sympathies he made it clear that Escrivá's canonization could not take place under his pontificate but would have to wait for his successor. Which leaves the issue very open indeed, since even in the best hypothesis (from the point of view of Opus interests) a new pope with such deep "elective affinities" with them cannot simply be taken for granted.

2. What did happen two years after Msgr. Escrivá's beatification was not his canonization, but a major event for the internal organization of Opus Dei: the sudden death of its leader, Msgr. del Portillo, in March 1994. This might well mean the end of an era in Opus history, and the beginning of a new one.

As a matter of fact, in the evolution of new movements and organizations the death of the "founder" is very often followed by a period of continuity in which the leadership is taken over by a close collaborator, belonging to the same generation, and formally or informally appointed by the founder himself before his death. This was clearly the case when Escrivá died in 1975; nobody doubted for a moment that Álvaro del Portillo would become his successor. So if any dramatic changes are to occur, they are much more likely to take place at the disappearance of the founding generation, symbolized in this case by Portillo's death.

According to Opus rules, when the prelate dies the general vicar of the organization assumes leadership for the interim and convenes a congress specifically devoted to the election of a new prelate. That is exactly the situation at the moment of writing these pages. The present general vicar is yet another Spaniard, Javier Echevarría, and he is said to be the most obvious candidate. Born in 1932 in Madrid, he entered Opus Dei at the age of sixteen, received his training in law and canon law in Rome, became a priest in 1955, and has spent his whole life in Opus central headquarters in Rome, permanently near Escrivá and Portillo. Apparently the typical apparatchik, he is scarcely known outside the organization; moreover, he has hardly published anything besides some laudatory comments on Escrivá's teachings. Trying to anticipate the significance of his hypothetical election would therefore be, at this stage, little more than an exercise in guessing and/or gossiping. Moreover, the fact of his being the most obvious candidate does not preclude some other possibilities. The only thing that can be taken for granted is that there will be a prelate from a different generation than Escrivá's. And given the long-standing efforts of Opus Dei to disentangle itself from its very Spanish origins in order to appear a truly international movement, one of the central issues at stake in the election of Msgr. Portillo's successor could well be, for a cupola still largely Spanish, the dilemma of choosing for the first time a non-Spanish prelate.

3. The whole concluding chapter of this book is an attempt to understand and to explain the paradoxical combination of traditionalism and modernity in the Opus Dei. A word should be added on this topic given the fact that, as frequently happens, several authors have been researching along the same lines simultaneously from different perspectives. For obvious reasons, the so-called fundamentalist movements have become in recent years a major subject of interest. And in many cases this same combination of traditionalism and modernity has been pointed out as one outstanding feature of most of these movements. Let us take three examples: The huge Chicago project under the editorship of Martin Marty, *Fundamentalisms,* is a truly encyclopedic endeavor that includes an article on Opus Dei. Gilles Kepel, in *La Revanche de Dieu: Chrétiens, juifs et musulmans à la reconquête du monde* (Paris: Seuil, 1991), compares the strong fundamentalist trends at work in significant sectors of Islam, Judaism, Protestantism, and Roman Catholicism. Kepel analyzes at length the Italian Comunione e Liberazione movement but does not say a word about Opus Dei. And the British sociologist David Martin, in connection with the Boston Institute for the Study of Economic Culture, has undertaken studies on the important expansion of Protestant fundamentalists in Latin America (*Tongues of Fire: The Explosion of Protestantism in Latin America,* Oxford: Blackwell, 1990).

All these works stress one and the same essential feature: the new fundamentalist movements do not necessarily appear as countermodernizing movements. In some aspects they even seem to act as the "carriers" of modernization. This would oblige a reexamination of the too easy correlation usually established between modernization and decline of religiosity. To refer to my own words, at the beginning of Chapter 15, what characterizes these movements is rather the acceptance and the use of modern means and modern techniques . . . together with the rejection of the underlying dominant values usually associated with the use of such means.

Seen from this more general perspective, then, Opus Dei would not be a unique case in the contemporary religious scene. Opus Dei members would rather be the carriers in the Roman Catholic Church of a more general and universal syndrome. This more general and universal syndrome is, precisely, the paradoxical combination of traditionalism and modernity.

4. I would content, however, that the Weberian distinction between an ethic of conviction and an ethic of responsibility is the best analytical tool at our disposal for arriving at an understanding of this paradox. In this sense, not only do I not feel obliged to add anything to the argument of the last chapter of the book, but rather I dare suggest that this Weberian frame of reference could be fruitfully applied to the analysis of other similar movements as well.

Nevertheless, at a different level there is indeed something to add. Since the original version of this book was written, two important official Vatican documents have appeared which have a direct bearing on the moral issues discussed in our conclusive chapter: the New Catechism of the Roman Catholic Church at the end of 1992, and the encyclical *Veritatis Splendor* in the fall of 1993.

(*a*) "Opus Dei's traditionalism is particularly noteworthy in those areas where it is guided by a strict ethics of conviction and not by the criteria of an ethics of responsibility. This traditionalism is shared by some of the official positions of the Roman Catholic Church and constitutes a good illustration of the difficulties of the insertion of Catholicism in contemporary Western societies, in spite of the efforts at adaptation of the Second Vatican Council, much opposed by significant sectors of the Roman Catholic Church—Opus Dei among them—without apparent success at the time, but in reality with notorious efficacy."

Let us resume at this concluding sentence of the last chapter of the book, and elaborate it a bit further in the light of the recent developments brought about by the two Vatican documents.

(*b*) Soon after the publication of the New Catechism several scholars pointed out that "the big question raised by the catechism is that of its true connection with the Vatican Council, whose heir it claims to be" (Jean-Claude Eslin, "Situation d'un catéchisme," *Esprit,* February 1993; see also René Marlé, "Le Catéchisme de l'Église catholique," *Études,* December 1992). On the other side, the Swiss theologian F. Georges Cottier, one of those who worked on the preparation of *Veritatis Splendor,* declared to the French journal *La Croix* that moral theologians were the true intended audience of the encyclical. It has equally been observed that both the encyclical and the catechism make of neoscholastical theology the "traditional Catholic doctrine," crudely identified with *"the* teaching of the Church." Vatican II, however, signified the official acknowledgment by the Roman Catholic Church of a different kind of theological thinking as equally legitimate: the thomistic school of thought was not repudiated but clearly lost its old monopoly in most of the central documents promulgated by the council. At the same time, however, the old Constitutions of Opus Dei stipulated that all full members had to be trained in ecclesiastical studies "of a thomistic orientation." And this orientation seems to have come back to the fore, some thirty years after the Second Vatican Council.

(*c*) Obviously present in both documents, and more specifically in the New Catechism, are all the moral issues that have been considered here as they were approached by the Opus Dei "official" literature. And not only are they present, but they are approached in much the same way as in the Opus Dei texts: a more or less conflictual integration of the ethics of conviction and responsibility in the cases of war, the death penalty, conscientious objection, and so on, and a sharp rejection of an ethics of responsibility in all sexual and family matters. In a similar way, the second chapter of *Veritatis Splendor* contains a frontal attack on the proposals of those Catholic moralists who stress the importance of taking into account the consequences of actions as well as the actors' intentions, defending instead—in the name of "natural law"—an "objectivist" morality.

(*d*) Opus Dei authors would of course reply that this only proves to what extent their arguments are in accordance with the authorized teaching of the Church. A less ingenuous point of view might lead us to ask to what extent Opus Dei scholars have been influential, or even directly involved, in the elaboration of both the New Catechism and *Veritatis Splendor*—which would mean that the theological and moral standpoints of Opus Dei are not just in agreement with the official Roman doctrine, but that they are nowadays contributing to configure it.

Whether this should be interpreted as a promise for the future of the Catholic Church, or rather as a threat, is certainly not for me to say, but for the reader to decide.

Bibliography

This bibliography includes only the sources and documentation used in the preparation of this book. Both here and in the text of the book, the author's name is generally followed by the original year of publication. In cases where the edition consulted was not the first edition, an asterisk indicates the year of the edition to which the page numbers cited in the text refer. The works of an author whose name is preceded by an asterisk were classified in the text as "official literature" of Opus Dei.

*Actas (1957). *Actas del Congreso Nacional de Perfección y Apostolado*. Vol. 1: *Introducción histórica y sesiones comunes*. Madrid: Coculsa.

*Alonso, Luis (1982). "La vocación apostólica del cristiano en la enseñanza de Mons. Escrivá de Balaguer. In Pedro Rodríguez et al., *Mons. Josemaría Escrivá de Balaguer y el Opus Dei, en el 50 aniversario de su fundación*. Pamplona: Eunsa, *2d ed. 1985, pp. 229–92.

Álvarez Bolado, Alfonso (1986–1990). "Guerra civil y universo religioso. Fenomenología de una implicación." *Miscelánea Comillas*, no. 44:233–300; no. 45:417–505; no. 47:3–86; no. 48:35–97.

Anderson, Charles (1970). *The Political Economy of Modern Spain: Policy-Making in an Authoritarian System*. Madison: University of Wisconsin Press.

Andrade, Jaime de [Francisco Franco] (1945). *Raza*. Madrid: Numancia.

Aranguren, José Luis (1952). *Catolicismo y protestantismo como formas de existencia*. Madrid: Revista de Occidente.

——— (1962). "El futuro de la Universidad española." *Cuadernos Taurus*, no. 56. Reprinted in *El futuro de la Universidad y otras polémicas*. Madrid: Taurus, 1973, pp. 9–43.

——— (1965). "La spiritualité de l'Opus Dei." *Esprit*, no. 337:762–71.

——— (1969). *La crisis del catolicismo.* Madrid: Alianza.

——— (1978). *Contralectura del catolicismo.* Barcelona: Planeta.

Artigues, Daniel [Jean Bécarud] (1971). *El Opus Dei en España (1928–1962).* Paris: Ruedo Ibérico, 1st ed. 1968.

Astráin, Antonio (1901–1920). *Historia de la Compañía de Jesús en la Asistencia de España,* 7 vols. Madrid.

*Aubert, Jean-Marie (1982). "La santificación en el trabajo." In Pedro Rodríguez et al., *Mons. Josemaría Escrivá de Balaguer y el Opus Dei.* Pamplona: Eunsa, *2d ed. 1985, pp. 215–24.

Ayala, Ángel (1940). *Formación de selectos.* Madrid: Atenas. Reprinted in *Obras completas,* 2 vols. Madrid: Edica, *1947.

Azpiazu, Joaquín (1944). *La moral del hombre de negocios.* Madrid: Razón y Fe.

*Barco Ortega, José (1974). "La justificación del gobernante." *Nuestro Tiempo,* no. 243:20–34.

Bécarud, Jean (1977). *De la Regenta al Opus Dei.* Madrid: Taurus.

Bellah, Robert N. (1964). "Religious Evolution." *American Sociological Review* 29, no. 3:358–74.

Beltrán, Miguel (1977). *La élite burocrática española.* Barcelona: Ariel.

*Benito, Ángel (1967). "El Fundador del Opus Dei en Pamplona." *Nuestro Tiempo,* no. 162.

*——— (1968). "Conversaciones con monseñor Escrivá de Balaguer." *Nuestro Tiempo,* no. 174:633–40.

Berger, Peter L. (1963). *Invitation to Sociology.* New York: Doubleday. Catalan edition: *Invitació a la sociologia.* Barcelona: Herder, *1986.

——— (1979). *The Heretical Imperative.* New York: Doubleday.

——— (1986). *The Capitalist Revolution.* New York: Basic Books.

Berger, Peter L., and Luckmann, Thomas (*1966). *The Social Construction of Reality.* New York: Doubleday. Catalan edition: *La construcció social de la realitat.* Barcelona: Herder, *1988.

*Berglar, Peter (1983). *Opus Dei: Leben und Werk des Gründers Josemaría Escrivá.* Salzburg: Otto Müller. Castilian edition: *Opus Dei: Vida y obra del Fundador Josemaría Escrivá de Balaguer.* Madrid: Rialp, *4th ed. 1988.

*Bernal, Salvador (1976). *Mons. Josemaría Escrivá de Balaguer: Apuntes sobre la vida del Fundador del Opus Dei.* Madrid: Rialp, *6th ed. 1980.

Beyer, Jean (1954). *Les Instituts séculiers.* Paris: Desclée de Brouwer.

*Blank, Wilhelm, and Gómez Pérez, Rafael (1970). *Doctrina y vida: El Opus Dei en la Iglesia.* Madrid: Palabra.

*Bonani, Giampaolo (1971). "La labor apostólica de los socios del Opus Dei en la diócesis de Roma." *Nuestro Tiempo,* no. 207:58–80.

Bourdieu, Pierre (1980). *Questions de sociologie.* Paris: Minuit.

Brodrick, James (1934). *The Economic Morals of the Jesuits.* London: Oxford University Press.

Brugarola, Martín (1947). *La cristianización de las empresas.* Madrid: Fax.

*Cacho Viu, Vicente (1962). *La Institución Libre de Enseñanza.* Madrid: Rialp.

*Calvo Serer, Rafael (1947). "Una nueva generación española." *Arbor,* no. 24:333–49.

*——— (1953). "La Iglesia en la vida pública española desde 1936." *Arbor,* no. 91–92:289–324.

*——— (1972). *Franco frente al rey.* Paris: Ruedo Ibérico.

*——— (1973). *La dictadura de los franquistas.* Paris: Ruedo Ibérico.

*Canals, Salvador (1954). *Institutos seculares y estado de perfección.* Madrid: Rialp, *2d ed. 1961.

*——— (1960). *Los institutos seculares.* Madrid: Rialp.

*——— (1962). *Ascética meditada.* Madrid: Rialp, *19th ed. 1989.

Carandell, Lluís (1975). *Vida y milagros de Mons. Escrivá de Balaguer, Fundador del Opus Dei.* Barcelona: Laia.

*Cardona, Carlos (1988). "Camino, una lección de amor." In José Morales, ed., *Estudios sobre* Camino. Madrid: Rialp, *2d ed. 1989, pp. 173–79.

*——— (1990). *Ética del quehacer educativo.* Madrid: Rialp.

Cardús, Salvador, and Estruch, Joan (1986). "Teoria i provocació: Reflexions per a una epistemologia paradoxal."*Papers,* no. 26:69–104.

Carmona, Francisco J. (1991). *La socialización del liderazgo católico en Barcelona durante el primer franquismo.* Doctoral diss., Universitat Autònoma de Barcelona.

*Carreño, Pablo A. (1983). *Fundamentos de sociología.* Madrid: Rialp.

Carrero Blanco, Luis (1948). *La victoria del Cristo de Lepanto.* Madrid: Editora Nacional.

Casañas, Joan (1989). *El progressisme catòlic a Catalunya (1940–1980).* Barcelona: La Llar del Llibre.

Casanova, José V. (1982). *The Opus Dei Ethic and the Modernization of Spain.* Doctoral diss., New School for Social Research, New York.

——— (1982a). "The First Secular Institute: The Opus Dei as a Religious Movement-Organization." *Annual Review of the Social Sciences of Religion* 6:243–85.

——— (1983). "The Opus Dei Ethic, the Technocrats and the Modernization of Spain." *Social Science Information* 22, no. 1:27–50.

*Casciaro, José María (1964). "La 'Consecratio Mundi' en las Epístolas de la Cautividad de San Pablo." *Nuestro Tiempo,* no. 120:755–74.

*——— (1982). "La santificación del cristiano en medio del mundo." In Pedro Rodríguez et al., *Mons. Josemaría Escrivá de Balaguer y el Opus Dei.* Pamplona: Eunsa, *2d ed. 1985, pp. 109–71.

*Cejas, José Miguel (1988). "Testimonios sobre un clásico de la literatura espiritual." In José Morales, ed., *Estudios sobre* Camino. Madrid: Rialp, *2d ed. 1989, pp. 89–109.

Charentenay, Pierre de (1990). "La Compagnie de Jésus, un ordre géopolitique?" *Hérodote,* no. 56:67–80.

Chesterton, Gilbert K. (1911). *The Innocence of Father Brown.* New York: Oxford University Press, *1987. Catalan edition: *La innocència del Pare Brown.* Barcelona: Plaza & Janés, *1965.

*Cipriani Thorne, Juan Luis (1989). *Catecismo de doctrina social.* Madrid: Palabra.

Clavera, Joan, et al. (1978). *Capitalismo español de la autarquía a la estabilización (1939–1959).* Madrid: Edicusa.

Comín, Alfonso Carlos (1967). *España, ¿país de misión?* Barcelona: Nova Terra.

Cooper, Norman B. (1975). *Catholicism and the Franco Regime.* London: Sage.

*Coverdale, John F. (1964). "Una respuesta a Von Balthasar." *Nuestro Tiempo,* no. 117–18:488–98.

Coy, Juan José (1974). *Réquiem por el jesuitismo.* Salamanca: Sígueme.

Creac'h, Jean (1958). *Le Coeur et l'épée.* Paris: Plon.

Crozier, Brian (1967). *Franco.* Boston: Little, Brown. Castilian edition: *Franco: Historia y biografía.* Madrid: Magisterio Español, *1969.

Dalmau, Bernabé (1975). "Reflexions ben intencionades sobre l'Opus Dei." *Serra d'Or,* September 1975.

Dalmau, Josep (1969). *Contrapunts al Camí de l'Opus Dei.* Barcelona: Pòrtic.

Díaz, Elías (1983). *Pensamiento español en la era de Franco (1939–1975)*. Madrid: Tecnos.

Díaz Salazar, Rafael (1981). *Iglesia, dictadura y democracia*. Madrid: HOAC.

Duch, Lluís (1976). "L'Opus Dei." In *Esperança cristiana i esforç humà*. Barcelona: Abadia de Montserrat, pp. 132–39.

Ebenstein, William (1960). *Church and State in Franco's Spain*. Princeton: Center of International Studies.

Encinas, Vicente M. (1964). "Una asociación llamada Opus Dei." *Colligite* 10, no. 37.

*Escrivá, José María (1934). *Consideraciones espirituales*. Cuenca: Imprenta Moderna.

*——— (1934a). *El Santo Rosario*. Madrid: Rialp, *25th ed. 1982. English edition: *Holy Rosary*. Dublin: Four Courts Press, 1979.

*——— (1939). *Camino*. Valencia: CID, *1939; Madrid: Rialp, *47th ed. 1988. Catalan edition: *Camí*. Barcelona: Atlàntida, *1955. English edition: *The Way*. London and New York: Scepter Press, *1953, *1987.

*——— (1944). *La Abadesa de Las Huelgas*. Madrid: Luz, *1944; Rialp, *2d ed. 1974.

*Escrivá de Balaguer, José María (1949). *La Constitución Apostólica Provida Mater Ecclesia y el Opus Dei*. Madrid: Boletín de la ACNP.

*Escrivá de Balaguer, Josemaría (1968). *Conversaciones con Mons. Escrivá de Balaguer*. Madrid: Rialp, *17th ed. 1989. English edition: *Conversations with Monsignor Escrivá de Balaguer*. Dublin: Scepter, 1970.

*——— (1973). *Es Cristo que pasa*. Madrid: Rialp, *15th ed. 1978. English edition: *Christ Is Passing By*. Manila: Sinag-Tala, 1974.

*——— (1975). "De la mano de Dios." In Pedro Rodríguez et al., *Mons. Josemaría Escrivá de Balaguer y el Opus Dei*. Pamplona: Eunsa, *2d ed. 1985, pp. 23–30.

*——— (1977). *Amigos de Dios*. Madrid: Rialp, *6th ed. 1980. English edition: *Friends of God*. London: Scepter, 1981.

*——— (1986). *Surco*. Madrid: Rialp.

*——— (1987). *Forja*. Madrid: Rialp. English edition: *The Forge*. London: Scepter, 1988.

Estradé, Antoni (1986). "Paradoxes: Apunts per a una epistemologia paradoxal en sociologia." *Papers*, no. 26:47–67.

Estruch, Joan (1972). *La innovación religiosa*. Barcelona: Ariel.

———, ed. (1984). *L'ètica protestant i l'esperit del capitalisme*, by Max Weber. Barcelona: Edicions 62, pp. 7–17.

Études (1963–1966). *Études sur les Instituts Séculiers*, 3 vols. Paris: Desclée de Brouwer.

Fanfani, Amintore (1934). *Cattolicesimo e protestantesimo nella formazione storica del capitalismo*. Milan: Vita e Pensiero, *2d ed. 1944.

Felzmann, Vladimir (1983). "Why I Left Opus Dei." *The Tablet*, March 26, 1983.

*Fernández Areal, Manuel (1970). *La política católica en España*. Barcelona: Dopesa.

Fernández de Castro, Ignacio (1968). *De las Cortes de Cádiz al Plan de Desarrollo*. Paris: Ruedo Ibérico.

Fesquet, Henri (1967). *Diario del Concilio*. Barcelona: Nova Terra.

*Fontán, Antonio (1961). *Los católicos en la universidad española actual*. Madrid: Rialp.

*——— (1962). "La doctrina social de la Iglesia y la actuación temporal de los cristianos." *Nuestro Tiempo*, no. 93.

Franco Salgado, Francisco (1976). *Mis conversaciones privadas con Franco*. Barcelona: Planeta.

*Fuenmayor, Amadeo de (1962). "La propiedad privada y su función social." *Nuestro Tiempo,* no. 93.

*——— (1983). "La erección del Opus Dei en prelatura personal." In Pedro Rodríguez et al., *Mons. Josemaría Escrivá de Balaguer y el Opus Dei.* Pamplona: Eunsa, *2d ed. 1985, pp. 439–65.

*Fuenmayor, Amadeo de, Gómez-Iglesias, Valentín, and Illanes, José Luis (1989). *El itinerario jurídico del Opus Dei.* Pamplona: Eunsa.

*Fuentes, Antonio (1988). *El sentido cristiano de la riqueza.* Madrid: Rialp.

Fuentes, Julio (1986). "Los cruzados de Wojtyla." *Cambio 16,* no. 737.

Gallo, Max (1969). *Histoire de l'Espagne franquiste.* Paris: Laffont.

*García de Haro, Ramón (1962). "La remuneración del trabajo." *Nuestro Tiempo,* no. 93.

García Escudero, José María (1956). *Catolicismo de fronteras adentro.* Madrid: Euramérica.

——— (1975). *Historia política de las dos Españas.* Madrid: Editora Nacional.

*García Hoz, Víctor (1941). *Pedagogía de la lucha ascética.* Madrid: CSIC.

*——— (1945). Book review: *Santo Rosario,* by D. José María Escrivá. *Arbor,* no. 9:594.

*——— (1976). "La educación en Mons. Escrivá de Balaguer." *Nuestro Tiempo,* no. 264:5–22.

*——— (1988). "Sobre la pedagogía de la lucha ascética en Camino." In José Morales, ed., *Estudios sobre* Camino. Madrid: Rialp, *2d ed. 1989, pp. 181–211.

García Molleda, María Dolores (1966). *Los reformadores de la España contemporánea.* Madrid: CSIC.

*García Suárez, A. (1970). "Existencia secular cristiana." *Scripta Theologica,* no. 2.

García Villoslada, Ricardo, ed. (1979). *Historia de la Iglesia en España,* vol. 5: *La Iglesia en la España contemporánea (1808–1975).* Madrid: BAC.

Giacomo, Maurizio di (1987). *Opus Dei.* Naples: Pironti.

Gilder, George (1990). "El altruismo en la empresa." In Carlos Llano et al., *La vertiente humana del trabajo en la empresa.* Madrid: Rialp, pp. 59–74.

Ginzburg, Carlo (1979). "Spie. Radici di un paradigma indiziario." In A. Gargani, ed., *Crisi della ragione.* Turin: Einaudi.

——— (1981). *Indagine su Piero.* Turin: Einaudi.

Goffman, Erving (1961). *Asylums.* New York: Doubleday.

*Gómez Pérez, Rafael (1976). *Política y religión en el régimen de Franco.* Barcelona: Dopesa.

*——— (1976a). *La minoría cristiana.* Madrid: Rialp.

*——— (1978). "Opus Dei 1928–1978: un camino que sigue." *Nuestro Tiempo,* no. 293:19–32.

*——— (1978a). "Filosofía de la liberación sexual." *Nuestro Tiempo,* no. 283:5–38.

*——— (1980). *Problemas morales de la existencia humana.* Madrid: Magisterio Español, *4th ed. 1987.

*——— (1986). *El franquismo y la Iglesia.* Madrid: Rialp.

*——— (1986a). "El Opus Dei en el conjunto de la Iglesia." *Cambio 16,* no. 737.

*——— (1987). *Introduccion a la ética social.* Madrid: Rialp, *3d ed. 1989.

*——— (1990). *Ética empresarial: teoría y casos.* Madrid: Rialp.

*Gondrand, François (1982). *Au pas de Dieu: Josemaría Escrivá de Balaguer, fondateur de l'Opus Dei.* Paris: France Empire. Castilian edition: *Al paso de Dios: Josemaría Escrivá de Balaguer, fundador del Opus Dei.* Madrid: Rialp, *4th ed. 1985.

González Estefani, José María (1978). *Creo en la historia: Del nacionalcatolicismo a la contracultura.* Bilbao: Desclée.

González González, Manuel Jesús (1979). *La economía política del franquismo (1940–1970): Dirigismo, mercado y planificación.* Madrid: Tecnos.

González Ruiz, José María (1964). "El Opus Dei, hijo de su tiempo." *Signo,* no. 1280.

Grootaers, Jan (1981). *De Vatican II à Jean-Paul II: Le Grand tournant de l'Église catholique.* Paris: Centurion.

Guichard, Alain (1974). *Les Jésuites.* Paris: Grasset. Castilian edition: *Los jesuitas.* Barcelona: Dopesa, *1974.

*Gutiérrez Ríos, Enrique (1970). *José María Albareda, una época de la cultura española.* Madrid: CSIC.

Hebblethwaite, Peter (1983). "Opus Dei: Lifting the Veil of Mystery." *National Catholic Reporter,* May 27, 1983.

*Helming, Dennis M. (1985). *Footprints in the Snow: A Pictorial Biography of Josemaría Escrivá, the Founder of Opus Dei.* New York: Scepter. Castilian edition: *Huellas en la nieve: Biografía ilustrada de Josemaría Escrivá de Balaguer, Fundador del Opus Dei.* Madrid: Palabra, *1987.

Hermet, Guy (1974). "Reflexiones sobre las funciones políticas del catolicismo en los regímenes autoritarios contemporáneos." *Sistema,* no. 4:23–34.

——— (1980–1981). *Los católicos en la España franquista,* 2 vols. Madrid: Siglo XXI, *1985, *1986.

*Hernández Garnica, José María (1957). *Perfección y laicado.* Madrid: Rialp.

*Herranz, Julián (1957). "El Opus Dei y la política." *Nuestro Tiempo,* no. 34:385–402.

*——— (1962). "El Opus Dei." *Nuestro Tiempo,* no. 97:5–30.

*——— (1964). "La evolución de los institutos seculares." *Ius Canonicum,* no. 4:303–33.

*Herrera Jaramillo, Francisco José (1984). *El derecho a la vida y el aborto.* Pamplona: Eunsa.

Hertel, Peter (1990). *"Ich verspreche euch den Himmel": Geistlicher Anspruch, gesellschaftliche Ziele und kirchliche Bedeutung des Opus Dei.* Düsseldorf: Patmos, 1st ed. 1985.

Hofmann, Paul (1985). *Anatomy of the Vatican.* London: Hale.

Howe, Richard Herbert (1978). "Max Weber's Elective Affinities: Sociology Within the Bounds of Pure Reason." *American Journal of Sociology* 84, no. 2:366–85.

*Huerta, Félix (1965). "El Instituto de Estudios Superiores de la Empresa." *Nuestro Tiempo,* no. 133–34, 136.

Hutch, Richard A. (1991). *Religious Leadership.* New York: Peter Lang.

*Ibáñez Langlois, José María (1985). *Teología de la liberación y lucha de clases.* Madrid: Palabra.

*Illanes, José Luis (1980). *La santificación del trabajo.* Madrid: Palabra, 1st ed. 1966, *9th ed. 1981.

*——— (1982). "Dos de octubre de 1928: alcance y significación de una fecha." In Pedro Rodríguez et al., *Mons. Josemaría Escrivá de Balaguer y el Opus Dei.* Pamplona: Eunsa, *2d ed. 1985, pp. 65–107.

*——— (1984). *Mundo y santidad.* Madrid: Rialp.

Iturralde, Juan de (1955–1965). *El catolicismo y la cruzada de Franco,* 3 vols. Vienna: Egui-Indarra.

*Ivars Moreno, Antonio (1962). "Nueva visión de la empresa." *Nuestro Tiempo,* no. 96.

*——— (1969). *Ideario para gerentes de empresa.* Madrid: Rialp.

*Ivars Moreno, Antonio, et al. (1971). "La empresa en los años setenta." *Nuestro Tiempo,* no. 208.

Jardiel Poncela, Eva (1974). *¿Por qué no es usted del Opus Dei?* Madrid. [volume of interviews]

Kolvenbach, Peter H. (1990). *Fedeli a Dio e all'uomo.* Milan: Paoline. Castilian edition: *Fieles a Dios y al hombre.* Madrid: Paulinas, *1991.

Lacouture, Jean (1991). *Jésuites.* Paris: Seuil.

Laín Entralgo, Pedro (1976). *Descargo de conciencia (1930–1960).* Barcelona: Barral.

Lannon, Frances (1987). *Privilege, Persecution and Prophecy: The Catholic Church in Spain, 1875–1975.* Oxford: Clarendon Press.

*Le Tourneau, Dominique (1984). *L'Opus Dei.* Paris: PUF, *2d ed. 1985. Castilian edition: *El Opus Dei.* Barcelona: Oikos-Tau, 1986.

Le Vaillant, Yvon (1971). *Sainte Maffia: Le Dossier de l'Opus Dei.* Paris: Mercure de France.

Linz, Juan J., and Miguel, Amando de (1966). *Los empresarios ante el poder público: El liderazgo y los grupos de intereses en el empresariado español.* Madrid: Instituto de Estudios Políticos.

——— (1968). *La élite funcionarial española ante la reforma administrativa.* Madrid: Centro de Estudios Sociales.

*Llano, Carlos, et al. (1990). *La vertiente humana del trabajo en la empresa.* Madrid: Rialp.

*López Amo, Angel (1952). *El poder político y la libertad: La monarquía de la reforma social.* Madrid: Rialp.

*López Rodó, Laureano (1964). "España 1964: Realidades y perspectivas del desarrollo económico." *Arbor,* no. 219.

*——— (1970). *Política y desarrollo.* Madrid: Aguilar.

*——— (1977). *La larga marcha hacia la monarquía.* Barcelona: Noguer.

*——— (1990–1991). *Memorias,* vols. 1 and 2. Barcelona: Plaza & Janés.

*Lucas Marín, Antonio (1797). *Introducción a la sociología.* Pamplona: Eunsa, *2d ed. 1982.

Lüthy, Herbert (1965). *Le Passé présent.* Monaco: Rocher.

Maeztu, Ramiro de (1957). *El sentido reverencial del dinero.* Madrid: Editora Nacional, 1st ed. 1933.

——— (1958). *Defensa del espíritu.* Madrid: Rialp.

Magister, Sandro (1986). "Santa Facciatosta"; "Il Polipo di Dio"; "Amore e cilicio"; "Primo, obbedire." *L'Espresso,* March 2, 9, 16; April 6.

Marquina Barrio, Antonio (1983). *La diplomacia vaticana y la España de Franco (1936–1945).* Madrid: CSIC.

Martí Gómez, Josep, and Ramoneda, Josep (1976). *Calvo Serer: el exilio y el reino.* Barcelona: Laia.

Martín-Sánchez Juliá, Fernando (1954). *Ideas claras.* Madrid: BAC, *1960.

*Martinell, Francisco, ed. (1970). *Cristianos corrientes: Textos sobre el Opus Dei.* Madrid: Rialp, 5th ed. 1974.

Martínez Alier, Joan (1978). "Notas sobre el franquismo." *Papers,* no. 8:27–51.

*Masllorens, Jorge (1968). "La empresa, realidad social." *Nuestro Tiempo,* no. 174:537–51.

*Mateo Seco, Lucas F. (1985). "Obras de Mons. Escrivá de Balaguer y estudios sobre el Opus Dei." In Pedro Rodríguez et al., *Mons. Josemaría Escrivá de Balaguer y el Opus Dei.* Pamplona: Eunsa, pp. 469–572.

Miguel, Amando de, and Linz, Juan J. (1964). "Nivel de estudios del empresariado español." *Arbor,* no. 219.

Mills, C. Wright (*1959). *The Sociological Imagination.* New York: Oxford University Press. Catalan edition: *La imaginació sociològica.* Barcelona: Herder, *1987.

Mitscherlich, Alexander, and Mitscherlich, Margarethe (1967). *Die Unfähigkeit zu Trauern.* Munich: Piper.

Moll, Xavier (1991). *De la Companyia de Jesús a l'Opus Dei.* Unpublished manuscript.

*Moncada, Alberto (1962). "El principio de subsidiariedad del Estado." *Nuestro Tiempo,* no. 93.

Moncada, Alberto (1974). *El Opus Dei: Una interpretación.* Madrid: Indice.

——— (1977). *Los hijos del Padre.* Barcelona: Argos-Vergara.

——— (1982). *Los españoles y su fe.* Madrid: Penthalon.

——— (1986). "Sociología del Opus Dei." *Cambio 16,* no. 737.

——— (1987). *Historia oral del Opus Dei.* Barcelona: Plaza & Janés.

——— (1990). "Sectas católicas: el Opus Dei." Madrid: XII Congreso de Sociología.

*Morales, José, ed. (1988). *Estudios sobre Camino.* Madrid: Rialp, 2d ed. 1989.

Moreno, María Angustias (1976). *El Opus Dei: Anexo a una historia.* Barcelona: Planeta.

——— (1978). *La otra cara del Opus Dei.* Barcelona: Planeta.

Moya, Carlos (1975). *El poder económico en España.* Madrid: Tucar.

——— (1984). "Las élites del poder económico en España (1939–1970)." In *Señas de Leviatán: Estado nacional y sociedad industrial: España 1936–1980.* Madrid: Alianza, pp. 64–154.

*Navarro Rubio, Mariano (1987). *Sobre el trabajo.* Madrid: Palabra.

Neuner, Peter (1991). "Die Warnungen sind berechtigt: Zur Diskussion über den 'katholischen Fundamentalismus.'" *Herder Korrespondenz,* no. 9:422–27.

Nicolás Cabo, Juan Martín de (1969). *La formación universitaria para la empresa.* Barcelona: Ariel.

*Ocáriz, Fernando (1982). "La filiación divina, realidad central en la vida y en la enseñanza de Mons. Escrivá de Balaguer." In Pedro Rodríguez et al., *Mons. Josemaría Escrivá de Balaguer y el Opus Dei.* Pamplona: Eunsa, *2d ed. 1985, pp. 173–214.

*Orlandis, José (1959). *La vocación cristiana del hombre de hoy.* Madrid: Rialp, *3d ed. 1973.

*——— (1962). "Hombre y sociedad en la Encíclica Mater et Magistra." *Nuestro Tiempo,* no. 93.

*——— (1965). "Una espiritualidad laical y secular." In Francisco Martinell, ed., *Cristianos corrientes: Textos sobre el Opus Dei.* Madrid: Rialp, *5th ed. 1974, pp. 29–65.

*——— (1967). *La crisis de la Universidad en España.* Madrid: Rialp.

*——— (1974). "¿Qué es ser católico?" *Nuestro Tiempo,* no. 246.

*Ors, Álvaro d' (1974). "Cuatro precursores de la ciencia española de nuestro tiempo." *Nuestro Tiempo,* no. 241:9–26.

Ortiz, Tomás (1982). *L'Opus Dei: Les nouveaux croisés.* Paris: Edimaf.

Ortuño, Manuel (1963). "Opus Dei." *Cuadernos Americanos,* January–February 1963.

Pániker, Salvador (1969). *Conversaciones en Madrid.* Barcelona: Kairós. [volume of interviews]

*Panikkar, Raimundo (1944). "Visión de síntesis del universo." *Arbor,* no. 1:9–60. Reprinted in *Humanismo y cruz.* Madrid: Rialp, 1963.

Pasamar, Gonzalo, et al. (1985). *Arbor,* no. 479–80:13–137. [monograph]

Pérez Díaz, Victor (1991). *The Church and Religion in Contemporary Spain.* Madrid: Instituto Juan March.

*Pérez Embid, Florentino (1949). "Ante la nueva actualidad del problema de España." *Arbor,* no. 45–46.

*——— (1952). "Breve historia de la revista *Arbor.*" *Arbor,* no. 75:305–16.

*——— (1953). *Ambiciones españolas.* Madrid: Editora Nacional.

*——— (1955). *Nosotros los cristianos.* Madrid: Rialp.

*——— (1956). *En la brecha.* Madrid: Rialp.

*——— (1961). "¿Qué es el Opus Dei?" *Vida Mundial,* April 29, 1961. [interview]

*——— (1963). "Monseñor José María Escrivá de Balaguer y Albás, Fundador del Opus Dei, primer Instituto Secular." In *Forjadores del Mundo Contemporáneo,* vol. 4. Barcelona: Planeta.

*Pérez López, Juan A. (1990). "El sentido de los conflictos éticos originados por el entorno en que opera la empresa." In Carlos Llano et al., *La vertiente humana del trabajo en la empresa.* Madrid: Rialp, pp. 33–58.

Pérez Vilariño, José, and Schoenherr, Richard A. (1990). "La religión organizada en España." In *España: Sociedad y política.* Madrid: Espasa-Calpe, pp. 449–69.

*Perosanz, José Miguel (1975). *Iglesia en tiempo de crisis.* Barcelona: Dopesa.

Petschen, Santiago (1977). *La Iglesia en la España de Franco.* Madrid: Sedmay.

Pinilla de las Heras, Esteban (1967). *Los empresarios y el desarrollo capitalista.* Barcelona: Península.

Piñol, Josep Maria (1969). *¿Nuevos caminos en la Iglesia?* Barcelona: Península.

*Polo, Leonardo (1990). "Ricos y pobres: Igualdad y desigualdad." In Carlos Llano et al., *La vertiente humana del trabajo en la empresa.* Madrid: Rialp, pp. 75–143.

*Portillo, Álvaro del (1966). "Le laïc dans l'Église et dans le monde." *La Table Ronde,* April 1966.

*——— (1976). "Monseñor Escrivá de Balaguer, instrumento de Dios." In Álvaro del Portillo, Francisco Ponz, and Gonzalo Herranz, *En memoria de Mons. Josemaría Escrivá de Balaguer.* Pamplona: Eunsa, pp. 17–60.

*——— (1978). "El camino del Opus Dei." In Pedro Rodríguez et al., *Mons. Josemaría Escrivá de Balaguer y el Opus Dei.* Pamplona: Eunsa, *2d ed. 1985, pp. 35–55.

*——— (1981). *Fieles y laicos en la Iglesia.* Pamplona: Eunsa, 1st ed. 1969.

*——— (1988). "Significado teológico-espiritual de *Camino.*" In José Morales, ed., *Estudios sobre* Camino. Madrid: Rialp, *2d ed. 1989, pp. 45–56.

*Portillo, Álvaro del, Ponz, Francisco, and Herranz, Gonzalo (1976). *En memoria de Monseñor Josemaría Escrivá de Balaguer.* Pamplona: Eunsa.

Preston, Paul, ed. (1976). *Spain in Crisis: The Evolution and Decline of the Franco Regime.* London: Harvester Press.

Robertson, Hector M. (1933). *Aspects of the Rise of Economic Individualism: A Criticism of Max Weber and His School.* Cambridge: Cambridge University Press.

Rocca, Giancarlo (1985). *L'Opus Dei: appunti e documenti per una storia.* Rome: Paoline.

——— (1989). "L'Opus Dei visto dall'Opus Dei." *Claretianum,* no. 29:379–391. [book review of Fuenmayor et al., 1989]

*Rodríguez, Pedro (1965). "*Camino* y la espiritualidad del Opus Dei." *Teología Espiritual,* no. 9:213–45.

*——— (1971). "Sobre la espiritualidad del trabajo." *Nuestro Tiempo,* no. 201:5–34.

*——— (1977). "La economía de la salvación y la secularidad cristiana." *Scripta Theologica,* no. 9:9–123.

*——— (1986). *Vocación, trabajo, contemplación.* Pamplona: Eunsa, *2d ed. 1987.

*——— (1986a). *Iglesias particulares y Prelaturas personales.* Pamplona: Eunsa.

*Rodríguez, Pedro, et al. (1985). *Mons. Josemaría Escrivá de Balaguer y el Opus Dei en el 50 aniversario de su fundación.* Pamplona: Eunsa, 1st ed. 1982.
*Roegele, Otto (1963). "L'Opus Dei, légendes et réalités." *La Revue Nouvelle,* November 15, 1963.
*Roegele, Otto, et al. (1967). *Opus Dei. Für und Wider.* Osnabrück: Fromm.
Roldán, Santiago, and García Delgado, José Luis (1973). *La formación de la sociedad capitalista en España,* 2 vols. Madrid: Confederación Española de Cajas de Ahorros.
Ros, Jacinto, ed. (1975). *Trece economistas españoles ante la economía española.* Barcelona: Oikos-Tau.
Roth, Jürgen, and Ender, Berndt (1984). *Dunkelmänner der Macht.* Köln: Lamuv.
Ruiz Rico, Juan José (1977). *El papel político de la Iglesia en la España de Franco (1936–1971).* Madrid: Tecnos.
Sáez Alba, A. (1974). *La Asociación Católica Nacional de Propagandistas.* Paris: Ruedo Ibérico.
Sánchez Ruiz, Valentín (1935). *Catecismo social.* Madrid: Apostolado de la Prensa.
*Saranyana, Josep Ignasi (1988). "Cincuenta años de historia." In José Morales, ed., *Estudios sobre* Camino. Madrid: Rialp, *2d ed. 1989, pp. 59–65.
*Sastre, Ana (1989). *Tiempo de caminar: Semblanza de Monseñor Escrivá de Balaguer.* Madrid: Rialp, *2d ed. 1990.
Saunier, Jean (1973). *L'Opus Dei et les sociétés secrètes catholiques.* Paris: Grasset.
*Seco, Luis Ignacio (1986). *La herencia de Mons. Escrivá de Balaguer.* Madrid: Palabra.
Shaw, Russell (1982). "The Secret of Opus Dei." *Columbia Magazine,* March 1982.
Sopeña, Federico (1970). *Defensa de una generación.* Madrid: Taurus.
Spaemann, Robert (1982). *Moralische Grundbegriffe.* Munich: Beck. Castilian edition: *Ética: cuestiones fundamentales.* Pamplona: Eunsa, *2d ed. 1988.
Steigleder, Klaus (1983). *Das Opus Dei: Eine Innenansicht.* Zurich: Benziger. Italian edition: *L'Opus Dei, vista dall'interno.* Turin: Claudiana; with an introduction by Maurizio di Giacomo, "L'Opus Dei in Italia," pp. 13–92.
*Suárez, Federico (1979). "Las dos caras del silencio." *Nuestro Tiempo,* no. 297.
Suñer, Enrique (1937). *Los intelectuales y la tragedia española.* Burgos: Editorial Española.
Tamames, Ramón (1973). *La República. La era de Franco.* Madrid: Alianza-Alfaguara.
——— (1976). *Estructura económica de España.* Madrid: Guadarrama.
Tapia, María del Carmen (1983). "Good Housekeepers for Opus Dei." *National Catholic Reporter,* May 27, 1983, pp. 10–13.
Tawney, Richard H. (1926). *Religion and the Rise of Capitalism.* London: Murray; Harmondsworth: Penguin Books, *8th ed. 1969.
Terricabras, Josep M. (1986). "Sobre tres models de coneixement." *Papers,* no. 26:27–45.
*Thierry, Jean-Jacques (1973). *L'Opus Dei: Mythe et réalité.* Paris: Hachette.
*Torelló, Juan B. (1965). *La espiritualidad de los laicos.* Madrid: Rialp.
*——— (1974). *Psicoanálisis y confesión.* Madrid: Rialp.
*Torelló, Juan B., et al. (1975). *La vocación cristiana.* Madrid: Palabra.
Troeltsch, Ernst (1912). *Die Soziallehren der Christlichen Kirchen und Gruppen.* Tübingen: Mohr. English edition: *The Social Teaching of the Christian Churches,* 2 vols. Chicago: University of Chicago Press, 1976.
Tuininga, M. (1970). "Qu'est-ce que l'Opus Dei?" *Informations Catholiques Internationales,* no. 361:19–28.

Tuñón de Lara, Manuel (1968). *El hecho religioso en España*. Paris: Librairie du Globe.

Tusell, Javier (1984). *Franco y los católicos*. Madrid: Alianza.

*Ullastres, Alberto (1961). "El desarrollo económico y su planteamiento en España." *Arbor*, no. 50:8–21.

Urbina, Fernando, et al. (1977). *Iglesia y sociedad en España (1939–1975)*. Madrid: Popular.

*Urteaga, Jesús (1948). *El valor divino de lo humano*. Madrid: Rialp, *32nd ed. 1988.

*——— (1988). "El impacto de *Camino* en los años cuarenta." In José Morales, ed., *Estudios sobre* Camino. Madrid: Rialp, *2d ed. 1989, pp. 79–88.

*Valero, Antonio (1962). "Estructura de la empresa." *Nuestro Tiempo*, no. 93.

Valverde, Carlos (1979). "Los católicos y la cultura española." In *Historia de la Iglesia en España*, vol. 5. Madrid: BAC, pp. 475–573.

*Vázquez de Prada, Andrés (1983). *El fundador del Opus Dei*. Madrid: Rialp, *2d ed. 1984.

Velarde Fuertes, José (1967). *Sobre la decadencia económica de España*. Madrid: Tecnos.

Von Balthasar, Hans Urs (1963). "Integralismus." *Wort und Wahrheit*, no. 12:737–44.

——— (1964). "Friedliche Fragen an das Opus Dei." *Der Christliche Sonntag*, April 12, 1964, p. 117.

Wach, Joachim (1944). *Sociology of Religion*. Chicago: University of Chicago Press.

Walf, Knut (1989). "Fundamentalistische Strömungen in der katholischen Kirche." In Thomas Meyer, ed., *Fundamentalismus in der modernen Welt*. Frankfurt: Suhrkamp, pp. 248–62.

Walsh, Michael (1989). *The Secret World of Opus Dei*. London: Grafton Books. Castilian edition: *El mundo secreto del Opus Dei*. Barcelona: Plaza & Janés, *1990.

Weber, Max (1904–1905). *Die Protestantische Ethik und der Geist des Kapitalismus*. In *Gesammelte Aufsätze zur Religionssoziologie*, vol. 1. Tübingen: Mohr, 1920, *5th ed. 1963. Catalan edition: *L'ètica protestant i l'esperit del capitalisme*. Barcelona: Edicions 62, *1984. English edition: *The Protestant Ethic and the Spirit of Capitalism*. London: Allen & Unwin, 1930.

——— (1920). *Gesammelte Aufsätze zur Religionssoziologie*, vol. 1. Tübingen: Mohr, 1920, *5th ed. 1963. English edition: In Hans Gerth and C. Wright Mills, eds. *From Max Weber: Essays in Sociology*. New York: Oxford University Press, 1948.

——— (1921). *Politik als Beruf*. In *Gesammelte Politische Schriften*. Tübingen: Mohr, *2d ed. 1958. English edition: In Hans Gerth and C. Wright Mills, eds. *From Max Weber: Essays in Sociology*. New York: Oxford University Press, 1948.

——— (1922). *Wirtschaft und Gesellschaft*. Tübingen: Mohr, *5th ed. 1972. English edition: *Economy and Society*, 3 vols. New York: Bedminster Press, 1968.

*West, William J. (1987). *Opus Dei: Exploding a Myth*. Sydney: Little Hills Press. Castilian edition: *Opus Dei: Ficción y realidad*. Madrid: Rialp, *1989.

Woodrow, Alain (1984). *Les Jésuites*. Paris: Lattès.

Ynfante, Jesús (1970). *La Prodigiosa aventura del Opus Dei: Génesis de la Santa Mafia*. Paris: Ruedo Ibérico.

Zafra, José (1974). "¿Negativa al servicio militar?" *Nuestro Tiempo*, no. 241:62–72.

Zizola, Giancarlo (1985). *La restaurazione di papa Wojtyla*. Roma-Bari: Laterza.

Name Index

General Index